Microbial Toxins

Current Research and Future Trends

Edited by Thomas Proft

Department of Molecular Medicine & Pathology
Faculty of Medical and Health Sciences
University of Auckland
Auckland
New Zealand

Caister Academic Press

Copyright © 2009

Caister Academic Press
Norfolk, UK

www.caister.com

British Library Cataloguing-in-Publication Data
A catalogue record for this book is available from the British Library

ISBN: 978-1-904455-44-8

Printed and bound in Great Britain

Contents

Contributors

Jorge Alegre-Cebollada
Departamento de Bioquímica y Biología Molecular I
Facultad de Ciencias Químicas
Universidad Complutense de Madrid
Madrid
Spain

ja2544@columbia.edu

Elisa Álvarez-García
Departamento de Bioquímica y Biología Molecular I
Facultad de Ciencias Químicas
Universidad Complutense de Madrid
Madrid
Spain

elisa@bbm1.ucm.es

John C. Atherton
Nottingham Digestive Diseases Centre, Biomedical
Research Unit and Institute of Infection, Immunity
and Inflammation
Centre for Biomolecular Science
University of Nottingham
Nottingham
United Kingdom

john.atherton@nottingham.ac.uk

Kathleen M. Averette-Mirrashidi
Department of Microbiology, Immunology, and
Molecular Genetics
University of California at Los Angeles
Los Angeles, CA
USA

kaverett@ucla.edu

Kenneth A. Bradley
Department of Microbiology, Immunology, and
Molecular Genetics
University of California at Los Angeles
Los Angeles, CA
USA

kbradley@microbio.ucla.edu

Nelson Carreras-Sangrà
Departamento de Bioquímica y Biología Molecular I
Facultad de Ciencias Químicas
Universidad Complutense de Madrid
Madrid
Spain

nelson@bbm1.ucm.es

Timothy L. Cover
Department of Medicine and Department of
Microbiology and Immunology
Vanderbilt University School of Medicine and
Veterans Affairs Tennessee Valley Healthcare System
Nashville, TN
USA

timothy.l.cover@vanderbilt.edu

J. Ross Fitzgerald
The Roslin Institute and Centre for Infectious
Diseases
University of Edinburgh
Edinburgh
Scotland
UK

ross.fitzgerald@ed.ac.uk

John D. Fraser
School of Medical Sciences
Faculty of Medical and Health Sciences
University of Auckland
Auckland
New Zealand

jd.fraser@auckland.ac.nz

José G. Gavilanes
Departamento de Bioquímica y Biología Molecular I
Facultad de Ciencias Químicas
Universidad Complutense de Madrid
Madrid
Spain

ppgf@bbm1.ucm.es

Brett Geissler
Department of Microbiology–Immunology
Feinberg School of Medicine
Northwestern University
Chicago, IL
USA

b_geissler@northwestern.edu

Elías Herrero-Galán
Departamento de Bioquímica y Biología Molecular I
Facultad de Ciencias Químicas
Universidad Complutense de Madrid
Madrid
Spain

elias@bbm1.ucm.es

Roshan Kukreja
Botulinum Research Center
and Department of Chemistry and Biochemistry
University of Massachusetts Dartmouth
North Dartmouth, MA
USA

g_rkukreja@umassd.edu

Javier Lacadena
Departamento de Bioquímica y Biología Molecular I
Facultad de Ciencias Químicas
Universidad Complutense de Madrid
Madrid
Spain

javierl@bbm1.ucm.es

Ries J. Langley
The Maurice Wilkins Centre for Molecular
Biodiscovery & School of Medical Sciences
University of Auckland
Auckland
New Zealand

r.langley@auckland.ac.nz

Francisco J. Maldonado-Arocho
Department of Microbiology, Immunology, and
Molecular Genetics
University of California at Los Angeles
Los Angeles, CA
USA

fmaldona@ucla.edu

Álvaro Martínez del Pozo
Departamento de Bioquímica y Biología Molecular I
Facultad de Ciencias Químicas
Universidad Complutense de Madrid
Madrid
Spain

alvaro@bbm1.ucm.es

Mercedes Oñaderra
Departamento de Bioquímica y Biología Molecular I
Facultad de Ciencias Químicas
Universidad Complutense de Madrid
Madrid
Spain

mos@bbm1.ucm.es

Joachim H.C. Orth
Institut für Experimentelle und Klinische
Pharmakologie und Toxikologie
Freiburg
Germany

joachim.orth@pharmakol.uni-freiburg.de

Adrienne W. Paton
School of Molecular and Biomedical Science
University of Adelaide
Adelaide, SA
Australia

adrienne.paton@adelaide.edu.au

James C. Paton
School of Molecular and Biomedical Science
University of Adelaide
Adelaide, SA
Australia

james.paton@adelaide.edu.au

José R Penadés
Centro de Investigación y Tecnología Animal
Instituto Valenciano de Investigaciones Agrarias
(CITA-IVIA)
Castellón
Spain

penades_jos@gva.es

Thomas Proft
The Maurice Wilkins Centre for Molecular
Biodiscovery & School of Medical Sciences
University of Auckland
Auckland
New Zealand

t.proft@auckland.ac.nz

Karla J. F. Satchell
Department of Microbiology–Immunology
Feinberg School of Medicine
Northwestern University
Chicago, IL
USA

k-satchell@northwestern.edu

Bal Ram Singh
Botulinum Research Center
and Department of Chemistry and Biochemistry
University of Massachusetts Dartmouth
North Dartmouth, MA
USA

bsingh@umassd.edu

Preface

Since Pierre Roux's and Alexandre Yersin's discovery of the first bacterial protein toxin, diphtheria toxin in 1888, more than 350 toxins have been identified to date. In particular, over the last few decades, our understanding of microbial toxins has been greatly increased. This is mainly due to the extraordinary technical advances in various disciplines involved in toxin research, such as molecular microbiology, complete genome sequencing, protein crystallography and experimental animal models.

During the last decade, more than 60 novel toxins have been discovered, most of them by bioinformatic mining of recently completed bacterial genomes. For example, screening of streptococcal and staphylococcal genomes resulted in the identification of more than 20 new superantigens and the novel family of superantigen-like toxins (SSLs) comprising 14 members.

Genomic sequencing also led to the discovery of new pathogenicity islands and other mobile genetic elements and has increased our understanding of horizontal transfer of toxin genes.

Crystallographic structures of microbial toxins have given us a better insight into structure–function relationships and helped to address questions about host receptor recognition, host cell penetration and ligand interaction.

Importantly, the result of extensive research on microbial toxins has provided scientists with a 'toolkit' to decipher metabolic pathways in eukaryotic cells and mechanisms of pathogenicity.

The aim of this book is not to provide a broad overview of microbial toxins, but to highlight very recent achievements in toxin research. For example, microbial toxins such as subtilase cytotoxin and several immune evasion toxins of *Staphylococcus aureus* were only recently discovered and have not been extensively reviewed in the scientific literature. The nine chapters of this book are written by a panel of 25 international experts from Australia, Germany, New Zealand, Spain, the United Kingdom and the United States of America, who describe the latest insights from this rapidly expanding field of research.

Thomas Proft, PhD
Auckland, January 2009

Other Books of Interest

Caister Academic Press   www.caister.com

Toxins Encoded by Mobile Genetic Elements

1

José R. Penadés and J. Ross Fitzgerald

Abstract

The identification of accessory genetic elements (plasmids, bacteriophages, and 'pathogenicity islands') encoding virulence-associated genes has facilitated our efforts to understand the evolution of pathogenic microorganisms. In many cases, toxigenic bacteria including *Vibrio cholerae*, *Escherichia coli* and *Staphylococcus aureus* acquired virulence by acquisition of toxin genes carried in mobile genetic elements. In fact, mobile genetic elements have had a profound influence on the emergence of pathogenic clones of these bacteria. In order to trace the evolution of pathogens from their non-pathogenic progenitors, it is important to identify and characterize the genetic elements that mediate lateral acquisition of virulence genes. Understanding the evolutionary events that lead to the emergence of pathogenic clones may provide new approaches to the control of infectious diseases.

Core and adaptative genome

The idea that bacterial genomes within one single species can vary widely in gene content is not new. Determination of genome size by pulse-field gel electrophoresis in the 1980s and 1990s showed that representatives of the *Escherichia coli* ECOR collection had genomes which varied in size from 4.5 to 5.5 Mbp (Bergthorsson and Ochman, 1998). However, it was only with the advent of the genomic era that the phenomenon could be properly investigated. Not only was the genome size different, but a significant portion of each genome included genes not found in other strains.

The shared 'core' genome accounts for around only 40% of the total gene pool in *E. coli* and it is interrupted by multiple variable regions unique to some strains. These findings have contributed to a change in the concept of bacterial species in recent years (Konstantinidis and Tiedje, 2005a,b) which are more appropriately defined by the 'pan-genome', consisting of the core genome and an accessory genome consisting of partially shared and strain-specific genes found after examination of multiple strains of the species. The core genome is the essence of the phylogenetic unit of the species and is thought to be representative at various taxonomic levels (Ochman and Santos, 2005). The accessory (or adaptive) genome, on the other hand, includes key genes required for survival in a specific environment, and which are commonly linked to virulence, niche adaptation and resistance, typically reflecting the organism's predominant lifestyle or habitat.

The core genome is a good starting point for the identification of essential genes that might be useful as antibiotic targets and genes involved in host–pathogen interactions which are common to all pathogenic strains could be potential vaccine targets (Rappuoli and Nabel, 2001). These shared genes may also represent useful sequences for phylogenetic inference (Daubin *et al.*, 2003). A second consequence of the core–adaptive split is the identification of dispensable and non-dispensable genes. Normally, human pathogens contain relatively large genomes that include many paralogous gene families. The disruption of many of those genes and their deletion in species with a restricted lifestyle shows that they are non-essential (Pushker et al., 2004). For example, a simple model found in nature is given by obligate host-related bacteria (typically symbionts) with a minimalist genome composition: The term 'minimal genome' has been used to describe the set of genes that are required for a self-sustainable cell (Mushegian and Koonin, 1996).

The mobile bacterial genome

In addition to the broadly conserved set of genes that are responsible for the basic cellular processes of growth and multiplication, bacterial species contain a second facultative set that is concerned with adaptations to environmental contingencies (Frost et al., 2005). It has become evident in recent years that the facultative genome includes a large variety of mobile genetic elements (MGEs), and that this 'mobile genome' may exceed 10% of the total for many species and is a key contributor to the plasticity of the bacterial genome. MGEs commonly contribute to antibiotic resistance and pathogenesis. All known classes of bacterial MGEs, including temperate phages, plasmids, transposons and chromosomal islands, have been shown to be involved in pathogenesis. Of particular importance, it is striking that the majority of the bacterial toxins that cause specific toxin-mediated diseases (toxinoses) are encoded by MGEs (Novick, 2003). This includes diseases such as diphtheria, dysentery, toxic shock syndrome, food poisoning, necrotizing pneumonia, scalded skin syndrome, botulism, haemolytic–uraemic syndrome or necrotizing fasciitis. This raises a number of important questions, including 'What are the evolutionary forces responsible for this association?' and 'What are the selective advantages for the bacterial host and the MGE?'.

In this chapter, we will analyse the role of the MGE in the dissemination of virulence toxin genes among bacterial pathogens. Additionally, we will analyse the role of the SOS response in the physiology of the MGE-encoded toxinoses.

Mobile genetic elements and their role in virulence

Plasmids

In 1887 Robert Koch published the results of several experiments demonstrating that the causative agent of anthrax was the rod-shaped bacterium *Bacillus anthracis*. Approximately 100 years later it was established that this bacterium harboured two plasmids that were required for its virulence properties (Little and Ivins, 1999).

Plasmids are autonomous replicons in bacterial cells. Their genetic constitution reflects their function as gene exchange machines. Thus, they always contain genes required for replication, stability, DNA transfer and establishment in recipient cells. In addition, they carry genes with adaptive functions and others of unknown function. Classically, plasmids are covalently closed, circular double-stranded DNA molecules, but linear double-stranded DNA plasmids have been found in an increasing number of bacterial species (Frost et al., 2005). The role of plasmids in bacterial virulence is linked to the fact that toxins are often encoded by plasmids, which can be easily transferred to a recipient strain by several different mechanisms, including natural transformation, conjugation or transduction. Bacterial transformation is the process by which bacterial cells take up naked DNA molecules. If the foreign DNA has an origin of replication recognized by the host cell DNA polymerases, as the plasmids have, the bacteria will replicate the foreign DNA along with their own. Conjugation is defined as the unidirectional transfer of genetic information between cells by cell-to-cell contact. This latter requirement for contact distinguishes conjugation from transduction and transformation. The term 'unidirectional' refers to the fact that a copy of the plasmid is transferred from one cell, termed the 'donor', to another cell, termed the 'recipient'. One additional significance of the conjugation lies in the fact that many plasmids can also affect the transfer of chromosomal DNA, as exemplified by the high frequency of recombination (Hfr) mode of F plasmids and by the chromosome mobilization ability (Cma) of plasmids in *Streptomyces* species (Frost et al., 2005). Such conjugative elements integrate into the host genome and transfer large sections of the chromosome, along with parts of the conjugative element, into recipient cells. Finally, transduction is defined as the transfer of genetic information between cells through the mediation of a virus (phage) particle. It therefore does not require cell to cell contact.

Bacteriophages

Phages are the most abundant and the most rapidly replicating life forms on earth and their genetic diversity is enormous (Brussow and Hendrix, 2002; Hendrix, 2003). Commonly used as tools in genetic engineering, they have gained new attention for their potential for use in antibacterial therapy and in nanotechnology (Fischetti, 2001). The genomes of phages can be composed of either single- or double-stranded DNA or RNA and can range in size from a few to several 100 kb. Their characteristic essential genes encode replicases, phage components involved in hijacking the host cell replicative machinery, and proteins that package DNA in a protein coat (capsid). Virulent bacteriophages replicate vigorously and characteristically lyse the host bacteria. Temperate bacteriophages have an alternative, quiescent, non-lytic growth mode called lysogeny. A phage in the lysogenic state is called a prophage. In most known cases of lysogeny, the phage genome integrates into the bacterial chromosome and replicates with it but in a few cases the phage genome replicates autonomously as a circular or linear plasmid. Some prophages alter the phenotype of the host bacterium. These are called converting phages, and the process is known as lysogenic conversion. For example, if the prophage encodes a toxin, then the bacterium will be lysogenically converted for toxin production (positive lysogenic conversion). Alternatively, when lysogenization results in the loss of a particular phenotype, it is referred to as negative lysogenic conversion. For example, in *S. aureus*, there are at least two examples of negative lysogenic conversion resulting from direct integration of the phage into the gene sequence which interrupts β-haemolysin and lipase expression, respectively (Coleman *et al.*, 1986, 1989; Ye and Lee, 1989).

Phages are also involved in the transfer of bacterial genes that they do not directly encode. During the lytic phase, chromosomal or plasmid DNA can be accidentally packaged into phage heads. This DNA can then be transferred to a recipient strain, and incorporated into the bacterial genome. Horizontal transmission of a gene by this process is called generalized transduction, which occurs a low frequency. Finally, recombination between prophages and other mobile elements that reside in the same bacterial host contributes to the well-documented mosaic structure of phages (Hendrix *et al.*, 1999).

Transposable elements

Transposons are segments of DNA that have the capacity to move between locations in the genome by a process known as transposition. Unlike other process that reorganize DNA, transposition does not require extensive areas of homology between the transposon and its integration site. Transposable elements differ from phages in lacking a virus life cycle and from plasmids in being unable to replicate autonomously. The simplest transposable elements are insertion sequences or IS elements. An IS element is a short DNA sequence (around 1000–1500 bp in length) containing only the genes for those enzymes required for its transposition and bounded at both ends by identical or very similar sequences of nucleotides in reverse orientation knows as inverted repeats. The repeats are typically 15 to 25 bp long and vary in sequence so that each IS has its own characteristic inverted repeat. Between the inverted repeats is a gene that codes for an enzyme called transposase (and sometimes a gene for a resolvase). This enzyme is required for transposition as it recognizes the ends of the IS element. In addition to the genes involved in the transposition process, some transposable elements also contain genes involved in virulence or antibiotic resistance. These elements are called composite transposons consisting of a central region containing the additional genes flanked on both sides by IS elements that are identical or very similar in sequence. Many composite transposon are more simple in organization. They are bounded by short inverted repeats, and the coding region contains both transposition genes and the extra genes. It is believed that composite transposons are formed when two IS elements flank a central chromosomal region containing one or more genes.

The process of transposition in prokaryotes involves a series of events, including self-replication and recombinational processes. Typically, the original transposon remains at the parental site on the chromosome, while a replicated copy inserts at the 5 to 9 bp target DNA in a process known as replicative transposition which results in duplication of the target site.

Transposable elements contribute to a variety of effects on the host bacterium, including gene mutations (when the element is inserted in the middle of the gene) or rearrangements (deletions) of genetic material (between two of these elements). Additionally, transposons can affect the expression of the flanking genes situated near its insertion site, since some of these elements contain promoters.

Transposons are frequently associated with other MGEs, such as plasmids or chromosomal islands, and in so doing they use the transfer capacities of the MGEs. Transposons are important elements for the spread virulence factors (toxin genes) among bacterial species.

Chromosomal and pathogenicity islands

The chromosomal islands are the most recently identified of the MGEs and consequently are somewhat less well defined than the others. They were initially recognized as discrete chromosomal segments that contained virulence genes, lacked essential genes, and represented important genomic differences between closely related organisms that differed in pathogenicity and were accordingly labelled 'pathogenicity islands' (PTIs) (Hacker and Kaper, 2000; Dobrindt et al., 2000). PTIs, however, are a subset of a much broader family of inserted units, the genomic or chromosomal islands (Hacker and Kaper, 2000). The key feature of these elements is, of course, their transferability. However, not all laterally transferred DNA represents a genomic island, and the distinction is not always obvious. It seems important to have a clear view of what constitutes a genomic island as opposed to an accidentally transferred segment of DNA. It is suggested that the primary criterion must be evidence of active mobility.

The minimum requirements for mobility are (i) a site-specific recombination function; (ii) a pair of flanking repeats upon which it can act; and (iii) a means of intercell transfer, which need not be encoded by the island. To be considered a chromosomal island, a genetic unit must either possess one or more of these functions or it must possess vestiges of such functions. Alternatively,

it must closely resemble an established genomic island. Well-characterized mobile islands include the staphylococcal PTIs (SaPIs) (Novick and Subedi, 2007), the symbiosis island of *Rhizobia* (Sullivan and Ronson, 1998) and genomic islands of enteric bacteria known as CONSTINS (conjugative, self-transmissible, integrating elements) (Hochhut and Waldor, 1999), also known as ICE (integrative and conjugative elements) (Schubert et al., 2004) which possess a conjugation system as well as integration–excision capability. Interestingly, the latter have been known for many years as *incJ* plasmids (see (Novick, 1969)), but their true nature has been appreciated only recently. Many other genomic islands encode functional site-specific recombinases catalysing their excision and re-insertion including the SCCmec islands of staphylococci (Katayama et al., 2000), several islands of *E. coli* (Hochhut et al., 2006), and the island of *Shigella flexneri* (Sakellaris et al., 2004). The PAPI-1 island of *Pseudomonas aeruginosa* is the first of the large pathogenicity islands of Gram-negative bacteria for which inter-strain transfer has been demonstrated (Qiu et al., 2006).

The extent of the role of pathogenicity islands in bacterial virulence has only been realized since the genomic era. In most bacteria, PAIs encode many different functions, which largely depend on the environmental context in which the bacterium lives. The genetic repertoire found within PAIs can be functionally divided into several groups, including genes encoding adhesins, invasins, iron uptake systems, pore-forming toxins, proteins causing apoptosis, superantigens, secreted lipases, secreted proteases, O antigens, proteins transported by type I, III, IV and V secretion systems, and genes involved in antibiotic resistance. A detailed description of these PIs can be found in recent reviews (Schmidt and Hensel, 2004; Gal-Mor and Finlay, 2006).

Why MGEs encode exotoxins and other virulence factors

If a gene confers some benefit to the host bacterium, but an encoding prophage does not otherwise confer any additional advantage to the same bacterium, then over time we would expect evolution to favour deletion of the prophage sequences accompanying a given gene (Lawrence *et al.*, 2001). These deletions result in gene stabilization within a bacterial genome. However, in most cases, the MGE containing the virulence factor is fully functional, indicating that this functionality is beneficial for the bacteria.

From the bacterial perspective, MGE-encoded proteins represent a reservoir of additional (non-core) genes that enable populations to respond and adapt to new environmental conditions. This is the case for the pathogenicity islands of *Staphylococcus aureus* (SaPIs). SaPIs have been implicated in the pathogenesis and evolution of *S. aureus* including adaptation to new hosts (Novick and Subedi, 2007). From bovine *S. aureus* isolates, two different SaPIs have been characterized (Fitzgerald *et al.*, 2001; Ubeda *et al.*, 2003). In addition to a central core region involved in replication and transfer of these elements (Ubeda *et al.*, 2008), both islands share the same integrase, implying that both islands compete to integrate at the same bacterial att_B site. SaPIbov1, identified in an isolate associated with clinical mastitis carried the toxin genes *tst*, *sel* and *sec*. In contrast, SaPIbov2, identified in an isolate from subclinical mastitis, substituted the toxin genes with a transposon containing *bap*, a gene involved in biofilm formation (Cucarella *et al.*, 2004).

In addition to the horizontal gene transfer between different strains of the same species, toxins encoded by MGEs can be transferred between different species. A recent example of interspecies dissemination involved the transfer of vancomycin resistance between *Enterococcus faecalis* and *S. aureus*. A clinical isolate of *S. aureus* with high-level resistance to vancomycin was isolated in June 2002. This isolate harboured a 57.9-kb multiresistance conjugative plasmid within which Tn1546 (*vanA*) was integrated. Additional elements on the plasmid encoded resistance to trimethoprim (*dfr*A), β-lactams (*bla*Z), aminoglycosides (*aac*A–*aph*D), and disinfectants (*qac*C). Genetic analyses suggested that the long-anticipated transfer of vancomycin resistance to a methicillin-resistant *S. aureus* occurred *in vivo* by interspecies transfer of Tn1546 from a co-isolate of *E. faecalis* (Weigel *et al.*, 2003).

Additionally, it is important to note that the evolution of MGEs is driven by selective forces that operate to maintain the functionality of these elements, including maintenance of their capabilities to replicate and be transferred. Accordingly, environmental stimuli such as DNA damaging agents that induce the bacterial SOS response compromising the viability of the cells activate the transfer of MGEs. Therefore, the association of virulence factors with 'functional' elements will increase their presence in the bacterial pool genome. Furthermore, virulence genes encoded by PIs or phages may withstand environmental exposure better than those encoded by bacterial chromosomes. For example, Muniesa *et al.* found that Stx2-encoding phages persisted in river water longer and were more resistant to chlorination and pasteurization than Stx2-encoding bacteria, suggesting that phage particles serve as a more durable environmental reservoir of the stx_2 genes than bacteria (Muniesa *et al.*, 1999).

Role of SOS induction

The SOS system, first described and thoroughly studied in *Escherichia coli* (Walker, 1984), is a global response designed to guarantee cell survival when massive DNA damage is introduced and the normal DNA replication of the bacterial cell is disturbed. This network is the prototypic cell cycle check-point control and DNA repair system and because of this, a detailed picture of the signal transduction pathway that regulates this response is required. A central part of the SOS response is the de-repression of more than 40 genes under the direct and indirect transcriptional control of the RecA and LexA proteins, which are also member of this regulon (Khil and Camerini-Otero, 2002). The LexA protein is the repressor of the system through its specific binding to the regulatory motifs present in the promoter region of the SOS genes. This regulatory motif, commonly known as the LexA box, has in *E. coli* the $CTGTN_8ACAG$ motif as the consensus sequence (Walker, 1984).

The signal transduction pathway leading to an SOS response ensues when RecA protein binds to single stranded DNA (ssDNA), which can be created by processing of DNA damage or stalled replication (Little and Mount, 1982; Walker, 1984; Miller *et al.*, 2004). This binding activates an otherwise dormant co-protease activity of RecA, which facilitates the proteolytic self-cleavage of the LexA repressor (Little, 1991) to trigger the expression of DNA repair genes. Once DNA lesions have been repaired, RecA ceases to be activated and non-cleaved LexA protein returns to its normal levels, repressing again the transcription of the SOS genes. The LexA gene is widespread among bacteria and is present in most phylogenetic groups for which different monophyletic LexA-binding motifs have been described including the GAACN7GTTC motif of Gram-positive bacteria.

The importance of the SOS response in the biology of the MGEs has emerged in recent years. Activated RecA also facilitates the auto-cleavage of phage repressors, which maintain the lysogenic state of temperate bacteriophages (Little, 1993).

Although the frequency of environmental SOS induction by typical DNA-damaging agents, such as radiation and certain chemicals, is probably not very high (though there does not seem to be much literature on this issue), a class of SOS-inducing compounds, the fluoroquinolone antibiotics, is in wide use with serious clinical implications. For example, Shiga toxin is encoded by a prophage and its gene is SOS-induced along with rest of the phage genome (Zhang et al., 2000). Treatment of the Shiga-toxin dependent haemolytic–uraemic syndrome (caused by E. coli H57:0157) with fluoroquinolones increases disease severity, sometimes fatally, as well as amplifying the population of phages encoding Shiga toxin (Muniesa et al., 1999; Wong et al., 2000; Zhang et al., 2000). Similarly, we demonstrated that the fluoroquinolone antibiotic ciprofloxacin induced staphylococcal prophages and any co-resident SaPI whose replication is controlled by the resident prophage which is strongly predicted to promote spread of the SaPI (Ubeda et al., 2005). Although there is no information on whether SOS induction increases the production of any SaPI-encoded superantigen, potentially worsening the associated clinical condition (e.g., TSS), it is highly likely that the increase in gene dosage will have this effect, whether or not transcription of the superantigen gene is under direct SOS induction control.

A third example of clinically adverse SOS induction by antibiotics is that of the integrating conjugative elements (ICEs), mobile elements that are transferred by conjugation and integrate site-specifically into the recipient chromosome. The prototype, SXT, is a ~100-kilobase V. cholerae ICE that carries resistance to chloramphenicol, sulphamethoxazole, trimethoprim and streptomycin (Beaber et al., 2004). SXT-related elements were not detected in V. cholerae before 1993 but are now present in almost all clinical V. cholerae isolates from Asia. ICEs related to SXT are also present in several other bacterial species and encode a variety of antibiotic and heavy metal resistance genes. Beaber and co-workers have shown that an SXT-encoded protein, SetR, represses SXT transfer and that this repression is relieved by SOS induction, stimulating transfer, with the implication that this has contributed to the dissemination of antibiotic resistance in Asian V. cholerae in recent years (Beaber et al., 2004). Thus, therapeutic agents can promote the spread of antibiotic resistance genes as well as virulence determinants.

It is suggested, in conclusion, that SOS induction by antibiotics of clinically important processes, such as virulence gene expression and transfer and antibiotic resistance transfer is likely to be of much wider importance than was previously thought.

Bacteriophage-encoding toxins

Many key virulence factors, including diphtheria toxin, Shiga toxin, cholera toxin, and several staphylococcal toxins, are encoded by genes found in the genomes of lysogenic bacteriophages (Wagner and Waldor, 2002; Brussow et al., 2004; Waldor and Friedman, 2005). In addition, recent studies with Shiga toxin (Stx)-encoding phages in E. coli and staphylococcal enterotoxin-encoding phages in S. aureus revealed that prophage induction can provide a mechanism to control toxin production. Here we provide a brief description of some of these processes that will not be described in other chapters:

Shiga toxin (Stx)-encoding phages

Shiga toxin producing Escherichia coli (STEC) are considered food- and water-borne pathogens, although person-to-person transmission has also been reported. STEC cause bloody diarrhoea and haemolytic–uraemic syndrome and can lead to severe complications. One of the best-known serotypes is O157:H7, which is the most common virulent serogroup in the USA and Canada, while other serotypes are frequently reported in European outbreaks. STEC are normally found in cattle, goat and sheep, which act as the reservoir, releasing STEC in their faeces.

Shiga toxins are the main virulence factors made by STEC, and are homologous to the toxin made by *Shigella dysenteriae* serotype I (O'Brien *et al.*, 1984). Currently two Stx have been described, Stx1 and Stx2 in addition to their variants (Jaeger and Acheson, 2000) which are only found in certain reservoirs only (Fraser *et al.*, 2004). Generally, Shiga toxins are characterized by their hexameric conformation, comprising five B subunits, which allow toxin attachment to their enterocyte receptor; Gb3, which is also present in endothelial cells of glomerular capillaries (Hoey *et al.*, 2002), and one A subunit, which is catalytically active and blocks translation of mRNA to protein, leading to cell death (Fraser *et al.*, 2004).

Shiga toxins are encoded in the genome of temperate phages (Schmidt, 2001) considered members of the lambdoid family since they share a common genome arrangement that conserves the relative positions of the genes with similar activities and associated regulatory signals (Campbell, 1994). Although Shiga toxin-encoding bacteriophages are a heterogeneous group, in terms of morphology and their genetic organization (Muniesa *et al.*, 2004; Serra-Moreno *et al.*, 2007), *stx* genes location is conserved, being found next to the lytic genes and downstream of the *Q* anti-terminator (O'Brien *et al.*, 1984). Therefore, Shiga toxin production is linked to the induction or progression of the phage lytic cycle, after activation of the SOS response in the bacterial host. Several studies have showed that treatment of STEC with SOS-inducer molecules, including mitomycin-C or quinolone antibiotics, increased Stx expression (Acheson and Donohue-Rolfe, 1989; Zhang *et al.*, 2000) and treatment of human STEC infection with bacteriophage-inducing antibiotics, such as fluoroquinolones, may have significant adverse clinical consequences and enhance the spread of virulence factors *in vivo*. After phage induction, the toxin is released through cell lysis (Wagner *et al.*, 1999). Following the lytic burst, Stx phages act as vectors in the *stx* horizontal transmission to other *Enterobacteriaceae* (Acheson *et al.*, 1998; Schmidt *et al.*, 1999).

In addition to the direct control of toxin expression by the phage, previous studies with STEC O157:H7 strains isolated from a single outbreak showed that isolates harbouring two different Stx prophages produced less toxin than strains from the same clone carrying only one Stx prophage (Muniesa *et al.*, 2003), suggesting that the expression of phage genes may be regulated when other temperate phages are present. In a recent study, Serra-Moreno and Muniesa demonstrated that the CI repressors of both prophages operating *in trans* could regulate the reduced production of the Stx toxin when a bacterium contains two prophages (Serra-Moreno *et al.*, 2008). The authors hypothesized that, although the sequence of the *cI* genes of the phages studied differed, the CI protein conformation was conserved. Consequently, the presence of more than one prophage in the host chromosome could be regarded as a mechanism to allow genetic retention in the cell, by reducing the activation of lytic cycle and hence the pathogenicity of the strains (Serra-Moreno *et al.*, 2008).

Staphylococcal phages

Bacteriophages of *S. aureus* encode many clinically relevant virulence factors, including toxins. For example, a recently emerged community-acquired epidemic strain is responsible for rapidly progressive, fatal diseases including necrotizing pneumonia, severe sepsis and necrotizing fasciitis. In addition to novel methicillin resistance genetic cassettes, these strains harbour a phage encoding Panton–Valentine leukocidin (PVL) which has been controversially implicated in the increased virulence of these clones (Labandeira-Rey *et al.*, 2007). Other known phage-encoded virulence factors include exfoliative toxin type A, staphylokinase, and staphylococcal enterotoxins (Yamaguchi *et al.*, 2000; Baba *et al.*, 2002; Sumby and Waldor, 2003).

Although the presence of virulence factors in different *S. aureus* phages was reported two decades ago by Betley and Mekalanos, who demonstrated that the gene for staphylococcal enterotoxin A (*sea*, formerly *entA*) is carried by related temperate bacteriophages. (Betley and Mekalanos, 1985), it is not well known whether the life cycles of the phages influence the expression of the virulence genes and thus *S. aureus* pathogenicity. In a pioneer study, and using as a model φSa3ms, a lysogenic bacteriophage encoding the staphylococcal enterotoxins SEA, SEG, and SEK and the fibrinolytic enzyme staphylokinase (Sak), Sumby and Waldor demonstrated that upon φSa3ms prophage induction, transcription of all four virulence factors were greatly increased (Sumby and Waldor, 2003). Interestingly, while the increase in *sea* and *sak* transcription was a result of read-through transcription from upstream latent phage promoters and an increase in phage copy number, the majority of the *seg2* and *sek2* transcripts were shown to initiate from the upstream phage *cI* promoter and hence were regulated by factors influencing *cI* transcription (Sumby and Waldor, 2003), suggesting that the production of phage-encoded virulence factors in *S. aureus* may be regulated by processes that govern lysogeny.

In addition to this work, several studies in *S. aureus* have described an antibiotic-induced SOS response that affects virulence by modulating phages. For example, Goerke and co-workers analysed the effects of ciprofloxacin and trimethoprim on phage induction and expression of phage-encoded virulence factors. Treatment of lysogens with subinhibitory concentrations of either antibiotic resulted in replication of phages in the bacterial host linked to elevated expression of the phage-encoded virulence genes, chiefly due to the activation of latent phage promoters (Goerke *et al.*, 2006).

The collective conclusion from these and other studies is that phages, together with DNA damaging agents that induce the phages, can promote enhanced horizontal gene transfer of phage-encoded toxins, as well as the expression of these factors, which can alter the course of a clinical infection.

Plasmids encoding toxins

Several well-known diseases are caused by toxin-producing clostridia; for example, gas gangrene and necrotic enteritis are caused by *Clostridium perfringens*, diarrhoea and pseudomembranous colitis are caused by *Clostridium difficile*, tetanus disease is caused by *Clostridium tetani*, and food-borne botulism is caused by *Clostridium botulinum*. The last two organisms produce the most powerful neurotoxins known to mankind, the tetanus (TeTX) and the botulinum toxin (BoNT), respectively (Bruggemann, 2005). Whereas the *tetX* gene can only be found on large plasmids in C. *tetani* (see below), different toxinotypes of BoNT exist (A–G), the genes of which are located on the chromosome (C. *botulinum* A, B, E and F; *Clostridium butyricum*), on plasmids (*Clostridium argentinense*) or on bacteriophages (C. *botulinum* C and D) (Raffestin *et al.*, 2004; Bruggemann, 2005). Some strains can produce a mixture of two BoNTs and many type A strains contain cryptic or silent BoNT type B genes. There is strong evidence that the BoNT gene loci are located on (degenerated) mobile genetic elements, which accounts for their presence on chromosomes, plasmids and phages as well as their probable transfer among different clostridial strains in evolutionary history (Dineen *et al.*, 2003).

As mentioned, The TeTx is encoded on a large plasmid and to date has only been found in toxigenic strains of C. *tetani* (Finn *et al.*, 1984). Its genome sequence has been determined for the Massachussetts derivative strain E88. The plasmid pE88 is 74 kb in size with a low G+C content of only 24.5% (Bruggemann, 2003; Bruggemann *et al.*, 2005). It encodes 61 genes, which cover 67% of the plasmid sequence. An additional virulence factor, a collagenase similar to the k toxin of C. *perfringens* is encoded on pE88. The only other gene of known function is *tetR*, which encodes a positive regulator of TeTx and is located directly upstream of *tetX*. Additional regulatory genes found on pE88 include three sigma factor-like proteins (CTP05, CTP10 and CTP11) and a two-component system with unknown regulatory function (CTP21 and CTP22). A large portion of genes on pE88 code for transport proteins: five multi-subunit ATP-binding cassette (ABC) transporters can be found, some of which show highest homology to peptide transporters responsible for bacteriocin efflux (Bruggemann *et al.*, 2003).

It has been shown that non-toxigenic C. *tetani* strains either completely lack plasmids or contain plasmids of different sizes. Recently, the plasmid of C. *tetani* strain E4222, a non-toxigenic variant, was sequenced (Bruggemann,

2005). It is 47 kb in size with a G+C content of 28.8% (slightly higher than that of pE88) encoding 63 genes, 30% of which show similarity to bacteriophage-related genes but no homology with plasmid pE88 was detected. Finally, it is important to remark that although the role of the plasmids in the *C. tetani* virulence is well established, no further studies have reported the mechanisms involved in the their horizontal transfer nor the environmental stimuli that trigger this process.

Pathogenicity islands encoding toxins

The term pathogenicity island was coined by Hacker *et al.* to describe two large unstable regions on the chromosome of uropathogenic *E. coli* (Blum *et al.*, 1994). Currently, this term is commonly used to describe regions in the genomes of certain pathogens that are absent is clonally related non-pathogenic strains. PIs have not only relevance as a repertoire of virulence factors, including toxins, but also changed our way of thinking about the evolution of bacterial pathogenicity.

SaPIs

Staphylococcus aureus pathogenicity islands are a family of related 15–17 kb mobile genetic elements that commonly carry genes for superantigen toxins and other virulence factors. SaPIs were the first pathogenicity islands described for any Gram-positive species and the first pathogenicity islands for which mobility has been demonstrated (Lindsay *et al.*, 1998; Ubeda *et al.*, 2005). They are, at present, probably the best characterized of any of the bacterial pathogenicity islands (Ubeda *et al.*, 2007a; Ubeda *et al.*, 2007b; Ubeda *et al.*, 2008). Complete sequences are known for 16 SaPIs, which form a highly coherent family with conserved functional and genetic organization. The SaPIs are biologically analogous to coliphage P4 (Lindqvist *et al.*, 1993) and to *Sulfolobus* plasmid pSSVx (Arnold *et al.*, 1999). The key feature of their mobility and spread is the induction by certain phages of their excision, replication and efficient encapsidation into specific small-headed phage-like infectious particles (Lindsay *et al.*, 1998; Ruzin *et al.*, 2001; Ubeda *et al.*, 2005). This sequence of events is referred to as the SaPI

excision-replication-packaging (ERP) cycle and is the result of an intimate interaction between phage and SaPI genomes, in which SaPIs divert key phage functions for its own ends. Since the SaPI uses phage proteins for particle formation and packaging (Tallent *et al.*, 2007; Tormo *et al.*, 2008), it initiates its own replication cycle only when an active prophage provides the necessary inducing/derepressing functions (Ubeda *et al.*, 2008). SaPIs encode an excisionase (*xis*), a site-specific integrase and proteins that initiate replication from its own specific replication origin (Ubeda *et al.*, 2007a). Other proteins are involved in remodelling the phage capsid to fit its smaller genome, and a terminase small subunit is encoded that recognizes its specific *pac* site, enabling the packaging of SaPI DNA via the phage terminase large subunit (LTS) and portal protein (Ubeda *et al.*, 2007b). The phage is responsible for lysis. SaPIs are identifiable by several universal features including a directly repeated *att* site core, a specific integrase, a terminase small subunit (but never a large subunit or portal protein), a replication initiator protein and an adjacent replication origin, and a key site of divergent transcription. SaPIs are very widespread among the staphylococci, and using these criteria we have very recently identified them in several other Gram-positive species (unpublished results).

Clinical significance

Staphylococcal superantigens have been found, thus far, only in association with mobile genetic elements, including SaPIs, temperate phages and plasmids. The biological basis for this is unclear, but is consistent with the observation that superantigens in other organisms are also carried by mobile genetic elements. The toxin most frequently encoded by SaPIs is toxic shock syndrome toxin-1 (TSST-1), which has been found, thus far, associated only with SaPIs and are exclusively responsible for menstrual TSS. Similarly, SEB has been found exclusively in association with SaPIs, which are responsible for a considerable fraction of non-menstrual TSS and food poisoning cases. SEB is also lethal in aerosol form and is listed as a Select Agent by the Department of Homeland Security (USA). Notably, TSST-1 appears to have a role in pathogenicity independently of toxic shock – it contributes importantly to morbidity and mortality in mice, which are not susceptible to TSS. Many of the SaPIs encode two or three superantigen toxins. Therefore, SaPIs may be regarded as major players in staphylococcal pathogenesis, perhaps in addition to the superantigenicity of their toxins. In this context SaPIbov2 contains *bap*, the gene for a ~240 kDa adhesin that is involved in biofilm formation and has a significant role in bovine mastitis (Cucarella *et al.*, 2001; Cucarella *et al.*, 2002; Cucarella *et al.*, 2004).

Expression of the toxin genes is increased during SaPI replication owing to the increase in gene dosage (unpublished data). Toxin production could also be specifically induced during SaPI replication if the gene is driven by a replication-induced promoter, as is the case with the phage-coded Shiga toxin (Zhang *et al.*, 2000). SOS induction caused by fluoroquinolones, β-lactams and probably other antibiotics therefore has the unfortunate side effect of inducing toxin synthesis, with adverse clinical consequences (Zhang *et al.*, 2000), as well as the production of phage or SaPI particles (Ubeda *et al.*, 2005; Maiques *et al.*, 2006), which would promote transfer of the toxin genes.

As described below, the vast majority of *S. aureus* strains carry one or more SaPIs, so that the SaPIs and their mobility constitute a significant feature of staphylococcal virulence. The occurrence of similar or identical SaPIs in unrelated strains suggests strongly that transfer occurs under natural circumstances as well as in the laboratory. An improved understanding of SaPI biology, genetics and dissemination will greatly aid in the epidemiological tracking and control of superantigen- and other SaPI-mediated diseases.

Open questions

Only recently the significant link between mobile genetics elements, genome diversity, and bacterial pathogenesis has been fully appreciated. The renewed interest in the biology of the MGEs has been a consequence of the complete genome sequencing of multiple and varied bacterial genomes. The results of comparative bacteria genomics, along with the discovery of novel MGEs, have allowed understanding of the role of these elements in bacterial physiology. Moreover, we are only just beginning to understand the complex relationship between the MGE's life cycle, the environmental condition that induces their transfer, and their role in the physiology of the recipient strain. Continued investigation of the biology of these elements will reveal much about the role that the MGEs play in bacterial pathogenesis.

Web resources

http://mml.sjtu.edu.cn/MobilomeFINDER/
For identification of bacterial strains rich in novel genetic material and for high-throughput genomic island discovery.
http://www.ispb.org/
International Society for Plasmid Biology and other Mobile Genetic Elements
http://kementari.bioinformatics.vt.edu/cgi-bin/islander.cgi
Islander Database of Genomic Islands
http://www.genomesonline.org/gold.cgi
Genomes online database

References

Acheson, D.W., and Donohue-Rolfe, A. (1989). Cancer-associated hemolytic uremic syndrome: a possible role of mitomycin in relation to Shiga-like toxins. J. Clin. Oncol. 7, 1943.

Acheson, D.W., Reidl, J., Zhang, X., Keusch, G.T., Mekalanos, J.J., and Waldor, M.K. (1998). In vivo transduction with Shiga toxin 1-encoding phage. Infect. Immun. 66, 4496–4498.

Arnold, H.P., She, Q., Phan, H., Stedman, K., Prangishvili, D., Holz, I., Kristjansson, J. K., Garrett, R., and Zillig, W. (1999). The genetic element pSSVx of the extremely thermophilic crenarchaeon *Sulfolobus* is a hybrid between a plasmid and a virus. Mol. Microbiol. 34, 217–226.

Baba, T., Takeuchi, F., Kuroda, M., Yuzawa, H., Aoki, K., Oguchi, A., Nagai, Y., Iwama, N., Asano, K., Naimi, T., *et al.* (2002). Genome and virulence determinants of high virulence community-acquired MRSA. Lancet 359, 1819–1827.

Beaber, J.W., Hochhut, B., and Waldor, M.K. (2004). SOS response promotes horizontal dissemination of antibiotic resistance genes. Nature 427, 72–74.

Bergthorsson, U., and Ochman, H. (1998). Distribution of chromosome length variation in natural isolates of *Escherichia coli*. Mol. Biol. Evol. 15, 6–16.

Betley, M.J., and Mekalanos, J.J. (1985). Staphylococcal enterotoxin A is encoded by phage. Science 229, 185–187.

Blum, G., Ott, M., Lischewski, A., Ritter, A., Imrich, H., Tschape, H., and Hacker, J. (1994). Excision of large DNA regions termed pathogenicity islands from tRNA-specific loci in the chromosome of an *Escherichia coli* wild-type pathogen. Infect. Immun. 62, 606–614.

Bruggemann, H. (2005). Genomics of clostridial pathogens: implication of extrachromosomal elements in pathogenicity. Curr. Opin. Microbiol. *8*, 601–605.

Bruggemann, H., Baumer, S., Fricke, W.F., Wiezer, A., Liesegang, H., Decker, I., Herzberg, C., Martinez-Arias, R., Merkl, R., Henne, A., and Gottschalk, G. (2003). The genome sequence of *Clostridium tetani*, the causative agent of tetanus disease. Proc. Natl. Acad. Sci. USA *100*, 1316–1321.

Brussow, H., Canchaya, C., and Hardt, W.D. (2004). Phages and the evolution of bacterial pathogens: from genomic rearrangements to lysogenic conversion. Microbiol. Mol. Biol. Rev. *68*, 560–602.

Brussow, H., and Hendrix, R.W. (2002). Phage genomics: small is beautiful. Cell *108*, 13–16.

Campbell, A. (1994). Comparative molecular biology of lambdoid phages. Annu. Rev. Microbiol. *48*, 193–222.

Coleman, D.C., Arbuthnott, J.P., Pomeroy, H.M., and Birkbeck, T.H. (1986). Cloning and expression in *Escherichia coli* and *Staphylococcus aureus* of the beta-lysin determinant from *Staphylococcus aureus*: evidence that bacteriophage conversion of beta-lysin activity is caused by insertional inactivation of the beta-lysin determinant. Microb. Pathog. *1*, 549–564.

Coleman, D.C., Sullivan, D.J., Russell, R.J., Arbuthnott, J.P., Carey, B.F., and Pomeroy, H.M. (1989). *Staphylococcus aureus* bacteriophages mediating the simultaneous lysogenic conversion of beta-lysin, staphylokinase and enterotoxin A: molecular mechanism of triple conversion. J. Gen. Microbiol. *135*, 1679–1697.

Cucarella, C., Solano, C., Valle, J., Amorena, B., Lasa, I., and Penades, J.R. (2001). Bap, a *Staphylococcus aureus* surface protein involved in biofilm formation. J. Bacteriol. *183*, 2888–2896.

Cucarella, C., Tormo, M. A., Knecht, E., Amorena, B., Lasa, I., Foster, T.J., and Penades, J.R. (2002). Expression of the Biofilm-Associated Protein Interferes with Host Protein Receptors of *Staphylococcus aureus* and Alters the Infective Process. Infect. Immun. *70*, 3180–3186.

Cucarella, C., Tormo, M.A., Ubeda, C., Trotonda, M.P., Monzon, M., Peris, C., Amorena, B., Lasa, I., and Penades, J.R. (2004). Role of Biofilm-Associated Protein Bap in the Pathogenesis of Bovine *Staphylococcus aureus*. Infect. Immun. *72*, 2177–2185.

Daubin, V., Moran, N.A., and Ochman, H. (2003). Phylogenetics and the cohesion of bacterial genomes. Science *301*, 829–832.

Dineen, S.S., Bradshaw, M., and Johnson, E.A. (2003). Neurotoxin gene clusters in *Clostridium botulinum* type A strains: sequence comparison and evolutionary implications. Curr. Microbiol. *46*, 345–352.

Dobrindt, U., Janke, B., Piechaczek, K., Nagy, G., Ziebuhr, W., Fischer, G., Schierhorn, A., Hecker, M., Blum-Oehler, G., and Hacker, J. (2000). Toxin genes on pathogenicity islands: impact for microbial evolution. Int. J. Med. Microbiol. *290*, 307–311.

Finn, C.W., Jr., Silver, R.P., Habig, W.H., Hardegree, M.C., Zon, G., and Garon, C.F. (1984). The structural gene for tetanus neurotoxin is on a plasmid. Science *224*, 881–884.

Fischetti, V.A. (2001). Phage antibacterials make a comeback. Nat. Biotechnol. *19*, 734–735.

Fitzgerald, J.R., Monday, S.R., Foster, T.J., Bohach, G.A., Hartigan, P.J., Meaney, W.J., and Smyth, C.J. (2001). Characterization of a putative pathogenicity island from bovine *Staphylococcus aureus* encoding multiple superantigens. J. Bacteriol. *183*, 63–70.

Fraser, M.E., Fujinaga, M., Cherney, M.M., Melton-Celsa, A.R., Twiddy, E.M., O'Brien, A.D., and James, M.N. (2004). Structure of Shiga toxin type 2 (Stx2) from *Escherichia coli* O157:H7. J. Biol. Chem. *279*, 27511–27517.

Frost, L.S., Leplae, R., Summers, A.O., and Toussaint, A. (2005). Mobile genetic elements: the agents of open source evolution. Nat. Rev. Microbiol. *3*, 722–732.

Gal-Mor, O., and Finlay, B.B. (2006). Pathogenicity islands: a molecular toolbox for bacterial virulence. Cell. Microbiol. *8*, 1707–1719.

Goerke, C., Koller, J., and Wolz, C. (2006). Ciprofloxacin and trimethoprim cause phage induction and virulence modulation in *Staphylococcus aureus*. Antimicrob. Agents Chemother. *50*, 171–177.

Hacker, J., and Kaper, J.B. (2000). Pathogenicity islands and the evolution of microbes. Annu. Rev. Microbiol. *54*, 641–679.

Hendrix, R.W. (2003). Bacteriophage genomics. Curr. Opin. Microbiol. *6*, 506–511.

Hendrix, R.W., Smith, M.C., Burns, R.N., Ford, M.E., and Hatfull, G.F. (1999). Evolutionary relationships among diverse bacteriophages and prophages: all the world's a phage. Proc. Natl. Acad. Sci. USA *96*, 2192–2197.

Hochhut, B., and Waldor, M.K. (1999). Site-specific integration of the conjugal *Vibrio cholerae* SXT element into *prfC*. Mol. Microbiol. *32*, 99–110.

Hochhut, B., Wilde, C., Balling, G., Middendorf, B., Dobrindt, U., Brzuszkiewicz, E., Gottschalk, G., Carniel, E., and Hacker, J. (2006). Role of pathogenicity island-associated integrases in the genome plasticity of uropathogenic *Escherichia coli* strain 536. Mol. Microbiol. *61*, 584–595.

Hoey, D.E., Currie, C., Else, R.W., Nutikka, A., Lingwood, C.A., Gally, D.L., and Smith, D.G. (2002). Expression of receptors for Verotoxin 1 from *Escherichia coli* O157 on bovine intestinal epithelium. J. Med. Microbiol. *51*, 143–149.

Jaeger, J.L., and Acheson, D.W. (2000). Shiga Toxin-Producing *Escherichia coli*. Curr. Infect. Dis. Rep. *2*, 61–67.

Katayama, Y., Ito, T., and Hiramatsu, K. (2000). A new class of genetic element, staphylococcus cassette chromosome mec, encodes methicillin resistance in *Staphylococcus aureus*. Antimicrob. Agents Chemother. *44*, 1549–1555.

Khil, P.P., and Camerini-Otero, R.D. (2002). Over 1000 genes are involved in the DNA damage response of *Escherichia coli*. Mol. Microbiol. *44*, 89–105.

Konstantinidis, K.T., and Tiedje, J.M. (2005a). Genomic insights that advance the species definition for prokaryotes. Proc. Natl. Acad. Sci. USA *102*, 2567–2572.

Konstantinidis, K.T., and Tiedje, J.M. (2005b). Towards a genome-based taxonomy for prokaryotes. J. Bacteriol. 187, 6258–6264.

Labandeira-Rey, M., Couzon, F., Boisset, S., Brown, E.L., Bes, M., Benito, Y., Barbu, E.M., Vazquez, V., Hook, M., Etienne, J., et al. (2007). Staphylococcus aureus Panton-Valentine leukocidin causes necrotizing pneumonia. Science 315, 1130–1133.

Lawrence, J.G., Hendrix, R.W., and Casjens, S. (2001). Where are the pseudogenes in bacterial genomes? Trends Microbiol. 9, 535–540.

Lindqvist, B.H., Deho, G., and Calendar, R. (1993). Mechanisms of genome propagation and helper exploitation by satellite phage P4. Microbiol. Rev. 57, 683–702.

Lindsay, J.A., Ruzin, A., Ross, H.F., Kurepina, N., and Novick, R.P. (1998). The gene for toxic shock toxin is carried by a family of mobile pathogenicity islands in Staphylococcus aureus. Mol. Microbiol. 29, 527–543.

Little, J.W. (1991). Mechanism of specific LexA cleavage: autodigestion and the role of RecA coprotease. Biochimie 73, 411–421.

Little, J.W. (1993). LexA cleavage and other self-processing reactions. J. Bacteriol. 175, 4943–4950.

Little, J.W., and Mount, D.W. (1982). The SOS regulatory system of Escherichia coli. Cell 29, 11–22.

Little, S.F., and Ivins, B.E. (1999). Molecular pathogenesis of Bacillus anthracis infection. Microbes Infect. 1, 131–139.

Maiques, E., Ubeda, C., Campoy, S., Salvador, N., Lasa, I., Novick, R.P., Barbe, J., and Penades, J.R. (2006). {beta}-lactam antibiotics induce the SOS response and horizontal transfer of virulence factors in Staphylococcus aureus. J. Bacteriol. 188, 2726–2729.

Miller, C., Thomsen, L.E., Gaggero, C., Mosseri, R., Ingmer, H., and Cohen, S. N. (2004). SOS response induction by beta-lactams and bacterial defense against antibiotic lethality. Science 305, 1629–1631.

Muniesa, M., Blanco, J.E., De Simon, M., Serra-Moreno, R., Blanch, A.R., and Jofre, J. (2004). Diversity of stx2 converting bacteriophages induced from Shiga-toxin-producing Escherichia coli strains isolated from cattle. Microbiology 150, 2959–2971.

Muniesa, M., de Simon, M., Prats, G., Ferrer, D., Panella, H., and Jofre, J. (2003). Shiga toxin 2-converting bacteriophages associated with clonal variability in Escherichia coli O157:H7 strains of human origin isolated from a single outbreak. Infect. Immun. 71, 4554–4562.

Muniesa, M., Lucena, F., and Jofre, J. (1999). Comparative survival of free Shiga toxin 2-encoding phages and Escherichia coli strains outside the gut. Appl. Environ. Microbiol. 65, 5615–5618.

Mushegian, A.R., and Koonin, E.V. (1996). A minimal gene set for cellular life derived by comparison of complete bacterial genomes. Proc. Natl. Acad. Sci. USA 93, 10268–10273.

Novick, R.P. (1969). Extrachromosomal inheritance in bacteria. Bacteriol. Rev. 33, 210–263.

Novick, R.P. (2003). Mobile genetic elements and bacterial toxinoses: the superantigen-encoding pathogenicity islands of Staphylococcus aureus. Plasmid 49, 93–105.

Novick, R.P., and Subedi, A. (2007). The SaPIs: mobile pathogenicity islands of Staphylococcus. Chem. Immunol. Allergy 93, 42–57.

O'Brien, A.D., Newland, J.W., Miller, S.F., Holmes, R.K., Smith, H.W., and Formal, S. B. (1984). Shiga-like toxin-converting phages from Escherichia coli strains that cause hemorrhagic colitis or infantile diarrhea. Science 226, 694–696.

Ochman, H., and Santos, S.R. (2005). Exploring microbial microevolution with microarrays. Infect. Genet. Evol. 5, 103–108.

Pushker, R., Mira, A., and Rodriguez-Valera, F. (2004). Comparative genomics of gene-family size in closely related bacteria. Genome Biol. 5, R27.

Qiu, X., Gurkar, A.U., and Lory, S. (2006). Interstrain transfer of the large pathogenicity island (PAPI-1) of Pseudomonas aeruginosa. Proc. Natl. Acad. Sci. USA 103, 19830–19835.

Raffestin, S., Marvaud, J.C., Cerrato, R., Dupuy, B., and Popoff, M.R. (2004). Organization and regulation of the neurotoxin genes in Clostridium botulinum and Clostridium tetani. Anaerobe 10, 93–100.

Rappuoli, R., and Nabel, G. (2001). Vaccines: ideal drugs for the 21st century? Curr. Opin. Investig. Drugs 2, 45–46.

Ruzin, A., Lindsay, J., and Novick, R.P. (2001). Molecular genetics of SaPI1 – a mobile pathogenicity island in Staphylococcus aureus. Mol. Microbiol. 41, 365–377.

Sakellaris, H., Luck, S.N., Al-Hasani, K., Rajakumar, K., Turner, S.A., and Adler, B. (2004). Regulated site-specific recombination of the she pathogenicity island of Shigella flexneri. Mol. Microbiol. 52, 1329–1336.

Schmidt, H. (2001). Shiga-toxin-converting bacteriophages. Res. Microbiol. 152, 687–695.

Schmidt, H., Bielaszewska, M., and Karch, H. (1999). Transduction of enteric Escherichia coli isolates with a derivative of Shiga toxin 2-encoding bacteriophage phi3538 isolated from Escherichia coli O157:H7. Appl. Environ. Microbiol. 65, 3855–3861.

Schmidt, H., and Hensel, M. (2004). Pathogenicity islands in bacterial pathogenesis. Clin. Microbiol. Rev. 17, 14–56.

Schubert, S., Dufke, S., Sorsa, J., and Heesemann, J. (2004). A novel integrative and conjugative element (ICE) of Escherichia coli: the putative progenitor of the Yersinia high-pathogenicity island. Mol. Microbiol. 51, 837–848.

Serra-Moreno, R., Jofre, J., and Muniesa, M. (2007). Insertion site occupancy by stx2 bacteriophages depends on the locus availability of the host strain chromosome. J. Bacteriol. 189, 6645–6654.

Serra-Moreno, R., Jofre, J., and Muniesa, M. (2008). The CI repressors of Shiga toxin-converting prophages are involved in co-infection of Escherichia coli strains, which causes a down regulation in the production of Shiga toxin 2. J. Bacteriol. (in press).

Sullivan, J.T., and Ronson, C.W. (1998). Evolution of rhizobia by acquisition of a 500-kb symbiosis island that integrates into a phe-tRNA gene. Proc. Natl. Acad. Sci. USA 95, 5145–5149.

Sumby, P., and Waldor, M.K. (2003). Transcription of the toxin genes present within the Staphylococcal phage phiSa3ms is intimately linked with the phage's life cycle. J. Bacteriol. *185*, 6841–6851.

Tallent, S.M., Langston, T.B., Moran, R.G., and Christie, G.E. (2007). Transducing particles of *Staphylococcus aureus* pathogenicity island SaPI1 are comprised of helper phage-encoded proteins. J. Bacteriol. *189*, 7520–7524.

Tormo, M.A., Ferrer, M.D., Maiques, E., Ubeda, C., Selva, L., Lasa, I., Calvete, J.J., Novick, R.P., and Penades, J.R. (2008). *Staphylococcus aureus* pathogenicity island DNA is packaged in particles composed of phage proteins. J. Bacteriol. *190*, 2434–2440.

Ubeda, C., Tormo, M.A., Cucarella, C., Trotonda, P., Foster, T.J., Lasa, I., and Penades, J.R. (2003). Sip, an integrase protein with excision, circularization and integration activities, defines a new family of mobile *Staphylococcus aureus* pathogenicity islands. Mol. Microbiol. *49*, 193–210.

Ubeda, C., Maiques, E., Knecht, E., Lasa, I., Novick, R.P., and Penades, J.R. (2005). Antibiotic-induced SOS response promotes horizontal dissemination of pathogenicity island-encoded virulence factors in staphylococci. Mol. Microbiol. *56*, 836–844.

Ubeda, C., Barry, P., Penades, J.R., and Novick, R.P. (2007a). A pathogenicity island replicon in *Staphylococcus aureus* replicates as an unstable plasmid. Proc. Natl. Acad. Sci. USA *104*, 14182–14188.

Ubeda, C., Maiques, E., Tormo, M.A., Campoy, S., Lasa, I., Barbe, J., Novick, R.P., and Penades, J.R. (2007b). SaPI operon I is required for SaPI packaging and is controlled by LexA. Mol. Microbiol. *65*, 41–50.

Ubeda, C., Maiques, E., Barry, P., Matthews, A., Tormo, M.A., Lasa, I., Novick, R.P., and Penades, J.R. (2008). SaPI mutations affecting replication and transfer and enabling autonomous replication in the absence of helper phage. Mol. Microbiol. *67*, 493–503.

Wagner, P.L., Acheson, D.W., and Waldor, M.K. (1999). Isogenic lysogens of diverse Shiga toxin 2-encoding bacteriophages produce markedly different amounts of Shiga toxin. Infect. Immun. *67*, 6710–6714.

Wagner, P.L., and Waldor, M.K. (2002). Bacteriophage control of bacterial virulence. Infect. Immun.*70*, 3985–3993.

Waldor, M.K., and Friedman, D.I. (2005). Phage regulatory circuits and virulence gene expression. Curr. Opin. Microbiol. *8*, 459–465.

Walker, G.C. (1984). Mutagenesis and inducible responses to deoxyribonucleic acid damage in *Escherichia coli*. Microbiol. Rev. *48*, 60–93.

Weigel, L.M., Clewell, D.B., Gill, S.R., Clark, N.C., McDougal, L.K., Flannagan, S.E., Kolonay, J.F., Shetty, J., Killgore, G.E., and Tenover, F.C. (2003). Genetic analysis of a high-level vancomycin-resistant isolate of *Staphylococcus aureus*. Science *302*, 1569–1571.

Wong, C.S., Jelacic, S., Habeeb, R.L., Watkins, S.L., and Tarr, P.I. (2000). The risk of the hemolytic-uremic syndrome after antibiotic treatment of *Escherichia coli* O157:H7 infections. N. Engl. J. Med. *342*, 1930–1936.

Yamaguchi, T., Hayashi, T., Takami, H., Nakasone, K., Ohnishi, M., Nakayama, K., Yamada, S., Komatsuzawa, H., and Sugai, M. (2000). Phage conversion of exfoliative toxin A production in *Staphylococcus aureus*. Mol. Microbiol. *38*, 694–705.

Ye, Z.H., and Lee, C.Y. (1989). Nucleotide sequence and genetic characterization of staphylococcal bacteriophage L54a *int* and *xis* genes. J. Bacteriol. *171*, 4146–4153.

Zhang, X., McDaniel, A.D., Wolf, L.E., Keusch, G.T., Waldor, M.K., and Acheson, D.W. (2000). Quinolone antibiotics induce Shiga toxin-encoding bacteriophages, toxin production, and death in mice. J. Infect. Dis. *181*, 664–670.

Botulinum Neurotoxins – Structure and Mechanism of Action

2

Roshan Kukreja and Bal Ram Singh

Abstract

Botulinum neurotoxins (BoNTs) are the most potent natural toxins known to humankind. The family of BoNTs comprises seven antigenically distinct serotypes (A to G) that are produced by various toxigenic strains of the spore-forming anaerobic bacterium *Clostridium botulinum*. They act as metalloproteinases that enter peripheral cholinergic nerve terminals and cleave proteins that are crucial components of the neuroexocytosis apparatus, causing a persistent but reversible inhibition of neurotransmitter release resulting in flaccid muscle paralysis.

Apart from being the sole causative agent of the deadly food poisoning disease, botulism, BoNTs pose a major biological warfare threat due to their extreme toxicity and easy production. Interestingly they also serve as powerful tools to treat an ever expanding list of medical conditions. A better understanding of the structure–function relationship of clostridial neurotoxins will not only help decipher their molecular mode of action but will also provide a greater understanding of the potential use of their individual domains in answering more fundamental questions of neuroexocytosis. It is also critical for designing effective specific inhibitors to counter botulism biothreat, and for the development of new therapeutics.

Journey of botulinum neurotoxins from food poisoning to Botox

Clostridial neurotoxins (CNTs) are produced by anaerobic, Gram-positive, spore forming bacteria of the genus *Clostridium*. They are the most toxic proteins known with mouse LD_{50} values in the 0.1–1 ng/kg range, and are solely responsible for the pathophysiology of botulism and tetanus (Schiavo et al., 2000). *Clostridium botulinum*, produces seven serotypes of botulinum neurotoxins (BoNTs) named A to G according to the order of their discovery (Sakaguchi, 1983). Botulinum neurotoxins are the causative agents of botulism, a severe neurological disease characterized by flaccid muscle paralysis, resulting from BoNT-mediated blockage of neurotransmitter release (Singh, 2006). Tetanus neurotoxins (TeNTs), produced by *Clostridium tetani*, cause tetanus, a disease characterized by spastic paralysis (Montecucco and Schiavo, 1995).

Botulism was first identified by Justinus Kerner in the early 19th century, when he linked deaths from food intoxication with a poison found in smoked sausages (Erbguth and Nauman, 1999). He had also proposed a variety of potential medical uses of botulinum toxin for movement disorders, hypersecretion of body fluids, ulcers, etc. (Ting and Freiman, 2004). The scientific parameters of the disease were uncovered in 1895 by Emile van Ermengem, who successfully isolated the bacterium and named it *Bacillus botulinus* which was renamed *Clostridium botulinum* in later years (van Ermengem, 1979; Cherington, 1998).

Botulism is one of the most terrifying paralysing diseases to afflict humankind. The clinical spectrum of botulism continues to expand and is today divided into six clinical categories:

1. Food-borne botulism results from ingestion of food containing pre-formed neurotoxin, produced by clostridial organisms that contaminate inadequately processed food. Classical food-borne botulism in humans is caused mainly by *C. botulinum* types A, B, E and rarely by type F neurotoxins.

2. Infant botulism is caused by the ingestion of spores that then germinate and produce toxin in the infant's gastrointestinal tract. The causative agents of infant botulism can be *C. botulinum* types A or B neurotoxins or *Clostridium butyricum* type E and *Clostridium baratti* type F neurotoxins.

3. Wound botulism arises as a consequence of toxin produced in wounds contaminated with the clostridial bacterium. Wound botulism is rare in humans and the causative agents are either type A or type B from group I, but is increasing significantly in recent years among drug users.

4. Hidden botulism is an adult variation of infant botulism.

5. Inadvertent botulism which is an unintended consequence of treatment with botulinum toxin A.

6. Intentional botulism which could result from contamination of air or food by terrorist groups (Hatheway, 1993; Merrison *et al.*, 2002; Cherington, 2004).

Clostridial neurotoxins act as metalloproteinases that enter peripheral cholinergic nerve terminals and cleave proteins that are crucial components of the neuroexocytosis apparatus, causing a persistent but reversible inhibition of neurotransmitter release. Tetanus results from the contamination of necrotic wounds with spores of *Clostridium tetani*, bacterial proliferation and production and release of toxin (Singh, 2000). While tetanus toxin affects interneurons in the spinal cord, botulinum toxin affects peripheral cholinergic transmission (Singh, 2000).

Because of their extreme toxicity, ease of production, and robust stability under adverse environmental conditions, BoNTs are on the top of the list of biological warfare threats (Caya *et al.*, 2004), and have been listed as Class A bioterror agents by the National Institute of Allergy and Infectious Diseases (NIAID) (http://www3.niaid.nih.gov/about/directors/congress/2007/Final_4–18–07_NIH_Bioshield_Testimony.pdf).

Paradoxically, because of their extreme neurospecificity, BoNTs are being exploited in the treatment of a myriad of neuromuscular disorders and for the removal of facial wrinkles (Johnson *et al.*, 1999; Rohrich *et al.*, 2003; Bigalke and Rummel, 2005; Dastoor *et al.*, 2007; Salti and Ghersetich, 2008).

Structure–function relationship of botulinum neurotoxins

Overview of modular architecture and mechanism of action of BoNTs

All botulinum neurotoxins are made in the bacterial cytosol and are released into the culture medium after bacterial lysis as a single, inactive polypeptide chain of 150 kDa which are subsequently cleaved by bacterial proteases at an exposed protein-sensitive loop. This generates the fully active neurotoxin, composed of a 100-kDa heavy chain (HC) and a 50-kDa light chain (LC) linked by both non-covalent protein–protein interactions and a conserved interchain S–S bond, whose integrity is essential for neurotoxicity (Singh, 2006) (Fig. 2.1). One exception is BoNT/E, which remains an intact polypeptide until it reaches the victim's gut where the proteases in the gut activate the toxin by splitting it into light and heavy chains (Aoki and Guyer, 2001). During intoxication process, the interchain bridge is reduced, and this is a prerequisite for the intracellular action of the toxins (Singh, 2006).

BoNTs belong to the A–B-type toxins wherein they contain two polypeptide chains referred to as A and B chains, which play distinct roles (Rossetto *et al.*, 2004). The B chain binds to the surface of the neuronal cells, commandeers the endocytotic pathway to facilitate internalization of the toxin, and then mediates translocation of the A chain into the cytoplasm. The A chain is an enzyme that executes the damage by modifying selected cellular target proteins.

BoNTs are folded into three distinct domains which are functionally related to their cell intoxication mechanism (Montecucco and

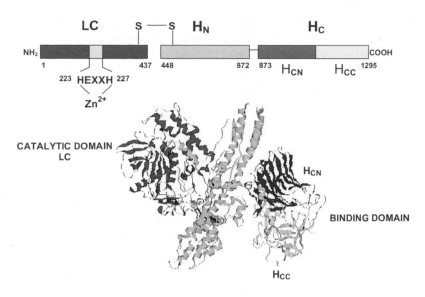

Figure 2.1 Top: Schematic diagram of different domains of BoNT/A. Bottom: Three-dimensional representation of the backbone of BoNT/A (Lacy *et al.*, 1998).

Schiavo, 1993). The intoxication by neurotoxins is proposed to be a multistep process which includes cell binding, internalization, translocation into the cytosol, and enzymatic modification of the cytosolic target (Montecucco *et al.*, 1994). The toxin first binds to the neuronal cell and is then internalized into the vesicles. To attack the targets in the cytosol the neurotoxin should cross the hydrophobic barrier of the vesicle membrane. It is proposed that the acidification of the vesicle lumen by a proton pumping ATPase leads to conformational changes in the toxin. The acidic conformation then exposes a hydrophobic area of the toxin molecule, creates an ion channel in the membrane, and inserts the LC into the cytosol. Once inside the cytosol, the light chain acts as a zinc-dependent endoprotease and selectively cleaves one of the three synaptic vesicle fusion proteins known as SNARE (soluble *N*-ethylmaleimide-sensitive factor attachment protein receptor) proteins. SNAREs are small conserved membrane proteins that include synaptobrevin on synaptic vesicles and synaptosomal associated protein of 25 kDa (SNAP-25), and syntaxin on the plasma membrane. SNAREs mediate the fusion of synaptic vesicles with the presynaptic membrane, so when any of the SNARE proteins in presynaptic nerve terminals are cleaved by the LC, neurotransmitter release

is inhibited (Rossetto *et al.*, 2001a; Davletov *et al.*, 2005).

The three-dimensional structures of BoNT/A and BoNT/B reveal three distinct 50-kDa domains: a N-terminal domain endowed with zinc-endopeptidase activity (LC); a membrane translocation domain (H_N) characterized by the presence of two 10-nm-long α-helices; and a binding domain (H_C), composed of two unique sub-domains (H_{CN} and H_{CC}) (Lacy *et al.*, 1998; Swaminathan and Eswaramoorthy, 2000) (Fig. 2.1). The binding domain is composed primarily of β-strands and is connected by one predominant α-helix. The catalytic domain LC comprises both α-helices and β-strands. The active site in BoNT/A is buried 20–24 Å deep, has negative surface charge, and is shielded from the solvent by the belt region from the translocation domain. The active site in BoNT/B is 16 Å deep and is not occluded by the belt region. The LC chain is a metalloprotease with a zinc atom in the centre of the active site coordinated via the two histidines and a glutamic acid residue of the highly conserved HEXXH zinc-binding motif, and by Glu262 in BoNT/A, a residue conserved among clostridial neurotoxins. Such structural organization is functionally related to the fact that the CNTs intoxicate neurons via a multistep mechanism.

Botulinum neurotoxins in complex form

Clostridial neurotoxins are synthesized in the bacterial cytosol without a leader sequence (Inoue *et al.*, 1996), which is consistent with the fact that they are released in the culture medium only after bacterial lysis. BoNTs are produced from *C. botulinum* in the form of multimeric complexes, with a set of non-toxic neurotoxin associated proteins (NAPs) coded for by genes adjacent to the neurotoxin gene (Fig. 2.2). These complexes are termed as progenitor toxins (Inoue *et al.*, 1996; Minton, 1995). A large non-toxic non-haemagglutinating protein coded for by a gene upstream to the BoNT gene, called the neurotoxin binding protein (NBP) is highly conserved in all serotypes of BoNTs (Minton, 1995; Fujita *et al.*, 1995). BoNTs produced with NAPs as complexes exist in three different forms: extra-large complex (LL complex, ~ 900 kDa, sediments at 19S), large (L complex ~ 500 kDa, sediments at 16S), and medium or M complex which sediments at 12S and is about 300 kDa. The M complex is the BoNT-NBP complex, L complex is composed of BoNT–NBP and several neurotoxin associated proteins (NAPs) with haemagglutinin activity, while the LL complex is believed to result from the association of two L complexes (Inoue *et al.*, 1996). Type A botulinum neurotoxin complex exists in three forms: LL, L, or M. Types B, C, D, and E exist in L and M forms. Type F exists only in M form, and type G exists only in L form (Fujita *et al.*, 1995;

Zhang and Singh, 1995). Some types of A and B neurotoxins that cause infant botulism, also exist in M form (Cordoba *et al.*, 1994).

The non-toxic neurotoxin-associated proteins in the complexes provide stability to the neurotoxin in adverse environmental conditions and during their passage through the low-pH environment of the gastrointestinal tract (Sugiyama, 1980; Sakaguchi, 1983; Schantz and Johnson, 1992). Such protection is important in the food poisoning activity of the neurotoxin. The oral toxicity of BoNT increases with incremental association of the neurotoxins with NAPs (Li *et al.*, 1998). BoNTs in their complex forms along with NAPs are known to be 10- to 100-fold more toxic than the non-complexed pure neurotoxin, through the oral route (Cheng *et al.*, 2008). In BoNT/A and BoNT/B, the larger forms of progenitor neurotoxins (19S and 16S) are more toxic by the oral route and more resistant to acid and pepsin than the smaller forms of the progenitor toxins (Schantz and Johnson, 1992; Fu *et al.*, 1998a; Sharma *et al.*, 1999). TeNT is produced as a single peptide of ~150 kDa, and is not known to form complexes with non-toxic proteins due to which it is not stable in the GI tract and thus is not a cause of food poisoning (Singh, 1999).

NAPs are also known to assist BoNT translocation across the intestinal mucosal layer (Fujinaga *et al.*, 1997, 2004). Certain carbohydrate groups that help the NAPs to anchor to the intestinal wall, have been identified (Inoue *et*

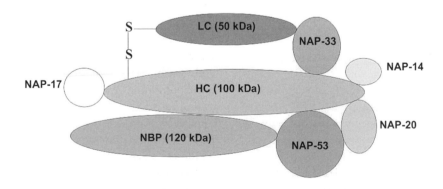

Figure 2.2 Schematic model of BoNT/A complex. The neurotoxin is composed of a 50-kDa LC and a 100-kDa HC linked by a disulphide bond. The neurotoxin is protected from the harsh environmental conditions of the GI tract by the neurotoxin binding protein NBP and 5 NAPs: NAP-53, NAP-33, NAP-20, NAP-17, and NAP14.

al., 2001; Nishikawa *et al.*, 2004). The progenitor neurotoxins are believed to dissociate into the neurotoxin and the non-toxic components (NAPs) in lymph after passing through the small intestine or in alkaline conditions (Li *et al.*, 1998).

Recently, NAPs have been shown to dramatically enhance the endopeptidase activity of BoNT/A (Cai *et al.*, 1999; Sharma and Singh, 2004). Because of the extreme toxicity and stability of BoNT in presence of NAPs, BoNT complexes top the list of biological warfare threats. Ironically, BoNT most commonly in its complex form is used as therapeutic agent to treat several muscular disorders and in cosmetic applications (Shukla and Sharma, 2005).

BoNTs A, B and E are responsible for majority of cases of human botulism (Singh *et al.*, 1995). Since much is known about the critical role of NAPs in the enhanced endopeptidase activity of BoNT/A complex, we recently reported a comparative study of the structure–function relationship of BoNT/A complex with the newly identified BoNT/E complex under different stages of protein unfolding using various spectroscopic techniques and biochemical assay (Kukreja and Singh, 2007).

Examination of the biological activity of BoNT/A and BoNT/E complexes revealed that they are enzymatically active under non-reducing conditions. BoNT/A complex under non-reducing conditions attains optimum activity at the physiological temperature of 37 °C whereas BoNT/E in its complex form under non-reducing conditions achieves maximum endopeptidase activity at 45 °C. One of the most intriguing features of the unique endopeptidase activity of botulinum neurotoxins is that it is observed only upon reduction of the disulphide bond that links the light and the heavy chain in the neurotoxin (Singh, 2000). However, observation of the optimal endopeptidase activity of BoNT/A complex at 37 °C under conditions where the disulphide bond linking the light and heavy chains remains intact, reinforces the hypothesis of a conformational alteration in BoNT/A upon its interaction with NAPs in the complex form making its active site readily accessible to the substrate (Cai *et al.*, 1999). This altered conformation of BoNT/A in the complex perhaps similar to the molten globule conformation of reduced BoNT/A or the PRIME conformation of the light chain of BoNT/A (Cai and Singh, 2001; Kukreja and Singh, 2005), at physiological temperature, may facilitate its favourable interaction with SNAP-25 leading to the optimal cleavage of the latter at this temperature even under conditions which keep the disulphide bond between the light and heavy chains of BoNT/A intact. We also observed that BoNT/E complex in non-reducing conditions undergoes conformational alterations in its polypeptide folding at 45 °C and it is at this temperature that it exhibits optimum activity. This may result in its favourable binding with the substrate leading to maximum cleavage of the latter (Kukreja and Singh, 2007).

This implication of a substantial functional role of NAPs in BoNT/A and BoNT/E complex requires further experimental evidence to provide a confirmed mechanism of the enhanced enzyme activity. At the minimum, these results suggest the substantial role of NAPs in the critical endopeptidase activity of BoNT, which may even have relevance to the biological activity of BoNT *in vivo*. While toxico-infection process of botulism assumes separation of the NAPs from BoNT before reaching the nerve cell, injection of BoNT/A complex for therapeutic or cosmetic use does not separate the two components. Furthermore, currently it is not known what role the toxin plays inside the bacterial cell or in the native ecologically conditions of *C. botulinum*. An activated form of the complex endopeptidase may have a critical role under those conditions.

The results also indicated that BoNT/A in its complex form is structurally more stable than BoNT/E complex against temperature whereas BoNT/E in its complex form is functionally better protected against temperature (Kukreja and Singh, 2007). Functional stability of the botulinum complexes against temperature plays a critical role in the survival of the agent in cooked food and in food-borne botulism.

A critical step in developing effective antidotes against botulinum endopeptidase requires a comprehensive knowledge of the molecular basis of its unique endopeptidase activity. Hence, intense chemical characterization of BoNTs providing structural information of the biologically active molecule seems to be crucial in understanding its toxic activity. Further more since BoNT/E targets the same intracellular substrate SNAP-25 as BoNT/A, an intense structural and biochemical analysis on BoNT/E provides an opportunity to compare with BoNT/A especially for its relevance to binding and recognition of SNAP-25.

Haemagglutinin-33 (Hn-33) is a 33-kDa component of NAPs, and forms the largest fraction of NAPs in the botulinum complex (Sharma and Singh, 2000). It has been shown to exhibit haemagglutinin activity (Fu et al., 1998a). The purified Hn-33 is found to be resistant to digestion by proteases such as trypsin, chymotrypsin, pepsin and subtilisin (Sharma and Singh, 1998). Hn-33 has been shown to enhance the endopeptidase activity of BoNT/A and BoNT/E (Sharma and Singh, 2004). Recently Hn-33 was found to bind to synaptotagmin, suggesting its possible role in the attachment of BoNT complexes to the nerve terminals. (Zhou et al., 2005).

Mode of action of botulinum neurotoxins

Passage of botulinum neurotoxins through the GI tract

The most common mechanism of botulism poisoning is through the oral ingestion of the toxin, which is found in food contaminated with *C. botulinum*. Hence BoNT must escape the GI system and reach the general circulation in route to the peripheral cholinergic nerve endings where it exerts neuroparalytic effects. One of the major reasons why the toxin is active by the oral route and can be absorbed into the general circulation, is because of its association with NAPs during its release from the bacterium in the form of a non-covalent mutimeric complex (Chen et al., 1998). The NAPs protect the toxin during its exposure to harsh conditions in the GI tract. Most proteins are broken down into peptides and amino acids in the stomach and small intestine during the process of digestion. However, BoNT in the complex form enters the stomach and withstands the acidic (pH 2) gastric juice containing the protease pepsin. The complex then enters the small intestine, where it encounters several more proteases (trypsin, chymotrypsin etc.) that func-

tion at pH 7 to 8. Botulinum toxin binds to the apical surface of epithelial cells of the gastrointestinal system, undergoes receptor-mediated endocytosis and transcytosis, and is thus carried from the lumen of the gut to interstitial fluid, and ultimately to the general circulation (Maksymowych and Simpson, 1998, 2004).

Penetration of the neurotoxin through epithelial cell barriers and its subsequent migration to cholinergic nerve terminals is the first essential step of botulinum intoxication. It was recently reported that BoNT/A crosses intestinal cell monolayers via receptor-mediated transcytosis. A double receptor model has been proposed wherein BoNT/A first interacts with GD_{1b} and GT_{1b} gangliosides on the intestinal cells and then with SV2 protein, which mediates BoNT/A transcytosis through these cells (Couesnon et al., 2008). The role of NAPs in transport of BoNT through epithelial cells is not very clear. Some reports suggest that NAPs assist in the binding of the toxin to the epithelial cells (Fujinaga et al., 2000; Inoue et al., 2001), whereas others suggest that the ligand binding moiety is present on the neurotoxin itself (Maksmowych et al., 1999; Ahsan et al., 2005).

It has been shown that type C progenitor toxin associates with high-molecular-weight glycoproteins, such as mucin, on the surface of human colon carcinoma HT-29 cells, thus permitting the uptake of progenitor toxin into cells via a clathrin and/or caveolae pathway (Uotsu et al., 2005). BoNT is also able to cross other epithelial layers of the respiratory system, indicating that botulism can also be acquired by toxin inhalation (Park and Simpson, 2003). The toxin from the general circulation eventually reaches the peripheral cholinergic nerve endings, which are its principal site of action and to which they bind specifically.

Neurospecific binding

The C-terminal domain of the heavy chain (H_{CC}) plays a major role in neurospecific binding. A double-receptor model that has been proposed suggests that BoNTs bind first to gangliosides on the neural membrane and then with a protein receptor through the H_C domain (Singh, 2000), (Fig. 2.3). Polysialogangliosides, particularly G_{T1b}, G_{D1b} are found to interact

A

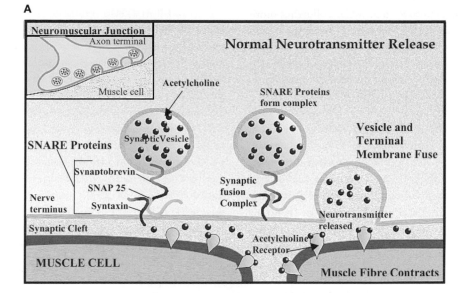

B

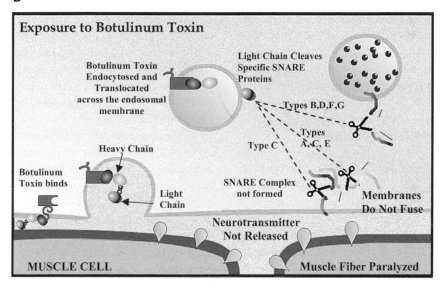

Figure 2.3 Schematic model of mode of action of botulinum neurotoxins. (A) Synaptic vesicles containing neurotransmitters dock and fuse with the plasma membrane through interaction of the SNARE proteins (VAMP, SNAP-25, and syntaxin). (B) Botulinum neurotoxin binds to the presynaptic membrane through gangliosides and a protein receptor followed by internalization into the endosomes via endocytosis. Following this the light chain is translocated across the membrane into the cytosol where it acts as a specific endopeptidase against either of the SNARE proteins. BoNTs cleave their substrates before the formation of SNARE complex.

with the receptor binding domains of BoNTs (Nishiki *et al.*, 1994, 1996; Rummel *et al.*, 2003, 2004). Since gangliosides are dispersed on the neuronal surface (Mirsky *et al.*, 1978), they are present in large excess compared to BoNT and are expected to diffuse laterally as fast as other membrane lipids. They would form an effective system to bind the neurotoxin and deliver it to its protein receptor (Montecucco *et al.*, 1986). Ganglioside binding sites have been identified for several serotypes of BoNTs (Swaminathan and Eswaramoorthy, 2000; Rummel *et al.*,

2004; Tsukamoto *et al.*, 2008). Following binding to the gangliosides, the membrane bound ganglioside–toxin complex moves to reach the toxin-specific receptor. Nishiki *et al.* have identified the synaptic vesicle protein synaptotagmin as a receptor for botulinum neurotoxin type B and G (Nishiki *et al.*, 1996). Synaptotagmin is a calcium sensor that couples calcium influx into the neuron to the fast phase of neurotransmitter release. The C-terminal portions of the heavy chains of BoNT/B and/G bind to the luminal domains of synaptotagmins I and II when they are exposed to the neuronal cell surface (Rummel *et al.*, 2004). Recently resolved crystal structures of the complex between BoNT/B and the luminal domain of synaptotagmin II show that the luminal domain which is largely unstructured in solution, folds into a helix, upon binding to BoNT/B. Synaptotagmin binds at the distal tip of the heavy chain subdomain of BoNT/B in a crevice on the surface (Jin *et al.*, 2006; Chai *et al.*, 2006). Mutagenesis of the synaptotagmin binding cleft revealed that synaptotagmin binding is more critical than ganglioside binding (Jin *et al.*, 2006). It is also suggested that BoNT/B and BoNT/G likely use the same strategy for receptor binding (Rummel *et al.*, 2007). Recently SV2 was identified as a receptor for botulinum neurotoxin type A (Dong *et al.*, 2006; Marhold *et al.*, 2006; Jahn, 2006). Among the SV2 isoforms, SV2c exhibited the most robust binding affinity for BoNT/A (Dong *et al.*, 2006; Marhold *et al.*, 2006). SV2 is a conserved membrane protein of synaptic vesicles that is structurally related to neurotransmitter transporters.

A dual receptor binding process could account for the extraordinary binding affinity and specificity of botulinum neurotoxins. The binding of toxin to gangliosides may bring the toxin in close proximity with the protein receptor, introduce conformational changes in the toxin to prime its binding with the receptor, or cause a conformational change at the binding site of the protein receptor to enhance its interaction (Singh, 2000). Recently isolated C-half of the C-terminus of the heavy chain of botulinum neurotoxin type A (HCQ) was shown to bind with synaptotagmin and gangliosides (Sharma *et al.*, 2006).

Internalization inside neurons

Since the light chains of CNTs block neuroexocytosis by acting in the cytosol, this toxin domain at least must reach the cell cytosol. BoNTs do not enter the cell directly from the plasma membrane, but are rather endocytosed inside acidic cellular components (Schiavo *et al.*, 2000) (Fig. 2.3). BoNTs enter the lumen of vesicular structures in a temperature and energy-dependent process (Dolly *et al.*, 1984). Once internalized, the vesicles fuse with other vesicles to generate endosomes with an acidic interior (Tycko and Maxfield, 1982). The acidic environment in the endosome is mandatory for intoxification to occur. Agents that interfere with intracellular vesicular acidification inhibit toxicity of the neurotoxins (Simpson *et al.*, 1994).

Translocation into neuronal cytosol

In order to reach the cytosol the light chain must cross the hydrophobic barrier of the endosomal membrane. The N-terminal region of the heavy chain (H_N) plays a critical role in mediating the translocation of the LC into the cytoplasm of the neuronal cell (Li and Singh, 1999). CNTs have to be exposed to a low pH for nerve intoxication to occur (Simpson *et al.*, 1994; Matteoli *et al.*, 1996). Acidic pH does not induce any direct activation of the toxin since the introduction of a non-acid treated LC into the cytosol is sufficient to block exocytosis (Bittner *et al.*, 1989; Penner *et al.*, 1986; Weller *et al.*, 1991). Low pH induces a conformational change in CNTs from a water-soluble 'neutral' structure to an 'acid' structure with surface-exposed hydrophobic segments, which enable the penetration of both the HC and LC in the hydrophobic core of the lipid bilayer (Cabiaux *et al.*, 1985; Montecucco *et al.*, 1989). Following the low pH-induced membrane insertion, BoNTs and TeNT form transmembrane ion channels in planar lipid bilayers (Hoch *et al.*, 1985; Shone *et al.*, 1987; Fu and Singh, 1999) and in PC12 membranes (Sheridan, 1998).

The membrane translocation domain of BoNT/A has a cylindrical shape determined by the presence of unusually long 10-nm-long helices (Lacy *et al.*, 1998) (Fig. 2.1). The overall structure of H_N of BoNT/A resembles that of other membrane translocating toxins, such as colicins and diphtheria toxin (Choe *et al.*, 1992;

Weiner *et al.*, 1997). The details of how the translocation domain changes conformation at acidic pH to form a pore and how it allows for the passage of a 50-kDa LC across the endosomal membrane are the least understood aspects of botulinum intoxication mechanism and warrant detailed investigation of the structural changes in LC that occur at low pH. Several models have been proposed for the translocation of BoNTs across the endosomal membrane. One of the hypothesis suggests that the LC of toxins unfold at low pH and permeate through the narrow transmembrane 'tunnel' formed by the HC (Hoch *et al.*, 1985). An alternative hypothesis is the 'cleft' model which assumes that the LC does not go directly through a channel, but rather, after membrane insertion, it is released from a cleft region formed by the hydrophilic protein–protein interface between L and H chain (Montecucco *et al.*, 1994). According to this hypothesis, the two toxin polypeptide chains are supposed to change their conformations at low pH such that both of them expose hydrophobic patches and interact with the hydrophobic core of the lipid bilayer. The toxin HC then forms a transmembrane hydrophilic cleft that nests the passage of the partially unfolded LC with its hydrophobic segments facing the lipids. The cytosolic neutral pH induces the LC to refold and regain its water-soluble neutral conformation, after reduction of the interchain disulphide bond (Lebeda and Singh, 1999). Recently we have shown that BoNT/A LC undergoes unique structural changes under low pH conditions, and adopts a molten globule state, exposing substantial number of hydrophobic groups. The flexibility of the molten globular structure, combined with retention of the secondary structure and exposure of hydrophobic domains could form a basis of specific interaction and translocation through the narrow membrane channel formed by the BoNT heavy chain (Cai *et al.*, 2006).

Structural changes observed at low pH leading to an elongated molten globule molecule are critical to support the hypothesis that LC passes through the heavy-chain channel under a refolded state, which renatures in the cytosol, and regains the endopeptidase activity. This observation is in conflict with the conclusions and perhaps the hypothesis of Koriazova and Montal, 2003, and suggest the BoNT/A LC may cross the HC channel through a different mechanism. Instead of a totally unfolded state (Koriazova and Montal, 2003), we have shown that the LC undergoes translocation in a molten globule conformation. The molten globule state in acidic conditions, with its secondary structure intact, can refold into an elongated structure (Fig. 2.4) that fits into a HC channel of 15 Å observed in electron micrographs of BoNT/B in two dimensional lipid arrays (Schmidt *et al.*, 1993), and may retain specific interactions with the HC and lipid bilayer through hydrophobic clusters as it passes through the HC membrane channel. Previously, it has been observed that low pH-induced structural changes allow the light chain to directly interact with the lipid bilayer which is consistent with our finding (Montecucco *et al.*, 1989). The acid-induced molten globule unfolding mechanism has been shown to be crucial to the translocation of bacterial toxins across cellular membranes (Bychkova *et al.*, 1988; vander Goot *et al.*, 1991; Zhao and London, 1986; Ren *et al.*, 1999). The lethal and oedema factors of anthrax toxin are also known to adopt a molten globule state at low pH during their translocation (Krantz *et al.*, 2004).

For BoNTs, a chaperone/channel activity of the heavy chain, driven by a pH gradient across the endosome has recently been suggested wherein the LC unfolds and enters the HC channel, which in turn prevents its aggregation during its translocation and releases the LC by reduction of the disulphide bond concomitant with its refolding in the cytosol (Koriazova and Montal, 2003; Fischer and Montal 2007a,b). It has also been proposed that the belt region of the HC translocation domain may act as a chaperone and as a pseudosubstrate inhibitor of the LC during its translocation across the endosomal membrane (Brunger *et al.*, 2007).

Zinc endopeptidase activity

After internalization into the cytosol, BoNTs specifically cleave neuronal proteins integral to vesicular trafficking and neurotransmitter release (Hanson *et al.*, 1997), (Fig. 2.3). BoNTs constitute a unique group of zinc proteases that contain the His-Glu-Xaa-Xaa-His zinc-binding motif of zinc endopeptidases (Schiavo *et al.*,

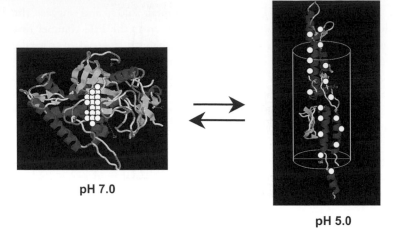

pH 7.0

pH 5.0

Figure 2.4 Schematic model representing low pH (pH 5.0) induced tertiary structural alteration in BoNT/A LC. The pH 7.0 folding (left panel) is derived from the X-ray crystallographic structure (Segelke *et al.*, 2004). The pH 5.0 folding (right) is a visual schematics drawn manually (no protein modelling used) to represent a stretched structure that fits within a cylindrical HC channel of 15 Å diameter, and is in accordance with experimental observation suggesting dramatic tertiary structural alterations while retaining virtually all its secondary structure content. The white balls represent the hydrophobic residues which are tightly packed in the native conformation (pH 7.0). At pH 5.0 BoNT/A LC adopts a molten globule conformation wherein the hydrophobic groups become exposed to the surface and probably aid in retaining specific interactions with the HC and lipid bilayer as the LC passes through the HC membrane channel during translocation (Cai *et al.*, 2006).

1992a), and recognize specific structural motifs and neuronal substrates referred to as SNAREs: (i) vesicle associated membrane protein (VAMP/synaptobrevin), (ii) syntaxin, and (iii) synaptosomal associated protein of 25 kDa (SNAP-25) (Burgoyne and Morgan, 1993; Pellegrini *et al.*, 1995; Vogel *et al.*, 1999; Graham *et al.*, 2002). These SNAREs reside on transport vesicles or on the target membranes (Schiavo *et al.*, 2000), (Fig. 2.5). Syntaxin and SNAP-25 bind with high affinity and undergo conformational changes enabling palmitoylation of SNAP-25 (Fasshauer, 1997; Veit, 2000). When synaptic vesicles come close to the plasma membrane, binding of VAMP to the syntaxin-SNAP25 complex, leads to the formation of a tight α-helical bundle which is a multi-protein 20S complex, together with NSF that possesses ATPase activity and probably drives vesicle fusion with the plasma membrane as shown in Fig. 2.5 (Banerjee *et al.*, 1996; Sutton *et al.*, 1998; Graham *et al.*, 2001). Formation of this complex is regulated by the Ca^{2+}-sensitive vesicular protein synaptotagmin (Südhof, 2004; Koh and Bellen, 2003). The core domains of SNARE proteins are mostly unstructured in so-

lution (Fasshauer *et al.*, 1997; Fiebig *et al.* 1999) but are entirely helical when the complex is formed (Sutton *et al.*, 1998). The helices formed by SNARE proteins are amphipathic and largely stabilized by hydrophobic interactions (Ossig *et al.*, 2000). When these proteins are assembled in the tight complex, the substrates are resistant to *in vitro* proteolysis by BoNTs (Pellegrini *et al.*, 1995) whereas free proteins are specifically and efficiently cleaved by BoNTs. The proteolysis of one SNARE protein results in a non-functional complex where the coupling between Ca^{2+} influx and fusion is disrupted (Humeau *et al.*, 2000). BoNT/B, BoNT/D, BoNT/F, BoNT/G and TeNT cleave VAMP at different single peptide bonds; BoNT/C cleaves both syntaxin and SNAP-25, two proteins of the presynaptic membrane; BoNT/A and BoNT/E cleave SNAP-25 at different sites within the COOH-terminus (Schiavo *et al.*, 2000). Strikingly, BoNT/B and TeNT cleave VAMP at the same peptide bond (Gln76–Phe77) and yet, cause opposite symptoms of botulism and tetanus, respectively, clearly demonstrating that the different symptoms of the two diseases derive from different sites of

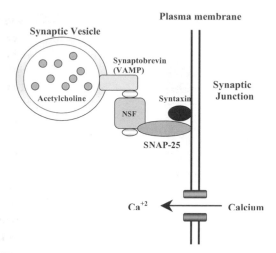

Figure 2.5 Schematic representation of the SNARE complex. The SNARE proteins synaptobrevin/VAMP, SNAP-25, and syntaxin form a complex with the ATPase NSF. The SNARE complex is involved in docking of the synaptic vesicles containing acetylcholine to the presynaptic membrane, where fusion and exocytosis of acetylcholine occurs (modified from Johnson, 1999).

intoxication rather than from different molecular mechanism of action of the two neurotoxins (Schiavo et al., 1992b).

The three SNAREs, VAMP, SNAP-25, and syntaxin possess a distinct three-dimensional motif (SNARE motif) that is required for specific proteolysis by the neurotoxins (Montecucco and Schiavo, 1995). Several lines of evidence suggest that botulinum neurotoxins recognize the tertiary rather than the primary structure of their proteolytic substrates at the level of the nine-residue long SNARE motif which is characterized by three carboxylate residues alternated with hydrophobic and hydrophilic residues, (Rosetto et al., 1994; Pellizari et al., 1996; Vaidyanathan et al., 1999). A double recognition model has been proposed involving interactions with the SNARE motif common to the three targets, and with the segment containing the peptide bond to be cleaved. Site-directed mutagenesis experiments suggest that the three negatively charged residues (Asp or Glu) of the motif play a major role in the recognition of specific targets by different neurotoxins (Pellizari et al., 1996). BoNTs differ with respect to the specific interaction with the SNARE motif (Rossetto et al., 2001a). The findings that only protein segments including at least one SNARE motif are cleaved by the toxins and that the motif is exposed at the protein surface clearly indicate the critical role of the SNARE motif in the specificity of action of botulinum neurotoxins.

Crystal v/s solution structure of BoNTs

Crystal structures of several substrate-free CNT proteases and their LC subunits have been determined including those of BoNT/A (Lacy et al., 1998; Segelke et al., 2004; Breidenbach and Brunger, 2004; Burnett et al., 2007), BoNT/B (Swaminathan and Eswaramoorthy, 2000; Hanon and Stevens, 2000), BoNT/E (Agarwal et al., 2004), and TeNT (Breidenbach and Brunger, 2005; Rao et al., 2005). These structures have focused on understanding the active site structure to explain the unique endopeptidase activity and to develop inhibitors. In spite of targeting different SNARE sites (with the exception of BoNT/B and TeNT), the LCs exhibit similar folding with differences primarily limited to surface features.

The striking overall similarity of LCs, especially their active sites provide little insight into their ability for target discrimination and none of the studies, so far, has provided a satisfactory explanation to the unique substrate requirements of BoNTs. For the endopeptidase activity of BoNT, there were two major chemical components identified at the beginning of the discovery of enzyme activity (Schiavo et al.,

1992a; Blasi *et al.*, 1993). These were: (i) the presence of Zn^{2+} in their active site and, (ii) reduction of the interchain disulphide bond between the heavy and light chains. Role of these two chemical components have been examined for the maintenance of the functional structure of BoNT, both in crystal and in solution. BoNT endopeptidase activity is observed only upon reduction of the disulphide bond between light and heavy chain (Singh, 2006). The data from X-ray crystal structure of BoNT/A and BoNT/B indicate that while similar in overall folding, they differ in key aspects of the depth and occlusion of the active sites, a key element expected to explain specificity and disulphide reduction-based activation of the endopeptidase activity of BoNTs (Lacy and Stevens, 1999, Singh, 2002). The BoNT crystal structure does not seem to undergo any noticeable change upon disulphide bond reduction, although spectroscopic analysis of BoNT in solution indicates dramatic changes in the polypeptide folding upon disulphide bond reduction (Cai and Singh, 2001). Zinc removal in BoNT/B showed no structural difference in the crystal structure (Swaminathan *et al.*, 2004), whereas removal of zinc in BoNT/A and BoNT/A LC causes dramatic changes in the tertiary structure folding (Fu *et al.*, 1998b; Li and Singh, 2000).

Thus there are substantial differences between structural elements of BoNTs in crystals and solutions. Structural characteristics in solution reveal a major role of the dynamic structure of BoNT in understanding the functioning of the molecule.

The latest three-dimensional (3-D) structure published with a double mutant of BoNT/A LC and a truncated form of SNAP-25 (residues 141–204) crystallized together provides the most promising information on the recognition and catalytic activity of BoNT/A endopeptidase (Breidenbach and Brunger, 2004). It provides a sound mechanism for BoNT/A endopeptidase specificity and selectivity towards cleavage of its substrate. According to this mechanism the catalytic efficiency of BoNT/A depends on an extensive array of exosites α and β (substrate binding sites remote from the active site), (Fig. 2.6). BoNT/A makes use of an extended extended enzyme–substrate interface to properly orient its substrate such that the scissile peptide bond is in close proximity with the catalytic centre of the enzyme. As seen in Fig. 2.6, many interactions that impart substrate specificity occur on the opposite side of the active site of the enzyme (α-exosite). The C-terminus of the substrate (β-exosite) introduces a conformational change in the active-site of the enzyme, probably

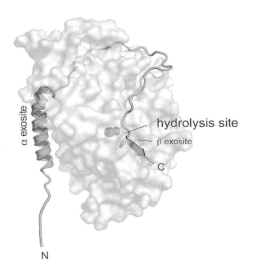

Figure 2.6 Structure of BoNT/A LC and SNAP-25 complex. BoNT/A (grey) forms an extended interface with the C-terminal domain of SNAP-25 (black). Multiple sites of enzyme-substrate interaction remote from the hydrolysis site extend around most of the circumference of the toxin, imparting the protease with exquisite specificity. SNAP-25 which is unstructured in solution, adopts a mix of α, β and extended conformations when it interacts with BoNT/A. (taken from Brunger *et al.*, 2008; published by Springer).

rendering the protease competent for catalysis. While the β-exosite 'does not confer much substrate specificity, its presence is vital for activity' (Breidenbach and Brunger, 2004). Such multisite binding strategy used by BoNT/A accounts for the extreme selectivity of this enzyme. The identification of novel substrate binding sites on the surface of BoNT/A (Fig. 2.6), presents new opportunities for the structure-based design of specific inhibitors of BoNT/A endopeptidase. Interestingly, this co-crystal structure does not show any direct interaction of the α-exosite of the light chain with the S4 SNARE motif of SNAP-25, demonstrated to be critical for the endopeptidase activity (Washbourne et al., 1997). Furthermore the co-crystal structure employed a truncated SNAP-25. Under normal physiological conditions, full-length SNAP-25 is the substrate and the optimal endopeptidase activity is observed only with full-length SNAP-25 (Vaidyanathan et al., 1999). Thus while the crystal structures provide useful information on the folding patterns of BoNTs, and on the interaction between BoNT/A light chain and SNAP-25, these do not take into consideration the role of dynamic structures in the functioning of the molecule. Structural characteristics in solution reveal a major role of dynamic structure of BoNT, which plays a critical role in the physiological interaction of BoNT/A light chain and SNAP-25. Crystal structures, therefore, may not be adequate either to explain the molecular basis of BoNT's action, or to design effective inhibitors.

Recently we have reported for the first time a prominently biologically active molten globule structure of the catalytic domain of BoNT/A and have also identified a novel conformation of this enzyme at the physiological temperature of 37 °C, and named it PRIME (pre-imminent molten globule enzyme) state (Kukreja and Singh, 2005). The PRIME state is dynamically flexible and is likely to facilitate specific interactions with its substrate, SNAP-25 for its optimum and selective enzymatic activity. Thus we suspect that the PRIME conformational state plays a very critical role in the biological function of BoNT/A, especially in its intracellular toxic action.

The schematic model of the temperature-induced PRIME conformational and molten globule states of BoNT/A light chain, as represented in Fig. 2.7, reflects how these states in BoNT/A light chain may facilitate binding and cleavage of SNAP-25. The intramolecular mobility of PRIME state is significantly higher than that of the native state which can be attributed to increased dynamics and expansion of the protein core, facilitating a maximum specific binding of SNAP-25, leading to its cleavage. The molten globule conformation observed at 50°C, on the other hand, has a considerably intact secondary structure, loose packing of side chains in the protein core and partial unfolding of loops. This structure also binds to SNAP-25 although to a lesser degree than the PRIME state, thus showing only about 61% of the optimum enzymatic activity (Fig. 2.7).

For decades, there has been increasing conviction that the biological function of protein can be interpreted only in terms of their 3-D structure, and ordered native structure generally obtained from X-ray crystallography has been accepted as the standard structure. Stable 3-D structures are the ultimate goals for most of the modern protein engineering efforts to generate proteins for specific functions. Paradoxically, many de novo synthesized proteins exist as molten-globules (Vamvaca et al., 2004), which are considered to be an obstacle to overcome in obtaining fully folded structures (Vamvaca et al., 2004; Bryson et al., 1995).

The static crystal structure of botulinum endopeptidase has been determined crystallographically (Segelke et al., 2004; Breidenbach and Brunger, 2004). However, it is important to adequately describe the dynamic properties of the enzyme in solution where it represents a manifold of conformational substrates. Conformational fluctuations resulting from the concerted motions of many atoms can push the unbound states of enzymes into conformations closely resembling the bound states thereby priming them to form complexes with the substrate. Hence every conformational state of the enzyme along the reaction pathway is likely to present a unique opportunity for interactions with drug molecules. Our observation of the PRIME conformation of BoNT/A LC with optimum enzymatic activity is the first observation of enzymatically active molten globule or molten

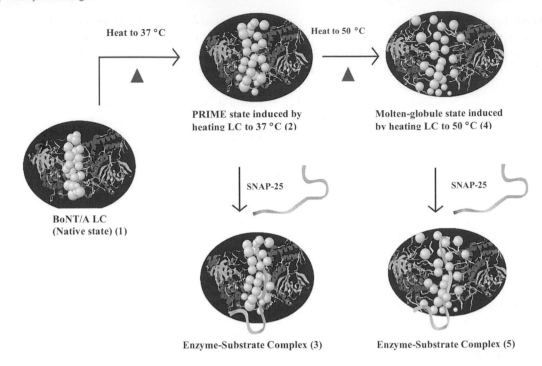

Heat to 37 °C

PRIME state induced by
heating LC to 37 °C (2)

Heat to 50 °C

Molten-globule state induced
by heating LC to 50 °C (4)

BoNT/A LC
(Native state) (1)

SNAP-25

SNAP-25

Enzyme-Substrate Complex (3)

Enzyme-Substrate Complex (5)

Figure 2.7 Schematic model representing the PRIME state and the molten globule state in BoNT/A LC facilitating binding of substrate SNAP-25. The grey balls represent the non-polar side chains which are tightly packed in the native conformation (1). Upon heating to 37°C there are significant alterations in the polypeptide folding and the protein core becomes slightly loosened compared to the native state and forms the PRIME state (2), facilitating binding of SNAP-25 to form the enzyme substrate complex (3). BoNT/A LC exhibits optimum activity in the PRIME state. Further heating of BoNT/A LC to 50°C leads to the formation of the molten globule (4), facilitating binding of SNAP-25 to form the enzyme substrate complex (5) although to a lesser degree than the PRIME state, thus exhibiting only about 61% of the optimum enzyme activity (Kukreja and Singh, 2005).

globule-like structure. The dynamic role of such structure is more significant in view of the extreme specificity of the endopeptidase towards its substrate, SNAP-25, and its tremendous utility in designing specific antidotes against botulism threats. An understanding of BoNT structure in physiological conditions is essential to provide detailed mechanism involved in its action.

These findings not only open new avenues to design effective diagnostics and antidotes against botulism, but also provide new information on enzymatically active molten globule or molten globule-like structures. The ability to obtain a stabilized and structurally defined molten globule provides a useful model for studying the folding and unfolding pathways of proteins. Structural analysis of protein intermediates or molten globules provides valuable information on biophysical and biological aspects of proteins.

The presence of the PRIME conformational state, although being reported for the first time in BoNT/A endopeptidase, may play a major role in the biological functions of wide variety of proteins.

Structural and functional role of zinc in botulinum neurotoxins

Light chains of clostridial neurotoxins are Zn^{2+} metalloproteases and contain the conserved HEXXH motif comprising hydrophobic amino acids. The X-ray crystallography structure of BoNT/A shows that His223 and His227 of the HEXXH motif directly coordinate the zinc and Glu224 coordinates through a water molecule. Glu262 has been identified as the fourth ligand that directly coordinates zinc, similar to thermolysin (Binz et al., 2002). The active site of botulinum neurotoxins has been extensively studied using site-directed mutagenesis and X-ray crystallography (Zhou et al., 1995; Li et al., 2000; Rossetto et al., 2001; Rigoni et al., 2001; Binz et al., 2002; Agarwal et al., 2004, 2005; Segelke et al., 2004; Breidenbach and Brunger, 2004). Although the crystal structures provide a great degree of insight in to the structures of proteins, experiments to establish the functional role of amino acid residues are essential to understand the mechanics of its biological function. The active site structure of BoNT/A is similar to that of thermolysin and identifies a primary sphere of residues essential to catalytic function. Glu224 in the HEXXH motif is known to be critical for the catalytic activity of BoNT/A LC because of its coordination with water which performs the hydrolytic reaction of proteolysis (Li et al., 2000). Site-directed mutagenesis experiments have revealed the essential roles in catalysis of the His residues in the Zn^{2+}-binding motif and the role of Glu262, which provides a negatively charged carboxylate moiety (Zhou et al., 1995; Rigoni et al., 2001).

Recently we showed that Glu224 residue of the active site motif of BoNT/A endopeptidase has strict topographical requirement for its role in the catalytic activity as moving it to position 225 in the HEXXH motif (resulting in HXEXH), resulted in complete loss of enzyme activity (Kukreja et al., 2007). Examining the role of Glu262 is particularly critical to understand the conformational motif for the recognition of specific substrate cleavage sites. Substitution of Glu262 with Asp resulted in a significant loss of enzyme activity revealing the essential role of the spatial positioning of the carboxyl group of Glu262 for BoNT/A endopeptidase activity. Replacement of Glu262 with Gln dramatically impaired the endopeptidase activity suggesting the pivotal role of the negatively charged carboxyl moiety Glu262 in the catalytic activity of BoNT/A endopeptidase. The loss of enzyme activity in Gln262 seems to result from a global conformational change introduced in BoNT/A light chain that most likely affects substrate binding not only at the cleavage site but also at secondary recognition sites of SNARE motifs.

The chemical characterization of the catalytic centre of botulinum neurotoxins is a prerequisite for the elucidation of the molecular basis of the mechanism of this family of metalloproteases and for the design of neurotoxin inhibitors based on active site structure.

In addition, there is a secondary layer of residues, less close to the zinc centre, in which Arg362 and Tyr365 (BoNT/A) have been shown to be directly involved in catalysis by providing transition state stabilization (Binz et al., 2002). Atomic absorption measurements of the zinc content of clostridial neurotoxins revealed that they contain one atom of zinc per molecule of toxin, bound to the LC (Schiavo et al., 1992a; Schiavo et al., 1992c; Schiavo et al., 1993; Fu et al., 1998b). The only exception is BoNT/C, which coordinates two atoms of zinc (Schiavo et al., 1994). Zinc in proteins can play a catalytic or structural role. (Vallee and Auld, 1990). The catalytic role of Zn^{2+} has been confirmed in the LC of BoNT/A (Li and Singh, 2000). Removal of Zn^{2+} from the LC of BoNT/A abolishes its endopeptidase activity and its replacement restores only 30% of the endopeptidase activity (Li and Singh, 2000). Biological activity of BoNT/A in PC12 cells is also abolished by removal of zinc and cannot be restored upon its replenishment (Fu et al., 1998b). Structural analysis of BoNT/A and its light chain suggested that removal of zinc results in an irreversible change in protein's tertiary structure (Li and Singh, 2000; Fu et al., 1998b). These results suggest that zinc plays both catalytic and structural role in BoNT/A.

Unique structural features of botulinum neurotoxins

The endopeptidase activity of clostridial neurotoxins is unique in several ways including the need of nicking and reduction of the interchain disulphide bond to activate their enzymatic activity, the selectivity of substrate and cleavage sites.

Nicking-mediated activation of botulinum neurotoxins

BoNTs are synthesized as single chain proteins and are subsequently cleaved by an endogenous protease (BoNT/A), or an exogenous protease (BoNT/E) or by both (BoNT/B) resulting in active di-chain neurotoxin. Proteolytic nicking is known to play an important role in maximizing the toxicity of BoNTs. Nicking of BoNT/E results in 100-fold increase in toxicity as revealed by mouse lethality test or neurotransmitter blockage in chromaffin and PC12 cells (Bittner et al., 1989; Lomneth et al., 1991), and by the cleavage of actin (DasGupta and Tepp, 1993). Proteolytic nicking experiments with single chain BoNT/A and BoNT/E indicated that formation of the di-chain neurotoxin involved a minimum of two steps around the nicking site. Ten residues [438]Ser–[447]Lys (Beecher and DasGupta, 1997) and three residues [419]Gly–[421]Arg (Anthravally and DasGupta, 1997) were truncated from BoNT/A and BoNT/E, respectively.

The nicked and unnicked BoNTs have the same binding characteristics for mouse brain synaptosomes (Kozaki and Sakaguchi, 1982; Evans et al., 1986). A similar interaction of nicked and unnicked BoNT with lipid membranes suggests that nicking is not essential for the binding of the toxin to the membrane (Evans et al., 1986). It is thus logical to speculate that nicking does not alter the binding site of BoNTs, although the possibility of conformational changes in other domains upon nicking cannot be ruled out. The spectroscopic profile of nicked BoNT/E is different from that of the unnicked form (Datta and DasGupta, 1988; Singh and DasGupta, 1990), indicating that structural change is indeed associated with nicking. Electrophoretic analysis and fluorescence quenching experiments have suggested that nicking mediated structural alterations introduce significant changes in the surface charge distribution of BoNT/E (Singh and Dasgupta, 1990). In contrast to the similarity in binding, a 100-fold increase of channel formation has been observed upon nicking of BoNT/E with trypsin, but was abolished by soybean trypsin inhibitor (Blaustein et al., 1988). It can be assumed that nicking may free the H_N domain from the LC for the insertion of the properly folded H_N into the membrane.

Nicking is also required for intracellular inhibitory activity. Single chain BoNT/E does not express toxicity when micro-injected intra-neurally in Aplysia (Poulain et al., 1989). The endopeptidase activity of BoNT/B and TeNT against their target protein, VAMP is observed only upon nicking of the neurotoxin and the reduction of interchain disulphide bond (Schiavo et al., 1992a). The endopeptidase activity of single-chain BoNT/E is dramatically increased upon nicking even under conditions where the interchain disulphide bond remains intact (Sharma and Singh, 1999). The structural difference between the nicked and unnicked toxins may account for the activation of their endopeptidase activity.

Recently it was been observed that clostridial neurotoxins undergo Tyr phosphorylation in vitro and in vivo resulting in higher enzymatic activity (Ferrer-Monteil et al., 1996). The higher activity of BoNTs after Tyr phosphorylation has been attributed to structural changes that accompany it (Encinar et al., 1998). Structural and thermal stabilization of BoNTs has been observed upon Tyr phosphorylation (Encinar et al., 1998; Blanes-Mira et al., 2001). One of the unique characteristics of BoNTs is the long-term paralytic effects exerted by these neurotoxins (Shone and Melling, 1992). This long lasting effect of BoNTs may be related to tyrosine phosphorylation, which is regulated by the intra-cellular signalling cascade. However, these results cannot rule out the possibility that one or more NAPs remain associated with BoNTs or their light chain subunits in the nerve terminal, which could enhance the endopeptidase activity of BoNTs and stabilize them. (Sharma et al., 1999; Sharma and Singh, 2004).

Role of interchain disulphide bond in the action of clostridial neurotoxins

Reduction of the disulphide bond between HC and LC of clostridial neurotoxins is a prerequisite for their toxicity (Stecher et al., 1989). Reduction was shown to be rate limiting step in BoNT-mediated blockage on neurotransmitter release in chromaffin cells (Erdal et al., 1995). Proteolysis of BoNT/A and BoNT/E by trypsin and pepsin revealed the presence of only one interchain disulphide bond between LC and HC (Krieglstein et al., 1994; Anthravally and DasGupta, 1997). Alkylation of free cysteine residues in BoNT/A after reduction of the disulphide bond does not influence the toxification by injection in Aplysia neurons (de Paiva et al., 1993). Hence it can be assumed that -SH groups themselves do not play a role in the intracellular toxicity of clostridial neurotoxins. However, reduction is required to restore the endopeptidase activity of BoNTs (Kurazano et al., 1992; Yamasaki et al., 1994; Schiavo et al., 2000). Experimental evidence suggests that the disulphide bond must be intact for the toxin to poison intact cells but it must be broken for the light chain to act catalytically in the cytosol (Schiavo et al., 1990; de Paiva et al., 1993).

The interchain disulphide bond also plays a role in the internalization and translocation step as suggested by the fact that the activity of BoNT/A is disrupted after reduction of the interchain disulphide bond at the nerve-diaphragm by bath-application toxin to the mouse phrenic motor end plates (de Paiva et al., 1993). Recently, a crucial role of the disulphide bond linking the light and heavy chains throughout the translocation of the LC through the HC channel was demonstrated, as the premature reduction of the disulphide bond after channel formation aborted translocation (Fischer and Montal, 2007b).

A study concerning the role of disulphide bond cleavage on the enhanced neurotoxin activity has revealed that BoNT/A is more resistant to cleavage by endoproteinase Glu-C than its reduced form and the isolated HC. These results suggest that disulphide cleavage of BoNT/A leads to an altered structure which might form a relatively active endopeptidase conformation (Beecher and DasGupta, 1997).

The X-ray crystallographic structure and the solution structure of BoNT/A provide clues to the possible role of disulphide reduction in the activation of BoNT/A endopeptidase activity. The active site in BoNT/A is buried 20–24 Å deep in the protein matrix and is partially shielded by a belt from the N-terminal domain of the heavy chain involved in the disulphide bond formation. The reduction of the disulphide bond presumably allows accessibility of the active site to the substrate (Lacy et al., 1998). The solution structural information of BoNT/A has indicated a more flexible structure upon reduction of disulphide bond (Cai et al., 1999; Cai and Singh, 2001). It has recently been shown that the activation mechanism of BoNT/A by reduction of the interchain disulphide bond involves a molten globule state under physiological temperature conditions (Cai and Singh, 2001). Taken together, these results suggest that the

structural change incorporated upon reduction of interchain disulphide bond plays a critical role in the disulphide-reduction mediated activation process. Although the overall crystal structures of BoNT/A and BoNT/B under non-reducing conditions are very similar, there is a difference in the belt region that extends from the translocation domain (Lacy et al., 1998; Swaminathan and Eswaramoorthy, 2000). In BoNT/A the catalytic site is partially covered by the belt enabling the zinc ion to be shielded from the environment. In BoNT/B, however, the belt region does not shield the zinc ion, thus making it completely accessible to substrate molecules (Swaminathan and Eswaramoorthy, 2000). Nevertheless, BoNT/B requires reduction of its disulphide bond between heavy and light chains to activate its endopeptidase activity (Schiavo et al., 1992b). Thus the disulphide bond in BoNT/B may play a crucial role in binding of the unstructured region of it substrate, synaptobrevin to the neurotoxin.

Recent evidence suggests that disulphide bond reduction of all BoNT serotypes produced conformational changes that diminished toxicity against intact cells but it produced conformational changes that led to exposure of the catalytic site only in BoNT serotypes A, C, and D. In case of BoNT/B,/E, and/F, the catalytic site was accessible even before disulphide bond reduction. Hence in these serotypes, disulphide bond reduction may govern something other than, or in addition to, the location of the belt segment (Simpson et al., 2004).

Substrate specificity and cleavage site selectivity

One of the most prominent unique characteristic features of BoNT is peptide bond cleavage selectivity for very specific and nearly exclusive substrates. BoNTs specifically proteolyse SNARE proteins (SNAP-25, synaptobrevin, and syntaxin) involved in neurotransmitter release process. Each of the seven BoNT serotypes either recognize an exclusive protein substrate or at least a selective cleavage site on the given substrate. No other cellular substrates are known to be proteolysed by BoNTs. No two BoNTs cleave at the same peptide bond, and except for BoNT/C1, no BoNT proteolyses more than one substrate. BoNT serotypes A and E cleave

SNAP-25 at the C-terminus, serotypes B, D, F, and G each cleave VAMP at different sites, and serotype C cleaves syntaxin and SNAP-25 (Singh, 2000). BoNT endopeptidases prefer the whole protein as substrate over short peptides encompassing the specific cleavage site, further enhancing substrate specificity. While peptide-based substrates, albeit relatively long peptides are cleaved by BoNT endopeptidases, cleavage is poor in terms of enzyme kinetics (Breidenbach and Brunger, 2004).

Therapeutic applications of botulinum neurotoxins

The first documented use of BoNT for the treatment of a neuromuscular disorder was in early 1980s, about 150 years after Kerner's initial observations of the potential use of BoNT as a therapeutic, when Dr. Alan Scott used local injection of minute doses to selectively inactivate muscle spasticity in strabismus (Scott, 1980). Following the success of series of clinical studies (Scott et al., 1985; Dutton and Buckley, 1988), the Food and drug Administration (FDA) in 1989, approved the use of BoNT/A for the treatment of strabismus, blepharospasm, and hemifacial spasm. Following this, the very lethal botulinum toxins, botulinum type A (marketed as BOTOX) and type B (marketed as Myobloc), have been extensively used for the treatment of cervical dystonia and other medical conditions including focal dystonias, spasticity, tremors, migraine and tension headaches, and hyperhidrosis (Brashear et al., 1999; Wheeler, 1998; Schulte-Mattler et al., 1999; Binder et al., 2000; Silberstein et al., 2000; Dressler et al., 2002; Sampaio et al., 2004; Farrugia and Nicholls, 2005). Botulinum neurotoxin is now widely considered as a pharmaceutical agent with multiple uses, and has been propelled into the public eye after it was widely reported to act as an anti-wrinkle drug for facial cosmetic enhancement (Bhidayasiri and Truong, 2005; Dayar and Mass, 2007; Salti and Ghersetich, 2008). The therapeutic utility of BoNTs has generated enormous interest, as evidenced by the popularity of 'Botox parties' for the treatment of facial wrinkles (Wenzel, 2004).

Incredibly, the list of conditions treated with botulinum toxin is expanding at a brisk rate. The remarkable therapeutic utility of botulinum toxin lies in its ability to specifically and potently inhibit involuntary muscle activity for an extended duration.

Recent reports have suggested that there are substantial systemic adverse reactions and respiratory compromise as a result of the use of types A and B botulinum neurotoxin complex based therapeutic and cosmetic products (Botox™ and Myobloc™), leading United States Food and Drug Administration to issue a medical advisory on their safety (http://www.fda.gov/cder/drug/early_comm/botulinium_toxins.htm). It is widely assumed that BoNT/A when injected for therapeutic purposes remains confined to the injection site. However, it was recently shown that BoNT/A spreads to different synapses in physiologically significant amounts via retrograde axonal transport and transcytosis to reach the CNS after peripheral administration (Antonucci et al., 2008), which may have serious clinical implications.

Concluding remarks and trends of future research

From a deadly food poison to frontline medicine, the discovery, understanding and utilization of botulinum neurotoxins has been an interesting story. Advancements in understanding the toxin structure have been valuable in deciphering the mechanisms of membrane exocytosis and to understand their therapeutic potential. These advances have provided us with molecular clues for further discoveries following the neurotoxin's use as a tool in neurobiology and cellular biology.

Novel uses can also be envisaged for BoNTs as carriers of biologically active molecules to particular areas of the cell or of the body. Many questions yet need to be answered about the molecular mode of action of clostridial neurotoxins. Understanding of the functionalities of the individual domains will be useful to develop medicines and research tools that can affect a multitude of cellular functions. Detailed study of the structure–function relationship of clostridial neurotoxins can not only help solve their molecular mode of action but will also provide a greater understanding of the potential use of individual domains to answer more fundamental questions of neuroexocytosis and for the development of new therapeutics.

Future research areas of botulinum toxins are very broad, and some of them are only indirectly related to the structure and mechanism of action. One such area is the purpose of the production of this toxin by the bacteria. Does the toxin have a function with the bacterial physiology or at least in its survival and reproduction? The fact that the endopeptidase activity of the toxin when released in the complex form containing the toxin and NAPs does not need reduction of the disulphide bond indicates that such an endopeptidase activity may be targeted towards another substrate present in organisms present within the ecological environment of *Clostridium botulinum*.

The presence of NAPs and its role in the protection and possible role in the translocation of the toxin across the gut wall is a unique example of a protein protecting another protein in the gastrointestinal system. This characteristic may be used to not only deliver botulinum toxin based therapeutics or vaccine orally but also use the system to deliver other drugs orally. More research is needed to understand the interactions between the toxin and its individual NAPs. Since NAPs and BoNT are synthesized polycistronically, regulated by a common regulatory protein, botR (Marvaud *et al.*, 1998), it is critical to examine the sequence of expression of these proteins and their assembly process to understand structural basis of their stability and functional activity.

Current use of botulinum neurotoxins types A and B mostly involve the complex form, and while the complex form provides stability of the product, it also puts larger load of protein to the immune system. While this has been reported as the reason why some patients become non-responders to botulinum neurotoxin-based drugs, the load on the immune system may also lead to some of the side effects. The interaction of NAPs and the toxin with immune system needs to be examined.

The existence of seven serotypes and further subtypes of BoNT has raised several issues, including their sequence homology, common or diverse set of specific receptors, efficiency of translocation into the neuronal cell, and selectivity of substrates. While significant progress has been made in identifying receptors for certain serotypes of BoNT, receptors for all the serotypes are not settled nor are their selectivities. Recent discovery that SV2, the receptor present at the neuronal cell, is also present at the epithelial cell (Couesnon *et al.*, 2008), raises a major question of why the toxin is selective to the nervous system.

The most unresolved question in botulinum structure function and mechanism of action remains the biochemical basis of the long lasting effects of botulinum neurotoxin, with the paralysis effects lasting several months. So far, research has focused on the sequestration of the light chain within the neuronal cell, selective ubiquitination of light chain, and stability of the cleaved SNAP-25. This issue requires research on signal transduction involved in the action of botulinum neurotoxin, including role of alternative signal transduction pathways (Ray *et al.*, 1993; Ishida *et al.*, 2004).

Understanding of structure–function and mechanism of action is critical for developing antidotes against botulism. There have been several discrepancies in the effectiveness of inhibitors developed so far against botulism under *in vitro*, *ex vivo*, and *in vivo* conditions, perhaps arising from the structural differences which might be encountered under these conditions. A better understanding of the structure–function relationship is critical to developing effective specific inhibitors to counter botulism biothreats.

Finally, expansion in the medical use of botulinum neurotoxin is leading to opportunities for developing innovative molecules with recombinant DNA technologies for targeting specific neuronal and non-neuronal cells (Chaddock *et al.*, 2004). This line of research is in its infancy, and is likely to attract many researchers from academic, government, and industrial laboratories.

Acknowledgements
This work was supported by a DoD/Army Contract No. W911NF-06-1-0095, a DARPA contract – W911NF-07-1-0623, and by the National Institutes of Health through the New England Center of Excellence for Biodefense (Grant AI057159-01).

References

Agarwal, R., Eswaramoorthy, S., Kumaran, D., Binz, T., and Swaminathan, S. (2004). Structural analysis of botulinum neurotoxin type E catalytic domain and its mutant Glu212→Gln reveals the pivotal role of the Glu212 carboxylate in the catalytic pathway. Biochemistry 43, 637–6644.

Agarwal, R., Binz, T., and Swaminathan, S. (2005). Analysis of active site residues of botulinum neurotoxin type E by mutational, functional and structural studies: Glu335Gln is an apoenzyme. Biochemistry 44, 8291–8302.

Ahsan, C.R., Hajnoczky, G., Maksymowych, A.B., and Simpson, L.L. (2005). Visualization of binding and transcytosis of botulinum toxin by human intestinal epithelial cells. J. Pharmacol. Exp. Ther. 315, 1028–1035.

Anthravally, B.S., and DasGupta, B.R. (1997). Covalent structure of botulinum neurotoxin type E: location of sulfydryl groups and disulfide bridges and identification of C-termini of light and heavy chain. J. Protein Chem. 16, 787–799.

Antonucci, F., Rossi, C., Gianfranceschi, L., Rossetto, O., and Caleo, M. (2008). Long-distance retrograde effects of botulinum neurotoxin A. J. Neurosci. 28, 3689–3696.

Aoki, K.R., and Guyer, B. (2001). Botulinum toxin type A and other botulinum toxin serotypes: a comparative review of biochemical and pharmacological actions. Eur. J. Neurol. 8, 21–29.

Baldwin, M.R., and Barbieri, J.T. (2007). Association of botulinum neurotoxin serotypes A and B with synaptic vesicle protein complexes. Biochemistry. 46, 3200–3210.

Banerjee, A., Kowalchyk, J.A., DasGupta, B.R., and Martin, T.F.J. (1996). SNAP-25 is required for a late postdocking step in Ca^{2+} dependent exocytosis. J. Biol. Chem. 271, 20227–20230.

Beecher, D.J., and DasGupta, B.R. (1997). Botulinum neurotoxin type A: limited proteolysis by endoproteinase Glu-C and alpha-chymotrypsin enhanced following reduction; identification of the cleaved sites and fragments. J. Protein. Chem. 16, 701–702.

Bhidayasiri, R., and Truong, D.D. (2005). Expanding use of botulinum toxin. J. Neurol. Sci. 235, 1–9.

Bigalke, H., and Rummel, A. (2005). Medical aspects of toxin weapons. Toxicology 214, 210–220.

Binder, W.J., Brin, M.F., Blitzer, A., Schuoenrock, L.D., and Pogoda, J.M. (2000). Botulinum toxin type A (BOTOX) for treatment of migraine headaches: an open-label study. Otolaryngol. Head Neck Surg. 123, 669–76.

Binz, T., Bade, S., Rummel, A., Kollewe, A., and Alves, J. (2002). Arg362 and Tyr 365 in botulinum neurotoxin type A light chain are involved in transition state stabilization, Biochemistry 41, 1717–1723.

Bittner, M.A., DasGupta, B.R., and Holz, R.W. (1989). Isolated light chains of botulinum neurotoxins inhibit exocytosis: studies in digitonin-permeabilized chromaffin cells. J. Biol. Chem. 264, 10354–10360.

Blanes-Mira, C., Ibanez, C., Fernandez-Ballester, G., Planells-Cases, R., Perez-Paya, E., and Ferrer-Montiel, A. (2001). Thermal stabilization of the catalytic domain of botulinum neurotoxin E by phosphorlyation of a single tyrosine residue. Biochemistry 40, 2234–2242.

Blasi, J., Chapman, E.R., Link, E., Binz, T., Yamasaki, S., De camilli, P., Südhof, T.C., Niemann, H., and Jahn, R. (1993). Botulinum neurotoxin A selectively cleaves the synaptic protein SNAP-25. Nature 365, 160–163.

Blaustein, R.O., Hoch, D.H., and DasGupta, B.R. (1988). Channels formed by botulinum type E neurotoxin in planar lipid bilayers. FASEB J. 2, A1750. (abstr).

Brashear, A., Lew, M.F., and Dykstra, D.D. (1999). Safety and efficacy of botulinum type B in type A responsive cervical dystonia. Neurology 53, 1439–1446.

Breidenbach, M.A., and Brunger, A.T. (2004). Substrate recognition strategy for botulinum neurotoxin serotype A. Nature 432, 925–929.

Breidenbach, M.A., and Brunger, A.T. (2005). 2.3 Å crystal structure of tetanus neurotoxin light chain. Biochemistry 44, 7450–7457.

Brunger, A.T., Jin, R., and Breidenbach, M.A. (2008). Highly specific interactions between botulinum neurotoxins and synaptic vesicle proteins. Cell Mol. Life Sci. 65, 2296–2306.

Brunger, A. T., Breidenbach, M.A., Rongsheng, J., Fischer, A., Santos, J.S., and Montal, M. (2007). Botulinum neurotoxin heavy chain belt as an intramolecular chaperone for the light chain. PLoS Pathogen. 3, 1191–1194.

Bryson, J.W., Betz, S.F., Lu, H.S., Suich, D.J., Zhou, H.X., O'Neil, K.T., and DeGrado, W.F. (1995). Protein design: a hierarchic approach. Science 270, 935–941.

Burgoyne, R.D., and Morgan, A. (1993). Regulated exocytosis. Biochem. J. 293, 305–316.

Burnett, J.C., Ruthel, G., Stegmann, C.M., Panchal, R.G., Nguyen, T.L., Hermone, A.R., Stafford, R.G., Lane, D.J., Kenny, T.A., McGrath, C.F., Wipf, P., Stahl, A.M., Schmidt, J.J., Gussio, R., Brunger, A.T., and Bavari, S. (2007). Inhibition of metalloprotease botulinum serotype A from a pseudo-peptide binding mode to a small molecule that is active in primary neurons. J. Biol. Chem. 282, 5004–5014.

Bychkova, V.E., Pain, R.H., and Ptitsyn, O.B. (1988). The "molten globule" state is involved in the translocation of proteins across membranes? FEBS Lett. 238, 231–234.

Cabiaux, V., Lorge, P., Falmagne, P., and Ruysschaert, J.M. (1985). Tetanus toxin induces fusion and aggregation of lipid vesicles containing phosphatidylionositol at low pH. Biochem. Biophys. Res. Commun. 128, 840–849.

Cai, S., and Singh, B.R. (2001). Role of the disulfide cleavage induced molten globule state of type A botulinum neurotoxin in its endopeptidase activity. Biochemistry 50, 15327–15333.

Cai, S., Sarkar, H.K., and Singh, B.R. (1999). Enhancement of the endopeptidase activity of botulinum neurotoxin by its associated proteins and dithiothreitol. Biochemistry 38, 6903–6910.

Cai, S., Kukreja, R., Shoesmith, S., Chang, T.W., and Singh, B.R. (2006). Botulinum neurotoxin light chain refolds at endosomal pH for its translocation. Protein J. 25, 455–62.

Caya, J.G., Agni, R., and Miller, J.E. (2004). Clostridium botulinum and the clinical laboratorian. Arch. Pathol. Lab. Med. 128, 653–662.

Chaddock, J.A., Purkiss, J.R., Alexander, F.C., Doward, S., Fooks, S.J., Friis, L.M., Hall, Y.H., Kirby, E.R., Leeds, N., Moulsdale, H.J., Dickenson, A., Green, G.M., Rahman, W., Suzuki, R., Duggan, M.J., Quinn, C.P., Shone, C.C., and Foster, K.A. (2004). Retargeted clostridial endopeptidases: inhibition of nociceptive neurotransmitter release in vitro, and antinociceptive activity in invivo models of pain. Mov. Disord. Suppl. 8, S42–47.

Chai, Q., Arndt, J.W., Dong, M., Tepp, W.H., Johnson, E.A., Chapman, E.R., and Stevens, R.C. (2006). Structural basis of cell surface receptor recognition by botulinum neurotoxin B. Nature. 444, 1096–1100.

Chen, F., Kuziemko, G.M., and Stevens, R.C. (1998). Biophysical characterization of the stability of the 150-kilodalton botulinum toxin, the nontoxic component, and the 900-kilodalton botulinum toxin complex species. Infect. Immun. 66, 2420–2425.

Cheng, L.W., Onisko, B., Johnson, E.A., Reader, J.R., Griffey, S.M., Larson, A.E., Tepp, W.H., Stanker, L.H., Brandon, D.L, and Carter, J.M. (2008). Effects of purification on the bioavailability of botulinum neurotoxin type A. Toxicology 249, 123–129.

Cherington, M. (1998). Clinical spectrum of botulism. Muscle and Nerve 21, 701–710.

Cherington, M. (2004). Botulism: update and review. Semin. Neurol. 24, 155–163.

Choe, S., Bennett, M.J., Fujii, G. Curmi, D.M., Kantardjieff, K.A., Collier, R.J., and Eisenberg, D. (1992). The crystal structure of diphtheria toxin. Nature 357, 216–222.

Cordoba, J.J., Collins, M.D., and East, K. (1994). Studies on the genes encoding for botulinum neurotoxin type A of Clostridium botulinum from a variety of sources. System Appl. Microbiol. 18, 13–22.

Couesnon, A., Pereira, Y., and Popoff, M.R. (2008). Receptor-mediated transcytosis of botulinum neurotoxin A through intestinal cell monolayers. Cell. Microbiol. 10, 375–387.

DasGupta, B.R., and Tepp, W. (1993). Protease activity of botulinum neurotoxin type E and its light chain: cleavage of Actin. Biochem. Biophy. Res. Commun. 190, 470–474.

Dastoor, S.F., Misch, C.E., and Wang, H.L. (2007). Botulinum toxin (Botox) to enhance facial macroesthetics: a literature review. J Oral Implantol. 33, 164–71.

Datta, A., and DasGupta, B.R. (1988). Circular dichroic and fluorescence spectroscopic study of the conformation of botulinum neurotoxin types A and E. Mol. Cell. Biochem. 79, 153–159.

Davletov, B., Bajohrs, M., and Binz, T. (2005). Beyond BOTOX: advantages and limitations of individual botulinum neurotoxins. Trends Neurosci. 28, 446–452.

Dayan, S.H., and Mass, C.S. (2007). Botulinum toxins for facial wrinkles: beyond glabellar lines. Facial Plast. Surg. Clin. North Am. 15, 41–49.

De Paiva, A., Poulain, B., Lawrence, G.W., Shone, C.C., Tauc, L., and Dolly, O.J. (1993). A role for the interchain disulfide or its participating thiols in internalization of botulinum neurotoxin A revealed by a toxin derivative that binds to ecto-acceptors and inhibits transmitter release intracellularly. J. Biol. Chem. 268, 20838–20844.

Dolly, O.J., Black, J., Williams, R.S., and Melling, J. (1984). Acceptors for botulinum neurotoxin reside on motor nerve terminals and mediate its internalization. Nature 307, 457–460.

Dong, M., Yeh, F., Tepp, W.H., Dean, C., Johnson, E.A., Janz, R., and Chapman, E.R. (2006). SV2 is the protein receptor for botulinum neurotoxin A. Science 312, 592–596.

Dresslar, D., Adib Saberi, F., and Benecke, R. (2002). Botulinum toxin type B for treatment of axillar hyperrhydrosis. J. Neurol. 249, 1729–1732.

Dutton, J.J., and Buckley, E.G. (1988). Long-term results and complications of botulinum A toxin in the treatment of blepharospasm. Ophthalmology 95, 1529–1534.

Encinar, J.A., Fernandez, A., Ferragut, J.A., Gonzalez-Ros, J.M., DasGupta, B.R., Montal, M., and Ferrer-Monteil, A. (1998). Structural stabilization of botulinum neurotoxins by tyrosine phosphorylation. FEBS Lett. 429, 78–82.

Erbguth, F.J., and Nauman, M. (1999). Historical aspects of botulinum toxin: Justinus Kerner (1786–1862) and the "sausage poison". Neurology 53, 1850–1853.

Erdal, E., Bartels, F., Binscheck, T., Erdmann, G., Frevert, J., Kistner, A., Weller, U., Wnere, J., and Bigalke, H. (1995). Processing of tetanus and botulinum A neurotoxins in isolated chromaffin cells. Naunyn-Schmeideberg's Arch. Pharmacol. 351, 67–78.

Evans, D.M., Williams, R.S., Shone, C.C., Hambleton, P., Melling, J., Dolly, O.J. (1986). Botulinum neurotoxin type B. Its purification, radioiodination and interaction with rat-brain synaptosomal memebranes. Eur. J. Biochem. 154, 409–416.

Farrugia, M.K. and Nicholls, E.A. (2005). Intradermal botulinum A toxin injection for axillary hyperhydrosis. J. Pediatr. Surg. 40, 1668–1669.

Fasshauer, D. Bruns, D., Shen, B., Jahn, R. and Brunger, A.T. (1997). A structural change occurs upon binding of syntaxin to SNAP-25. J. Biol. Chem. 272, 4582–4590.

Ferrer-Monteil, A.V., Canaves, J.M., DasGupta, B.R., Wilson, M.C., Montal, M. (1996). Tyrosine phosphorylation modulates the activity of clostridial neurotoxins. J. Biol. Chem. 271, 18322–18325.

Fiebig. K.M., Rice, L.M., Pollock, E., and Brunger, A.T. (1999). Folding intermediates of SNARE complex assembly. Nat. Struct. Biol. 6, 117–123.

Fischer, A., and Montal, M. (2007a). Single molecule detection of intermediates during botulinum neurotoxin translocation across membranes. Proc. Natl. Acad. Sci. USA 104, 10447–10452.

Fischer, A., and Montal, M. (2007b). Crucial role of the disulfide bridge between botulinum neurotoxin light and heavy chains in protease translocation across membranes. J. Biol. Chem. *282*, 29604–29611.

Fu, F.N., and Singh, B.R. (1999). Calcein permeability of liposomes mediated by type A botulinum neurotoxin and its light and heavy chains. J. Protein Chem. *18*, 701–707.

Fu, F.N., Sharma, S.K., and Singh, B.R. (1998a). A protease-resistant novel hemagglutinin purified from type A *Clostridium botulinum*. J. Protein Chem. *7*, 63–60.

Fu, F., Lomneth, R.B., Cai, S., and Singh, B.R. (1998b). Role of zinc in the structure and toxic activity of botulinum neurotoxin. Biochemistry *37*, 5267–5278.

Fujinaga, Y., Inoue, K., Watanabe, S., Yokota, K., Hirai, Y., Nagamachi, E., and Oguma, K. (1997). The hemagglutinin of *Clostridium botulinum* type C progenitor toxin plays an essential role in binding of toxin to the epithelial cells of guinea pig small intestine, leading to the efficient absorption of the toxin. Microbiology *143*, 3841–3847.

Fujinaga, Y., Inoue, K., Nomura, T., Sasaki, J., and Marvaud, J.C. (2000). Identification and characterization of functional subunits of *Clostridium botulinum* type A progenitor toxin involved in binding to intestinal microvilli and erythrocytes. FEBS Lett. *467*, 179–183.

Fujinaga, Y., Inoue, K., Watanabe, S., Sakaguchi, Y., Arimitsu, H., Lee, J.C., Jin, Y., Matsumura, T., Kabumoto, Y., Watanabe, T., and Ohyama, T. (2004). Molecular characterization of binding subcomponents of *Clostridium botulinum* type C progenitor toxin for intestinal epithelial cells and erythrocytes. Microbiology *150*, 1529–1538.

Fujita, R., Fujinaga, Y., Inoue, K., Nakajima, H., Kumon, H., and Oguma, K. (1995). Molecular characterization of two forms of non-toxic non-haemagglutinin components of *Clostridium botulinum* type A progenitor toxins. FEBS Lett. *376*, 41–44.

Graham, M.E., Washbourne, P., Wilson, M.C., and Burgoyne, R.D. (2001). SNAP-25 with mutations in the zero layer supports normal membrane fusion kinetics. J. Cell Sci. *114*, 4397–4405.

Graham, M.E., Washbourne, P., Wilson, M.C., and Burgoyne, R.D. (2002). Molecular analysis of SNAP-25 function in exocytosis. Ann. N.Y. Acad. Sci. *971*, 210–221.

Hanson, M.A., and Stevens, R.C. (2000). Co-crystal structure of synaptobrevin-II bound to botulinum neurotoxin type B at 2.0Å resolution. Nat. Struct. Biol. *7*, 687–692.

Hanson, P.I., Heuser, J.E., and Jahn, R. (1997). Neurotransmitter release – four years of SNARE complexes. Curr. Opin. Neurobiol. *7*, 310–315.

Hatheway, C.L. (1993). In Botulinum and Tetanus Neurotoxins.: Bacteriology and pathology of neurotoxigenic clostridia, DasGupta BR, ed. (Plenum Press, New York).

Hoch, D.H., Romero-Mira, M., Ehrlich, B.E., Finkelstein, A., DasGupta, B.R., and Simpson, L.L. (1985). Channels formed by botulinum, tetanus, and diphtheria toxins in planar lipid bilayers: relevance to translocation of proteins across membranes. Proc. Natl. Acad. Sci. USA *82*, 1692–1696.

Humeau, Y., Dossau, F., Grant, N.J., and Poulain, B. (2000). How botulinum and tetanus neurotoxins block neurotransmitter relaease? Biochimie *82*, 427–446.

Inoue, K., Fujinaga, Y., Watanabe, T., Ohyama, T., Takeshi, K., Moriishi, K., Nakajima, H., and Oguma, K. (1996). Molecular composition of *Clostridium botulinum* type A progenitor toxins. Infect. Immun. *64*, 1589–1594.

Inoue, K., Fujinaga, Y., Honke, K., Arimitsu, H., Mahmut, N., Sakaguchi, Y., Ohyama, T., Watanabe, T., Inoue, K., and Oguma, K. (2001). *Clostridium botulinum* type A haemagglutinin-positive progenitor toxin (HA(+)-PTX) binds to oligosaccharides containing Gal beta1–4GlcNAc through one subcomponent of haemagglutinin (HA1). Microbiology *147*, 811–819.

Ishida, H., Zhang, X., Erickson, K., and Ray, P. (2004). Botulinum toxin type A targets RhoB to inhibit lyso-phosphatidic acid-stimulated actin reorganization and acetylcholine release in nerve growth factor-treated PC12 cells. J. Pharmacol. Exp. Ther. *310*, 881–889.

Jahn, R. (2006). A neuronal receptor for botulinum toxin. Science *312*, 540–541.

Jin, R., Rummel, A., Binz, T., and Brunger, A.T. (2006). Botulinum neurotoxin B recognizes its receptor with high affinity and specificity. Nature. *444*, 1092–1095.

Johnson, E.A. (1999). Clostridial toxins as therapeutic agents: benefits of nature's most toxic proteins. Annu. Rev. Microbiol. *53*, 551–575.

Koh, T.W., and Bellen, H.J. (2003). Synaptotagmin I, a Ca^{2+} sensor for neurotransmitter release. Trends Neurosci. *26*, 413–422.

Koriazova, L.K., and Montal, M. (2003). Translocation of botulinum neurotoxin light chain protease through the heavy chain channel. Nat. Struct. Biol. *10*, 13–18.

Kozaki, S., and Sakaguchi, G. (1982). Binding of mouse brain synaptosomes of *Clostridium botulinum* type E derivative toxin before and after tryptic activation. Toxicon *20*, 841–846.

Krantz, B. A., Trivedi, A. D., Cunningham, K., Christensen, K.A., and Collier, R. J. (2004). Acid-induced unfolding of the amino-terminal domains of the lethal and edema factors of anthrax toxin. J. Mol. Biol. *344*, 739–756.

Krieglstein, K.G., DasGupta, B.R., Henschen, A.H. (1994). Covalent structure of botulinum neurotoxin type A: location of sulfhydryl groups, and disulfide bridges and identification of C-termini of light and heavy chains. J. Protein Chem. *13*, 49–57.

Kukreja, R., and Singh, B.R. (2005). Biologically active novel conformational state of botulinum, the most poisonous poison. J. Biol. Chem. *280*, 39346–39352.

Kukreja, R.V., and Singh, B.R. (2007). Comparative role of neurotoxin-associated proteins in the structural stability and endopeptidase activity of botulinum neurotoxin complex types A and E. Biochemistry *46*, 14316–14324.

Kukreja, R.V., Sharma, S., Cai, S., and Singh, B.R. (2007). Role of two active site Glu residues in the molecular action of botulinum neurotoxin endopeptidase. Biochim. Biophys. Acta *1774*, 213–22.

Kurazono, H., Mochida, S., Binz, T., Eisel, U., Quanz, M., Grebenstein, O., Wernars, K., Poulain, B., Tauc,

L., and Niemann, H. (1992). Minimal essential domains specifying toxicity of the light chains of tetanus and botulinum neurotoxin type A. J. Biol. Chem. *267*, 14721–14729.

Lacy, D.B., and Stevens, R.C. (1999). Sequence homology and structural analysis of the clostridial neurotoxins. J. Mol. Biol. *291*, 1091–1104.

Lacy, D.B., Tepp, W., Cohen, A.C., DasGupta, B.R., and Stevens, R.C. (1998). Crystal structure of botulinum neurotoxin type A and implications for toxicity. Nat. Struct. Biol. *5*, 898–902.

Lebeda, F.J., and Singh, B.R. (1999). Membrane channel activity and translocation of tetanus and botulinum neurotoxins. Toxin Rev. *18*, 45–76.

Li, B., Qian, X., Sarkar, H.K., and Singh, B.R. (1998). Molecular characterization of type E *Clostridium botulinum* and comparison to other types of Clostridium botulinum. Biochim. Biophys. Acta *1395*, 21–7.

Li, L., and Singh, B.R. (1999). Structure-function relationship of clostridial neurotoxins. J. Toxicol.-Toxin Rev. *18*, 95–112.

Li, L., and Singh, B.R. (2000). Role of zinc binding in type A botulinum neurotoxin light chain's toxic structure. Biochemistry *39*, 10581–10586.

Li, L., Binz, T., Niemann, H., and Singh, B.R. (2000). Probing the role of glutamate residue in the zinc-binding motif of type A botulinum neurotoxin light chain, Biochemistry *39*, 2399–2405.

Lomneth, R., Martin, T.F., DasGupta, B.R. (1991). Botulinum neurotoxin light chain inhibits neoepinephrine secretion in PC12 cells at an intracellular membranous or cytoskeletal site. J. Neurochem. *57*, 1413–1421.

Maksymowych, A.B., and Simpson, L.L. (1998). Binding and transcytosis of botulinum toxin by polarized human colon carcinoma cells. J. Biol. Chem. *273*, 21950–21957.

Maksymowch, A.B., and Simpson, L.L. (2004). Structural features of the botulinum neurotoxin molecule that govern binding and transcytosis across polarized human intestinal epithelial cells. J. Pharmacol. Exp. Ther. *310*, 633–641.

Maksymowych, A.B., Reinhard, M., Malizio, C.J., Goodnough, M.C., Johnson, E.A., and Simpson, L.L. (1999). Pure botulinum neurotoxin is absorbed from the stomach and small intestine and produces peripheral neuromuscular blockade. Infect. Immun. *67*, 4708–4712.

Marhold, S., Rummel, A., Bigalke, H., Davletov, B., and Binz, T. (2006). The synaptic vesicle protein SV2C mediates the uptake of botulinum neurotoxin A into phrenic nerves. FEBS Lett. *580*, 2011–2014.

Marvaud, J.C., Gilbert, M., Inoue, K., Fujinaga, Y., Oguma, K., and Popoff, M.R. (1998). botR/A is a positive regulator of botulinum neurotoxin and associated non-toxin protein genes in *Clostridium botulinum* A. Mol. Microbiol. *29*, 1009–1018.

Matteoli, M., Verderio, C., Rossetto, O., Lezzi, N., Coco, S., Schiavo, G., and Montecucco, C. (1996). Synaptic vesicle endocytosis mediates the entry of tetanus neurotoxin into hippocampal neurons. Proc. Natl. Acad. Sci. USA. *93*, 13310–13315.

Merrison, A.F., Chidley, K.E., Dunnett, J., and Sieradzan, K.A. (2002). Wound botulism associated with subcutaneous drug use. BMJ. *325*, 1020–1021.

Minton, N.P. (1995). Molecular genetics of clostridial neurotoxins. Curr. Top. Microbiol. Immunol. *195*, 161–194.

Mirsky, R., Wendon, L.M., Black, P., Stolkin, C., and Bray, D. (1978). Tetanus toxin: a cell surface marker for neurons in culture. Brain Res. *148*, 251–259.

Montecucco, C., and Schiavo, G. (1993). Tetanus and botulism neurotoxins: a new group of zinc proteases. Trends Biochem. Sci. *11*, 314–317.

Montecucco, C., and Schiavo, G. (1995). Structure and function of tetanus and botulinum neurotoxins. Q. Rev. Biophys. *28*, 423–472.

Montecucco, C., Schiavo, G., Brunner, J., Duflot, E., Boquet, P., Roa, M. (1986). Tetanus toxin is labeled with photoactivable phospholipids at low pH. Biochemistry *25*, 919–924.

Montecucco, C., Schiavo, G., Dasgupta, B. R. (1989). Effect of pH on the interaction of botulinum neurotoxins A, B and E with liposomes. Biochem. J. *259*, 47–53.

Montecucco, C., Papini, C., and Schiavo, G. (1994). Bacterial toxins penetrate cells via a four-step mechanism. FEBS Lett. *346*, 92–98.

Nishikawa, A., Uotsu, N., Arimitsu, H., Lee, J.C., Miura, Y., Fujinaga, Y., Nakada, H., Watanabe, T., Ohyama, T., Sakano, Y., and Oguma, K. (2004). The receptor and transporter for internalization of Clostridium botulinum type C progenitor toxin into HT-29 cells. Biochem Biophys. Res. Comm. *319*, 327–333.

Nishiki, T., Kamata, Y., Nemoto, Y., Omori, A., Ito, T. Takahashi, M., and Kozaki, S. (1994). Identification of protein receptor for *Clostridium botulinum* type B neurotoxin in rat brain synaptosomes. J. Biol. Chem. *269*, 10498 - 10503.

Nishiki, T., Tokuyama,Y., Kamata, Y., Nemoto, Y., Yoshida, A., Sato, K., Sekiguchi, M., Takahashi, M., and Kozaki, S. (1996). The high-affinity binding of *Clostridium botulinum* type B neurotoxin to synaptotagmin II associated with gangliosides GT_{1b}/GD_{1a}. FEBS Lett. *378*, 253–257.

Ossig, R., Schmitt, H.D., de Groot, B., Riedel, D., Keränen, S., Ronne, H., Grubmüller, H., and Jahn, R. (2000). Exocytosis requires asymmetry in the central layer of the SNARE complex. EMBO J. *19*, 6000–6010.

Park, J.B., and Simpson, L.L. (2003). Inhalation poisoning by botulinum toxin and inhalation vaccination with its heavy-chain component. Infect. Immun. *71*, 1147–1154.

Pellegrini, L.L., O'Connor, V., Lottspeich, F., and Betz, H. (1995). Clostridial neurotoxins compromise the stability of a low energy SNARE complex mediating. NSF activation of synaptic vesicle fusion. EMBO J. *14*, 4705–4713.

Pellizari, R., Rossetto, O., Lozzi, L., Giovedi, S., Johnson, E., Shone, C.C., and Montecucco, C. (1996). Structural determinants of the specificity for synaptic-vesicle associated membrane protein/synaptobrevin of tetanus and botulinum type B and G neurotoxins. J. Biol. Chem. *271*, 20353–20358.

Penner, R. Neher, E., and Dreyer, F. (1986). Intracellularly injected tetanus toxin inhibits exocytosis in bovine adrenal chromaffin cells. Nature 324, 76–78.

Poulain, B., Wadsworth, J.D., Maisey, E.A., Shone, C.C., Melling, J., Tauc, L., and Dolly, O.J. (1989). Inhibition of transmitter release by botulinum neurotoxin A, contribution of various fragments to the intoxication process. Eur. J. Biochem. 185, 197–203.

Rao, K.N. (2005). Structural analysis of the catalytic domain of tetanus neurotoxin. Toxicon 45, 929–939.

Ray P., Berman J.D., Middleton W., and Brendle J. (1993). Botulinum toxin inhibits arachidonic acid release associated with acetylcholine release from PC12 cells. J. Biol. Chem. 268, 11057–11064.

Ren, J., Kachel, K., Kim, H., Melenbaum, S.E., Collier, R.J., and London, E. (1999). Interaction of diphtheria toxin T domain with molten globule like proteins and its implications for translocation. Science 284, 955–957.

Rigoni, M., Caccin, P., Johnson, E.A., Montecucco, C., and Rossetto, O. (2001). Site-directed mutagenesis identifies active-site residues of the light chain of botulinum neurotoxin type A. Biochem. Biophys. Res. Commun. 288, 1231–1237.

Rohrich, R.J., Janis, J.E., Fagien, S., and Stuzin, J.M. (2003). Botulinum toxin: expanding role in medicine. Plastic Reconst. Surg. 112, 1S–3S.

Rossetto, O., Schiavo, G., Montecucco, C., Poulain, B., Deloye, F., Lozzi, L., and Shone, C.C. (1994). SNARE motif and neurotoxins. Nature 372, 415–416.

Rossetto, O., Seveso, M., Caccin, P., Schiavo, G., and Montecucco, C. (2001a). Tetanus and botulinum neurotoxins: turning bad guys into good by research. Toxicon. 39, 27–41.

Rossetto, O., Caccini, P., Rigoni, M., Tonello, F., Bortoletto, N., Stevens, R.C., and Montecucco, C. (2001b). Active site-mutagenesis of tetanus neurotoxin implicates TYR-375 and GLU-271 in metalloproteolytic activity. Toxicon 39, 1151–1159.

Rossetto, O., Rigoni, M., and Montecucco, C. (2004). Different mechanism of blockade of neuroexocytosis by presynaptic neurotoxins. Toxicology Lett. 149, 91–101.

Rummel, A., Bade, S., Alves, J., Bigalke, H., and Binz, T. (2003). Two carbohydrate binding sites in the H(CC)-domain of tetanus neurotoxin are required for toxicity. J. Mol. Biol. 326, 835–847.

Rummel, A., Mahrhold, S., Bigalke, H., and Binz, T. (2004). The HCC-domain of botulinum neurotoxins A and B exhibits a singular ganglioside binding site displaying serotype specific carbohydrate interaction. Mol. Microbiol. 51, 631–643.

Rummel, A., Eichner, T., Weil, T., Karnath, T., Gutcaits, A., Mahrhold, S., Sandhoff, K., Proia, R.L., Acharya, K.R., Bigalke, H., and Binz, T. (2007). Identification of the receptor binding site of botulinum neurotoxins B and G proves the double-receptor concept. Proc. Natl. Acad. Sci. USA 104, 359–364.

Sakaguchi, G. (1983). Clostridium botulinum toxins. Pharmac. Ther. 19, 165–194.

Salti, G. and Ghersetich, I. (2008). Advanced botulinum toxin techniques against wrinkles in the upper face. Clin Dermatol. 26, 182–91.

Sampaio, C., Costa, J., and Ferreira, J.J. (2004). Clinical comparability of marketed formulations of botulinum toxin. Mov Disord. 18, S129-S136.

Schantz, E.J. and Johnson, E.A. (1992). Properties and use of botulinum toxin and their microbial toxins in medicine. Microbiol. Rev. 56, 80–99.

Schiavo, G., Papini, E., Genna, G. Montecucco, C. (1990). An intact inter-chain disulfide bond is required for the neurotoxicity of tetanus neurotoxin. Infect. Immun. 58, 4136–4141.

Schiavo, G., Rossetto, O., Santucci, A., DasGupta, B.R., Montecucco, C. (1992a). Botulinum neurotoxins are zinc proteins. J. Biol. Chem. 267, 23479–23483.

Schiavo, G., Benfenati, F., Poulain, B., Rossetto, O., Polverino de Laureto, P., DasGupta, B.R., and Montecucco, C. (1992b). Tetanus and botulinum B neurotoxins block neurotransmitter release by proteolytic cleavage of synaptobrevin. Nature 359, 832–835.

Schiavo, G. Poulain, P., Rossetto, O., Benfenati, F., Tauc, L., and Montecucco, C. (1992c). Tetanus toxin is a zinc protein and its inhibition of neurotransmitter release and protease activity depends on zinc. EMBO J. 11, 3577–3583.

Schiavo, G., Shone, C.C., Rossetto, O., Alexander, F.C., Montecucco, C. (1993). Botulinum neurotoxin type F is a zinc endopeptidase specific for VAMP/synaptobrevin. J. Biol. Chem. 268, 11416–11519.

Schiavo, G., Rossetto, O., Montecucco, C. (1994). Clostridial neurotoxins as tools to investigate the molecular events of neurotransmitter release. Semin. Cell. Biol. 5, 221–229.

Schiavo, G., Matteoli, M., Montecucco, C. (2000). Neurotoxins affecting neuroexocytosis. Physiol. Rev. 80, 715–760.

Schmidt, M.F., Robinson, J.P., and DasGupta, B.R. (1993). Direct visualization of botulinum neurotoxin-induced channels in phospholipid vesicles. Nature 364, 827–830.

Schulte-Mattler, W.J., Wieser, T., and Zierz, S. (1999). Treatment of tension-type headache with botulinum toxin: a pilot study. Eur. J. Med. Res. 4, 183–186.

Scott, A.B. (1980). Botulinum toxin injection into extraocular muscles as an alternative to strabismus surgery. Ophthalmology 87, 1044–1049.

Scott, A.B., Kennedy, R.A., and Stubbs, H.A. (1985). Botulinum A toxin injection as a treatment for blepharospasm. Arch. Ophthalmol. 103, 347–350.

Segelke, B., Knapp, M., Kadkhodayan, S., Balhorn, R., and Rupp, B. (2004). Crystal structure of Clostridium botulinum neurotoxin protease in a product-bound state: Evidence for non-canonical zinc protease activity. Proc. Natl. Acad. Sci. USA 101, 6888–6893.

Sharma, S.K., and Singh, B.R. (1998). Haemagglutinin binding mediated protection of botulinum neurotoxin from proteolysis. J. Nat. Toxins 7, 239–253.

Sharma, S.K, and Singh, B.R. (1999). Nicking-mediated enhancement in the endopeptidase activity of type E botulinum neurotoxin. Protein Sci. 8 (Suppl. 1), 383M (abstr)

Sharma, S.K., and Singh, B.R. (2000). Immunological properties of Hn-33 purified from type A Clostridium botulinum. J. Nat. Toxins 9, 357–362.

Sharma, S., and Singh, B.R. (2004). Enhancement of the endopeptidase activity of purified neurotoxins A and E by an isolated component of the native neurotoxin associated proteins. Biochemistry 43, 4791–4798.

Sharma, S.K., Fu, F.N., and Singh, B.R. (1999). Molecular properties of a haemagglutinin purified from type A *Clostridium botulinum*. J. Protein. Chem. 18, 29–38.

Sharma, S., Zhou, Y., and Singh, B.R. (2006). Cloning, expression, and purification of C-terminal quarter of the heavy chain of botulinum neurotoxin type A. Protein Expression Purif. 45, 288–295.

Sheridan, R.E. (1998). Gating and permeability of ion channel produced by botulinum toxin types A and E in PC12 cells membrane. Toxicon 36, 703–717.

Shone, C.C., Hambleton, P., and Melling, J. (1987). A 50-kda fragment from NH2-terminus of the heavy chain subunit of *Clostridium botulinum* type A neurotoxin forms channels in lipid vesicles. Eur. J. Biochem. 167, 175–180.

Shone, C.C., and Melling, J. (1992). Inhibition of calcium-dependent release of noradrenaline from PC12 cells by botulinum neurotoxin type A: long term effects of the neurotoxin on intacT-cells. Eur. J. Biochem. 207, 1009–1016.

Shukla, H.D., and Sharma, S.K. (2005). *Clostridium botulinum*: A bug with beauty and weapon. Crit. Rev. Microbiol. 31, 11–18.

Silberstein, S., Mathew, N., Saper, J., and Jenkins, S. (2000). Botulinum toxin type A as a migraine preventive treatment. For the BOTOX Migraine Clinical Research Group. Headache 40, 445–450.

Simpson, L.L., Coffield, J.L., and Barkry, N. (1994). Inhibition of vacuolar adenosine triphosphate antagonizes the effects of clostridial neurotoxins but not phospholipase A2 neurotoxins. J. Pharmacol. Exp. Ther. 269, 256–269.

Simpson, L.L., Maksymowych, A.B., Park, J.B., and Bora, R.S. (2004). The role of the interchain disulfide bond in governing the pharmacological actions of botulinum toxin. J. Pharmacol. Exp. Ther. 308, 857–864.

Singh, B.R. (1999). Biomedical and toxico-chemical aspects of botulinum neurotoxins. J. Toxicol-Toxin Rev. 18, vii–x.

Singh, B.R. (2000). Intimate details of the most poisonous poison. Nat. Struct. Biol. 7, 617–619.

Singh, B.R. (2002). In Scientific and Therapeutic Aspects of Botulinum Toxin: Molecular basis of the unique endopeptidase activity of botulinum neurotoxin, M.F. Brin, J. Jankovic, and M. Hallet, eds., (Lippincott Williams and Wilkins, Philadelphia), pp. 75–88.

Singh, B. R. (2006). Botulinum neurotoxin structure, engineering, and novel cellular trafficking and targeting. Neurotoxicity Res. 9, 73–92.

Singh, B.R., and DasGupta, B.R. (1990). Conformational changes associated with the nicking and activation of botulinum neurotoxin type E. Biophys. Chem. 38, 123–130.

Singh, B.R., Li, B., and Read, R., (1995). Botulinum versus tetanus neurotoxins: why is botulinum neurotoxin a food poison but not tetanus? Toxicon 33, 1541–1547.

Stecher, B., Gratzl, M., and Ahnert-Hilger, G. (1989). Reductive chain separation of botulinum A toxin – a prerequisite to its inhibitory action on exocytosis in chromaffin cells. FEBS Lett. 248, 23–27.

Südhof, T.C. (2004). The synaptic vesicle cycle. Annu. Rev. Neurosci. 27, 509–547.

Sugiyama, H. (1980). *Clostridium botulinum* neurotoxins. Microbiol. Rev. 44, 419–448.

Sutton, R.B., Fasshauer, D., Jahn, R. and Brunger, A.T. (1998). Crystal structure of a SNARE complex involved in synaptic exocytosis at 2.4 Å resolution. Nature 395, 347–353.

Swaminathan, S., and Eswaramoorthy, S. (2000). Structural analysis of the catalytic and binding sites of *Clostridium botulinum* neurotoxin B. Nat. Struct. Biol. 7, 693–699.

Swaminathan, S., Eswaramoorthy, S., and Kumaran, D., (2004). Structure and enzymatic activity of botulinum neurotoxins. Mov. Disord. 19, S17-S22.

Ting, P.T., and Freiman, A. (2004). The story of *Clostridium botulinum*: from food poisoning to botox. Clin. Med. 4, 258–262.

Tsukamoto, K., Kozai, Y., Ihara, H., Kohda, T., Mukamoto, M., Tsuji, T. and Kozaki, S. (2008). Identification of the receptor-binding sites in the carboxy-terminal half of the heavy chain of botulinum neurotoxin types C and D. Microb. Pathog. 44, 484–493.

Tycko, B., and Maxfield, F.R. (1982). Rapid acidification of endocytic vesicles containing alpha 2-macroglobulin. Cell 28, 643–651.

Uotsu, N., Nishikawa, A., Watanabe, T., Ohyama, T., Tonozuka, T., Sakano, Y. and Oguma, K. (2005). Cell internalization and traffic pathway of *Clostridium botulinum* type C neurotoxin in HT-29 cells. Biochim Biophys Acta 1763, 120–128.

Vaidyanathan, V. V., Yoshino, K., Jahnz, M., Dorries, C., Bade, S., Nauenburg, S., Niemann, H., and Binz, T. (1999). Proteolysis of SNAP-25 isoforms by botulinum neurotoxin types A, C, and E: Domains and amino acid residues controlling the formation of enzyme-substrate complexes and cleavage. J. Neurochem. 72, 327–337.

Vallee, B.L., and Auld, D.S. (1990). Zinc coordination, function, and structure of zinc enzymes and other proteins. Biochemistry 29, 5647–5659.

Vamvaca, K., Vogeli B., Kast, P., Pervushin, K., and Hilvert, D. (2004). An enzymatic molten globule: Efficient coupling of folding and catalysis. Proc. Natl. Acad. Sci. USA 101, 12860–12864.

van der Goot, F.G., Gonnzalez-Manas, J.M., Lakey, J.H., and Pattus, F. (1991). A "molten-globule" membrane-insertion intermediate of the pore-forming domain of Colicin A. Nature 354, 408–410.

van Ermengem, E. (1979). Classics in infectious diseases. A new anaerobic bacillus and its relation to botulism. Rev. Infect. Dis. 1, 701–719.

Veit, M. (2000). Palmitoylation of the 25-kDa synaptosomal protein SNAP-25 *in vitro* occurs in the absence of an enzyme, but is stimulated by binding to syntaxin. Biochem. J. 345, 145–151.

Vogel, K., Pierre, C., and Roche, P.A. (1999). Targeting of SNAP-25 to membranes is mediated by its association with the target SNARE syntaxin. J. Biol. Chem. *275*, 2959–2965.

Washbourne, P., Pellizari, R., Baldini, G., Wilson, M.C., and Montecucco, C. (1997). Botulinum neurotoxin type A and E require the SNARE motif in SNAP-25 for proteolysis. FEBS Lett. *418*, 1–5.

Weiner, M., Freymann, D., Ghosh, P., Stroud, R.M. (1997). Crystal structure of Colicin 1a. Nature *385*, 461–464.

Weller, U., Dauzenroth, M.E., Gransel, M., Dreyer, F. (1991). Cooperative action of the light chain of tetanus neurotoxin and heavy chain of botulinum toxin type A on the transmitter release of mammalian motor and plates. Neurosci. Lett. *122*, 132–134.

Wenzel, R.G. (2004). Pharmacology of botulinum neurotoxin serotype A. Am. J. Health-Syst. Pharm. *61*, S5-S10.

Wheeler, A.H. (1998).Botulinum toxin A, adjunctive therapy for refractory headaches associated with pericranial muscle tension. Headache *38*, 468–471.

Yamasaki, S., Baumeister, A., Binz, T., Blasi, J., Link, E., Cornille, F., Roques, B., Fykse, E.M., Sudhof, T.C.,

Jahn, R., Niemann, H. (1994). Cleavage of members of the synaptobrevin/VAMP family by types D and F botulinal neurotoxins and tetanus toxin. J. Biol. Chem. *269*, 12764–12772.

Zhang, Z., and Singh, B.R. (1995). Protein Sci. *4 (suppl 2)*, 110.

Zhao, J.M., and London, E. (1986). Similarity of the conformation of diphtheria toxin at high temperature to that in the membrane-penetrating low-pH state. Proc. Natl Acad. Sci. USA *83*, 2002–2006.

Zhou, L., de Paiva, A., Liu, D., Aoki, R., and Dolly, O.J. (1995). Expression and purification of the light chain of botulinum neurotoxin A: a single mutation abolishes its cleavage of SNAP-25 and neurotoxicity after reconstitution with the heavy chain. Biochemistry *34*, 15175–15181.

Zhou, Y., Foss, S., Lindo, P., Sarkar, H., and Singh, B.R. (2005). Haemagglutinin-33 of type A botulinum neurotoxin complex binds with synaptotagmin II. FEBS J. *272*, 2717–2726.

Anthrax Toxin

3

Francisco J. Maldonado-Arocho, Kathleen M. Averette-Mirrashidi and Kenneth A. Bradley

Abstract

Bacillus anthracis produces two major virulence factors, a tripartite exotoxin referred to as anthrax toxin and an antiphagocytic capsule. These virulence factors mediate pathogen survival and, in the case of toxin, directly induce damage to the host. Two distinct enzymatic activities are associated with anthrax toxin, each encoded by a separate protein. The enzymatic subunits are lethal factor (LF), a zinc-dependent metalloproteinase, and oedema factor (EF), a calcium- and calmodulin-dependent adenylate cyclase (Leppla, 1982, 1984; Vitale *et al.*, 1998, 1999). LF and EF gain access to the host cytosol by binding to and translocating through a pore formed by the shared binding subunit, protective antigen (PA) (Blaustein *et al.*, 1989; Milne *et al.*, 1994; Zhang *et al.*, 2004a,b). The combination of LF and PA is called lethal toxin (LT), and this toxin inactivates MAPK signalling in the host. Oedema toxin (ET), formed by the combination of EF and PA, produces high cAMP levels in host cells. Early during infection, systemic toxin levels are low, and likely modulate the host immune response locally, thereby allowing for establishment of infection. Late in infection, toxin concentrations increase causing organ damage, vascular leakage, and ultimately death of the host. Indeed, both LT and ET are capable of inducing mortality in animal models when injected as purified toxins. This chapter will focus on current trends in LT and ET research aimed at understanding the mechanisms by which they affect the host and alter disease outcome.

Bacillus anthracis and anthrax

B. anthracis is a rod-shaped Gram-positive, non-motile, aerobic, bacterium that exists as either a vegetative bacillus or a dormant spore. The spore is found normally in the soil, is capable of remaining viable for decades and can withstand extreme environmental insults (Turnbull, 2002). The vegetative, actively dividing form is predominantly present in the context of a host, though vegetative bacilli are also likely present in the rhizosphere of plant roots (Saile and Koehler, 2006). This lifestyle has permitted *B. anthracis* a high degree of monomorphism, since the bacteria has very little chance of exposure to various environmental factors and mutagens that allow for species variation (Turnbull, 2002).

Anthrax is a zoonotic disease with the potential to affect all mammals, but is most commonly associated with grazing animals. The human manifestation of the disease commonly occurs by accidental contact with contaminated animals or animal products. Anthrax can occur in one of three forms, depending on the route of entry of the spores (Dixon *et al.*, 1999). Cutaneous anthrax, the most common of the three forms in humans, occurs when spores come in contact with lesions in the skin. Gastrointestinal anthrax presents with variable mortality rates, has traditionally been thought to be rare (though this may be due to underreporting) and occurs upon ingestion of contaminated meats (Sirisanthana and Brown, 2002; Beatty *et al.*, 2003). The third and most deadly form, inhalation anthrax, occurs following uptake of airborne spores into the lungs. This last form is of great concern given the stability of the spores and the ability to aerosolize spores for use as a biological weapon (Matsumoto, 2003).

Although *B. anthracis* is normally an extracellular pathogen, one current model, referred to as the Trojan horse model, maintains that this pathogen undergoes a short intracellular stage at the initiation of infection. This model is based on early experiments of inhalation anthrax that demonstrated trafficking of spores from the lumen of the lungs to the regional lymph nodes within phagocytic cells (Lincoln *et al.*, 1965). These cells were assumed to be alveolar macrophages, but early studies did not distinguish between macrophages and other phagocytic cells such as dendritic cells (DCs). According to the Trojan horse model, spores germinate to the vegetative form within the macrophage phagolysosome during migration to the draining lymph nodes (Guidi-Rontani, 2002; Sanz *et al.*, 2008). During germination, *B. anthracis* expresses virulence factors that promote lysis of the macrophage and release of the bacilli into the extracellular environment, thus initiating a systemic infection (Shafa *et al.*, 1966; Guidi-Rontani *et al.*, 1999; Banks *et al.*, 2005).

However, the specific details of the Trojan horse model have been called into question. Recent studies suggest that macrophages are very efficient at clearing infection and thus are less likely to serve as a transport mechanism for germinating spores (Cote *et al.*, 2004, 2006; Kang *et al.*, 2005; Hu *et al.*, 2006). Indeed, macrophage depletion in the lungs actually increases susceptibility to anthrax, indicating that alveolar macrophages are important for limiting *B. anthracis* and are not required for establishing infection (Cote *et al.*, 2004, 2006). Furthermore, internalization of spores has been observed in other cell types, including lung epithelial cells and DCs, suggesting that spores may have alternative mechanisms of disseminating (Brittingham *et al.*, 2005; Cleret *et al.*, 2007; Russell *et al.*, 2007, 2008). Second, the Trojan horse model stipulates that spores must be phagocytosed for germination and dissemination. Several studies now suggest that spores germinate without the need for phagocytosis and that spores germinate at the initial site of infection without a need to

be transported to the lymph nodes (Bischof *et al.*, 2007; Glomski *et al.*, 2007b; Sanz *et al.*, 2008). Regardless of the mode of infection and germination, once bacilli gain access to the lymphatic and circulatory systems they continue to divide and secrete toxins resulting in bacteraemia and toxaemia that lead to the ultimate demise of the host.

Survival of *B. anthracis* in the host is attributed to production of a capsule and the tripartite anthrax toxin. Biosynthetic genes for the capsule and structural genes for toxin are each encoded by one of two large extrachromosomal plasmids, pXO1 and pXO2. Strains lacking either of these two plasmids are attenuated. In fact, generation of plasmid-cured strains were the basis for the veterinary vaccines created by Louis Pasteur and Maxwell Sterne (Sterne, 1937; Mock and Fouet, 2001). The genes required for the synthesis of the poly- γ-D-glutamic acid capsule are encoded on the 96-kb plasmid, pXO2 (Green *et al.*, 1985). The 182-kb plasmid pXO1 carries the genes *pagA*, *lef* and *cya* that encode the toxin components protective antigen (PA), lethal factor (LF) and oedema factor (EF), respectively (Mikesell *et al.*, 1983; Okinaka *et al.*, 1999a,b).

Smith and Keppie were the first to demonstrate the presence of a toxin produced by *B. anthracis* (Smith and Keppie, 1954). Guinea pigs infected with *B. anthracis* could be treated with antibiotics to clear infection, but only if treatment was given early. A 'point of no return' was defined after which antibiotics could clear infection but disease progressed to death as if no treatment had been given. This point of no return was attributed to production of a toxin that was later identified as a combination of PA, LF and EF. While all three proteins assemble as a single toxic complex during infection, two distinct AB toxins can be generated in the laboratory from different pairwise combinations of these proteins. Anthrax toxin is unique among AB toxins in that a common binding B moiety, PA, can assemble with two catalytic A moieties, LF and EF. The combination of EF plus PA is referred to as oedema toxin (ET) while the combination of LF plus PA is referred to as lethal toxin (LT). Injection of LT induces many of the symptoms of anthrax, including vascular collapse and death (Beall *et al.*, 1962). Non-encapsulated *B.*

anthracis strains lacking either LF or PA are not lethal to mice (Pezard *et al.*, 1991). As a result, LT has historically received more attention than ET or capsule. It is worth noting that while strains lacking EF are virulent, these strains require a ~10-fold higher dose than wild type to induce mortality. This, as well as early work performing injection of the toxin components in different combinations, suggested that while LT may be the primary virulence factor in specific rodent models, ET also contributes to pathogenesis (Smith and Stoner, 1967; Pezard *et al.*, 1991).

Anthrax toxin

The existence of an exotoxin that contributes to pathology associated with *B. anthracis* infection was initially presented by Smith and Keppie, who showed that, in addition to the point of no return described above, injection of filtered plasma from infected guinea pigs into naïve animals induced anthrax-like symptoms including death (Smith and Keppie, 1954). Characterization of this exotoxin led to the isolation of three biochemical fractions subsequently identified as EF (fraction I), PA (fraction II) and LF (fraction III) (Beall *et al.*, 1962). Injection of ET (PA + EF) into skin produced oedema but was initially found to be non-toxic when injected intravenously. In contrast, LT (PA + LF) was toxic when injected intravenously but did not produce overt tissue oedema when injected intradermally (Beall *et al.*, 1962). Injecting EF and/or LF was non-toxic, demonstrating a requirement of PA for toxin activity, whereas injection of all three fractions was more toxic than either ET or LT alone (Beall *et al.*, 1962; Smith and Stoner, 1967).

In addition to identifying anthrax toxin, Smith and Keppie's original experiment also indicated that toxin production is restricted by environmental cues. Indeed, prior attempts to identify a toxin from *B. anthracis* grown *in vitro* had failed (King and Stein, 1950). Environmental triggers for toxin gene transcription are bicarbonate/CO_2 and a neutral pH (Bartkus and Leppla, 1989; Sirard *et al.*, 1994), which likely represent physiological cues that the bacterium is within an animal host. The transcriptional regulator *atxA*, present on pXO1, serves as a trans-acting positive regulator of the toxin gene expression and *atxA*

activity is dependent on CO_2 levels (Dai *et al.*, 1995). Expression of *atxA* itself is temperature dependent with higher temperatures inducing higher transcript levels (Dai and Koehler, 1997). Interestingly, while no sequence similarities exist in the promoter regions of AtxA-regulated genes, recent work suggests that the structural topology of the DNA of toxin gene promoter regions plays an important role in the control of toxin gene expression (Hadjifrangiskou and Koehler, 2008). The anthrax toxin promoter regions are associated with high intrinsic curvature hypothesized to play a role in recognition by the transcription machinery and/or trans-acting regulators such as AtxA.

A second regulator, AbrB, encoded by a chromosomal gene also contributes to toxin gene expression (Phillips and Strauch, 2002; Saile and Koehler, 2002). AbrB acts as a transition state regulator, suppressing post-exponential-phase gene expression during logarithmic phase of growth. During logarithmic phase growth, AbrB binds to specific sequences in the *atxA* promoter region, thereby repressing its expression and indirectly inhibiting toxin expression (Strauch *et al.*, 2005). AbrB levels begin to drop in mid-logarithmic phase of growth due to negative regulation by Spo0A (Saile and Koehler, 2002). In the post-exponential growth phase, the AbrB concentration is low leading to *atxA* expression. In addition to binding the promoter region of *atxA*, AbrB binds the promoter of *sigH*, a gene encoding the RNA polymerase sigma factor σ^H (Hadjifrangiskou *et al.*, 2007). Similar to σ^H of *B. subtilis*, *B. anthracis* σ^H is involved in post-exponential-phase gene regulation, including expression of genes involved in sporulation. Furthermore, σ^H positively controls *atxA* in an AbrB-independent manner.

Finally, a two-component system consisting of BrrA and BrrB contributes to regulation of toxin expression in *B. anthracis* (Vetter and Schlievert, 2007). Interestingly, BrrA–BrrB is also involved in regulation of metabolic processes. The current model is that BrrB acts as an environmental sensor that activates BrrA. Upon activation BrrA binds toxin promoters and up-regulates their expression. The involvement of the BrrA–BrrB two-component system suggests that in addition to the host-associated environmental cues, toxin expression is also tied to the metabolic state of the bacterium.

Protective antigen

Protective antigen was so named for giving rise to protective immunity in animals immunized with culture supernatants (Gladstone, 1946; Turnbull, 2002). This antigenic property is independent of the role of PA as a toxin, but the fact that antibody-based inactivation of toxin correlates with protective immunity highlights the critical role of toxin in disease. The presence of PA is necessary for binding and internalization of the two anthrax toxin catalytic moieties, EF and LF (Fig. 3.1). PA is a 735 amino acid, 83-kDa, protein encoded by the *pagA* gene on pXO1. The three-dimensional structure of monomeric PA obtained by X-ray crystallography reveals four structural domains (Petosa *et al.*, 1997). Domain 1 contains a protease cleavage site (RKKR) that allows for processing of PA into an N-terminal 20 kDa (PA_{20}) fragment and a C-terminal 63-kDa (PA_{63}) fragment (Klimpel *et al.*, 1992). Release of PA_{20} relieves a steric block to PA self-association and allows assembly of PA_{63} into a prepore complex (Petosa *et al.*, 1997). Current structural and biochemical data indicate that the ring consists of seven PA monomers (Petosa *et al.*, 1997). In addition, release of PA_{20} exposes a hydrophobic patch formed by domain 1', which consists of domain 1 residues that remain associated with PA_{63}. The binding sites for the catalytic A moieties are formed at the interface of two adjacent PA_{63} monomers and involve domain 1' residues from each monomer (Cunningham *et al.*, 2002). Proteolytic processing of PA is a necessary event for intoxication, as PA variants lacking protease recognition sites still bind host cell surfaces but are unable to internalize the catalytic moieties (Singh *et al.*, 1989; Klimpel *et al.*, 1992; Gordon *et al.*, 1995).

Domain 2 is involved in pore formation and receptor binding. It contains several long β-strands that line the inner lumen of the channel as well as a large flexible loop implicated in membrane insertion (Novak *et al.*, 1992; Petosa *et al.*, 1997; Benson *et al.*, 1998; Mourez *et al.*, 2003). Indeed, multiple lines of experimental evidence indicate that the PA prepore undergoes structural rearrangement to the pore form upon

lowering of pH, where the large flexible loop of domain 2 extends down, forming a 14-stranded transmembrane β-barrel (Petosa *et al.*, 1997; Miller *et al.*, 1999; Nassi *et al.*, 2002; Katayama *et al.*, 2008). Analysis of the co-crystal structure of PA with one of two identified anthrax toxin receptors (ANTXR2) reveals that domain 2 of PA directly interacts with this host-cell receptor (Lacy *et al.*, 2004a; Santelli *et al.*, 2004). ANTXR2 contact with PA domain 2 is proposed to prevent pore formation until the toxin complex is trafficked to an acidic environment (Miller *et al.*, 1999; Nassi *et al.*, 2002; Lacy *et al.*, 2004a; Santelli *et al.*, 2004). The finding that PA domain 2 makes contact with ANTXR2 is surprising, as all prior biochemical data indicates that receptor binding activity of PA is restricted to domain 4 (Little and Lowe, 1991; Brossier *et al.*, 1999; Singh *et al.*, 1999; Varughese *et al.*, 1999; Bradley *et al.*, 2001).

Domain 3, the smallest of the four PA domains, is involved in oligomer formation. Mutations in domain 3 abolish the ability of PA to oligomerize into the prepore form, thus blocking intoxication (Petosa *et al.*, 1997; Ahuja *et al.*, 2001; Mogridge *et al.*, 2001).

Domain 4 is composed of an Ig-like fold and is loosely associated with the other three domains (Petosa *et al.*, 1997). Mutational analysis demonstrates that domain 4 is necessary and sufficient for binding host cell surface receptors (Little and Lowe, 1991; Brossier *et al.*, 1999; Singh *et al.*, 1999; Varughese *et al.*, 1999; Bradley *et al.*, 2001). Specifically, removal of either the entire C-terminus or a small exposed loop of PA results in complete abrogation or reduced binding to cells, respectively (Brossier *et al.*, 1999; Varughese *et al.*, 1999). These studies demonstrate that an exposed 19-amino-acid loop located within domain 4 is involved in the binding of PA to ANTXR. The involvement of PA-domain 4 is further confirmed by co-crystal structure analysis of PA with ANTXR2 demonstrating that domain 4 interacts directly with receptor (Lacy *et al.*, 2004a; Santelli *et al.*, 2004).

Lethal factor

LF, the catalytic component of LT, is a 90-kDa protein encoded by the *lef* gene on pXO1 (Mikesell *et al.*, 1983). LF was named due to early

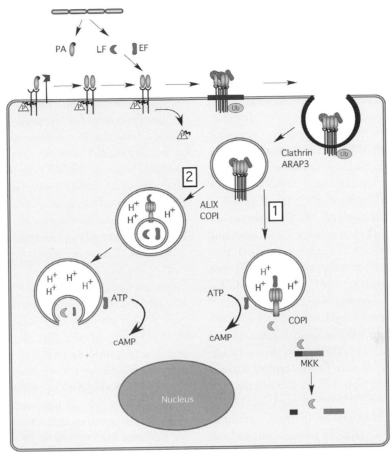

Figure 3.1 Cellular intoxication pathway for anthrax toxin. *Bacillus anthracis* secretes the tripartite anthrax toxin consisting of PA, EF, and LF. PA binds to one of two ANTXRs on the cell surface where it is cleaved by host furin-like proteases. Processing of PA allows oligomerization into a prepore complex that permits binding of the catalytic subunits, EF and LF, and leads to clustering of ANTXRs. Receptor clustering and removal of palmitate (P; triangle) from the cytosolic tail allows the receptor-bound toxin complex to move to lipid rafts. Entry into lipid rafts leads to ubiquitination (Ub; oval) of receptors and rapid clathrin- and ARAP3-dependent endocytosis of the toxin complex. Endocytosed toxin is trafficked to late endosomes (path 1) where a drop in pH converts the prepore to a pore leading to COPI-mediated translocation of the catalytic subunits. Alternatively (path 2), toxin is trafficked to early endosomes and sorted to ECV/MVBs in an ALIX- and COPI coatomer-dependent manner. A drop in pH leads to conversion of the PA prepore to a pore and translocation of the catalytic subunits into the lumen of the intraluminal vesicles. Toxin is then transported to late endosomes where back fusion of the intraluminal vesicles to late endosomes releases the catalytic subunits into the cytosol. Irrespective of the path used, LF is released into the cytosol and cleaves MKKs, whereas EF remains associated with the membrane vesicles and catalyses the conversion of ATP to cAMP.

work demonstrating that injection of LF and PA into experimental animals is capable of inducing many of the symptoms of anthrax, including death (Beall *et al.*, 1962). LF was the last of three anthrax toxin components to be discovered (factor III), and the last for which the mechanism of action was known (Beall *et al.*, 1962; Duesbery *et al.*, 1998).

Nearly 40 years passed from identification of LF to discovery of its enzymatic activity (Beall *et al.*, 1962; Duesbery *et al.*, 1998; Hammond and

Hanna, 1998; Vitale *et al.*, 1998; Vitale *et al.*, 1999). Initial insight came from identification of a small motif, with amino acid sequence HEXXH, at the C-terminus that is characteristic of a Zn^{2+} binding site of metalloproteases (Klimpel *et al.*, 1994). Substituting key residues in the putative catalytic site decreases Zn^{2+} binding and abolishes the ability of LT to kill macrophages. Protease inhibitors block LT action, further supporting the hypothesis that LF is a metalloprotease (Klimpel *et al.*, 1994; Hammond and Hanna,

1998). Final proof came with demonstration that LF cleaves the N-termini of several mitogen activated protein kinase (MAPK) kinases (MKKs), including MAPK and extracellular regulated kinase kinases (MEK) 1 and 2, and MKKs 3, 4, 6, and 7 (Duesbery *et al.*, 1998; Vitale *et al.*, 1998; Vitale *et al.*, 1999).

LF disrupts MAPK signalling pathways through cleavage of MKKs just upstream of or within MAPK docking domains, disrupting MKK-MAPK binding and MKK intrinsic kinase activity (Chopra *et al.*, 2003; Bardwell *et al.*, 2004). The N-terminal MKK sequence that LF recognizes and cleaves has a loose consensus consisting of a string of basic residues, followed by two hydrophobic residues separated by one amino acid (Turk *et al.*, 2004). LF cleaves MKKs before the second hydrophobic residues. A loose consensus sequence suggests that LF may cleave targets other than MKKs. Indeed, neuronal nitric oxide synthase (nNOS) is also cleaved by LF although the significance of this finding has yet to be determined (Kim *et al.*, 2008b). In the case of MEKs, a second LF-interacting region (LFIR) exists. This LFIR is present at the C-terminus of MEKs and is required for recognition and cleavage by LF (Chopra *et al.*, 2003). Therefore, additional specificity for MEKs exists, since disrupting the C-terminal LFIR abolishes LF-mediated MEK cleavage although the N-terminal consensus sequence is still intact.

Amino acid and X-ray crystallographic analysis of LF has revealed four structural domains. Domain I, residues 1–263, is homologous to domain I of EF and is necessary and sufficient for PA binding and for internalization of both catalytic subunits (Bragg and Robertson, 1989; Arora and Leppla, 1993). This domain, also called LF_N, can be fused to heterologous proteins or conjugated to nucleic acids to serve as a PA-dependent vehicle for cellular delivery (Arora *et al.*, 1992; Arora and Leppla, 1993; Milne *et al.*, 1995; Blanke *et al.*, 1996; Wesche *et al.*, 1998; Gaur *et al.*, 2002; Kim *et al.*, 2004). Domain II, residues 263–550, has structural similarities to the ADP-ribosylating toxin VIP-2 of *B. cereus*, but the active site has been mutated in LF (Pannifer *et al.*, 2001). Domain III is inserted within domain II and shares a hydrophobic surface with domain IV. Domain III excludes globular proteins from

accessing the LF active site, and thus LF binds linear or tail-like substrates in its defined proteolytic active site (Pannifer *et al.*, 2001). Domain IV contains the zinc-binding active site, which is structurally similar to the active site of the metalloprotease thermolysin. The crystal structure of LF has been solved in the presence of a peptide derived from the N-terminus of MEK2 as well as an optimized peptide substrate that was designed based on the MKK consensus sequence (Pannifer *et al.*, 2001; Turk *et al.*, 2004). These structures reveal a long, 40-Å groove made of domains II, III and IV that creates an interface for substrate binding (Pannifer *et al.*, 2001; Tonello *et al.*, 2004; Turk *et al.*, 2004).

Oedema factor

EF, an 89-kDa protein, is encoded by the *cya* gene present on the pXO1 plasmid (Mikesell *et al.*, 1983). The N-terminus of EF, which bears amino acid homology to the N-terminus of LF, mediates binding to PA, whereas the C-terminal portion of EF encodes the catalytic activity. Cellular effects of ET are inhibited by addition of excess LF, suggesting that EF and LF compete for the same binding site on PA (Leppla, 1982). Insight into the catalytic activity of EF first came from the observation that subdermal injection of ET induces oedema similar to that caused by cholera toxin (CT) (Beall *et al.*, 1962; Craig, 1965). This led to the finding that, like CT, ET increases cAMP levels in cultured cells, although this occurs via distinct mechanisms (Leppla, 1982). While CT triggers constitutive activation of host cell adenylate cyclase through ADP ribosylation of host $G_s\alpha$, EF is itself an adenylate cyclase. EF is secreted by *B. anthracis* in an inactive form, requiring the presence of host calmodulin (CaM) and Ca^{2+} for its activation (Leppla, 1982, 1984; Kumar *et al.*, 2002). Once bound to calmodulin, EF is a very potent adenylate cyclase that can induce up to a 5000-fold increase in intracellular cAMP levels, with differences in response varying between the cell types used (Leppla, 1982; Voth *et al.*, 2005; Maldonado-Arocho *et al.*, 2006).

Although EF shares very little structural homology with mammalian adenylate cyclases, it contains two regions with amino acid homology to consensus sequences for A-type ATP binding domains present in many ATP binding proteins (Xia and Storm, 1990). Two conserved lysines, Lys-346 and Lys-353, are predicted to interact with ATP. Mutation of either of these two residues completely abolishes enzymatic activity, suggesting that these two residues are important for catalytic activity by potentially interacting directly with ATP (Xia and Storm, 1990). More recently, structural analyses of EF, EF in complex with calmodulin, and EF in complex with ATP have been performed (Drum et al., 2002; Shen et al., 2004a; Shen et al., 2004b; Shen et al., 2005). EF contains three globular domains that constitute the catalytic portion of this enzyme (Drum et al., 2002). The catalytic active site lies at the interface of two of these domains, C_A (amino acids 294–349 and 490–622) and C_B (350–489). The third domain (660 – 880) is helical and is connected to the C_A domain by a linker. CaM binds between C_A and the helical domain, causing a conformational change in EF to accommodate CaM and lock it into position while at the same time forming a binding pocket for ATP (Drum et al., 2002). EF uses a two-metal-ion catalysis mechanism that is facilitated by a histidine at position 351 (Shen et al., 2005). The lack of structural similarity to mammalian enzymes combined with the use of a two-metal-ion catalysis mechanism prevalent in mammalian adenylate cyclases suggests a mechanism-driven convergent evolution for EF (Shen et al., 2005).

Binding and internalization of anthrax toxin

Cellular intoxication begins with the binding of PA to one of two proteinaceous receptors, ANTXR1 or ANTXR2 (Fig. 3.1) (Bradley et al., 2001; Scobie et al., 2003). Upon binding, PA is processed by host furin or furin-like proteases on the cell surface resulting in the release of PA_{20} and leaving PA_{63} bound to the receptor (Klimpel et al., 1992; Gordon et al., 1995; Gordon et al., 1997). Alternatively, PA may be processed prior to cell-surface binding by serum proteases (Ezzell and Abshire, 1992; Moayeri et al., 2007). Receptor-bound PA_{63} then self-associates, forming a heptameric ring-shaped prepore complex that binds up to three molecules of LF and/or EF in a competitive fashion (Cunningham et al., 2002;

Mogridge et al., 2002a; Mogridge et al., 2002b; Ren et al., 2004). This toxin–receptor complex is internalized and delivered to an acidic endosome that triggers a conformational change in PA, resulting in formation of a pore that translocates EF and LF to the host cell cytosol (Friedlander, 1986; Gordon et al., 1988; Milne and Collier, 1993; Abrami et al., 2004; Zhang et al., 2004a; Krantz et al., 2006).

Anthrax toxin receptors

PA interacts with host cells by binding directly to one of two related cell-surface receptors, anthrax toxin receptor 1 (ANTXR1) and anthrax toxin receptor 2 (ANTXR2) (Escuyer and Collier, 1991; Bradley et al., 2001; Scobie et al., 2003). Both receptors are type-I transmembrane proteins that possess an extracellular von Willebrand factor type A (VWA) domain, alternatively called an inserted (I) domain, which binds directly to PA. Additionally, both receptors bind extracellular matrix (ECM) proteins, including collagen and laminin (Bell et al., 2001; Bradley et al., 2003; Scobie et al., 2003; Nanda et al., 2004; Hotchkiss et al., 2005).

ANTXR1 was originally identified as one of nine tumour endothelial markers (TEMs), and was named TEM8 (St Croix et al., 2000). Of all the TEMs, ANTXR1 is the most highly conserved between mouse and human (Carson-Walter et al., 2001). ANTXR1 is expressed at increased levels in endothelial cells (ECs) associated with colorectal tumours, as well as the vasculature of developing mouse embryos (St Croix et al., 2000; Carson-Walter et al., 2001). This receptor is also expressed by epithelial cells lining the three sites of entry for B. anthracis, which may have an impact in the initiation of infection and disease pathogenesis (Bonuccelli et al., 2005). Consistent with a role in angiogenesis, expression of ANXTR1 is increased in ECs upon initiation of tube formation and in response to interleukin-1β (IL-1β) a cytokine known to induce angiogenesis, (Rmali et al., 2004; Hotchkiss et al., 2005). Furthermore, ANTXR1 expression is quickly lost as human umbilical vein ECs (HUVECs) age or de-differentiate in vitro. Finally, ANTXR1 binds to gelatin, collagen I and collagen VI, and expression of this receptor affects motility and cell spreading of ECs grown on ECM (Nanda *et al.*, 2004; Hotchkiss *et al.*, 2005; Werner *et al.*, 2006; Rogers *et al.*, 2007).

There are three known splice variants of ANTXR1, producing three protein isoforms (Bradley *et al.*, 2001). Splice variants 1 and 2 are both type 1 transmembrane proteins, differing only in the length of their cytoplasmic domains. Proteins encoded by splice variants 1 and 2 function as anthrax toxin receptors, consistent with the finding that the cytoplasmic tail of ANTXR1 is not necessary for toxin binding and internalization (Liu and Leppla, 2003). In contrast, the cytoplasmic domain does play a role in coupling the binding of extracellular ligands to actin rearrangements associated with ANTXR1-mediated cell spreading (Werner *et al.*, 2006). Splice variant 3 is a shorter version of ANTXR1, possessing the extracellular domain but no cytoplasmic or transmembrane domain. Splice variant 3, therefore, is unable to function as an anthrax toxin receptor.

ANTXR2 has parallel features to ANTXR1 including a type I transmembrane region and a VWA domain with 60% amino acid identity to the ANTXR1 VWA domain (Scobie *et al.*, 2003). ANTXR2 was originally identified as a gene up-regulated in HUVECs undergoing capillary formation in three-dimensional collagen matrices, and was named capillary morphogenesis gene 2 (CMG2) (Bell *et al.*, 2001). The VWA domain binds ECM components collagen IV and laminin (Bell *et al.*, 2001; Dowling *et al.*, 2003). This gene appears to be ubiquitously expressed in normal adult tissues (Scobie *et al.*, 2003).

At least four splice variants exist for ANTXR2. The first variant initially identified as CMG2 (ANTXR2^{386}; where superscript number refers to number of amino acids encoded by each variant) is predominantly found in the endoplasmic reticulum and does not appear to play a role in intoxication (Bell *et al.*, 2001; Scobie *et al.*, 2003). The second, ANTXR2^{488}, is the variant initially identified to mediate anthrax toxin binding and internalization (Scobie *et al.*, 2003). ANTXR2^{386} lacks a 100 amino acid region between the VWA domain and the transmembrane domain, which is conserved in ANTXR1 and may be important for receptor function. The third variant, ANTXR2^{489}, is identical to ANTXR2^{488} except for a 12-amino-acid C-terminal sequence, which it shares with ANTXR2^{386}. The fourth variant, ANTXR2^{322}, lacks the transmembrane domain and is believed to be secreted in a manner similar to ANTXR1 variant 3.

Mutations in ANTXR2 abrogate normal cell interaction with the ECM and have been linked to two autosomal recessive genetic disorders, juvenile hyaline fibromatosis (JHF) and infantile systemic hyalinosis (ISH), which are characterized by soft-tissue tumours (Dowling *et al.*, 2003; Hanks *et al.*, 2003). In both diseases, hyaline ECM-like material is deposited in various organs, and patients display an increased susceptibility to infection. Different mutations have been identified in ANTXR2 that correlate with severity of disease (Dowling *et al.*, 2003; Hanks *et al.*, 2003). Many of the mutations identified lead to abrogation of binding of ANTXR2 to laminin, with no noticeable differences in binding to either collagen I or collagen IV. This alteration is suggested to contribute to disease pathogenesis for both ISH and JHF (Dowling *et al.*, 2003). Interestingly, while ISH and JHF mutations that affect the VWA/I domain or transmembrane region of ANTXR2 completely abrogate its ability to function as an anthrax toxin receptor, mutations leading to a truncated cytosolic domain do not (Liu *et al.*, 2007). In fact, ANTXR2 harbouring a truncation in the cytosolic tail makes cells hypersensitive to anthrax toxin. These findings highlight differences in the role of ANTXR2 in normal physiology and anthrax pathology.

The discovery of the anthrax toxin receptors has allowed extensive study of the mechanism of PA binding to host cells. Structural and biochemical studies demonstrate that PA–ANTXR interactions resemble the manner by which some α-integrin chains bind their ligands (Bradley *et al.*, 2003; Lacy *et al.*, 2004b; Wigelsworth *et al.*, 2004). Similar to integrins, the two ANTXRs possess extracellular VWA/I domains responsible for protein–protein interactions. Nine of the 18 α-integrin chains contain I domains, which are required for binding of these integrins to their natural ligands. Also like integrins, the ANTXR I domains contain functional metal ion adhesion site (MIDAS) motifs. The MIDAS motif consists of the amino acid sequence DXSXS...T..D (where X is any amino acid), and it functions in mediating ligand binding by coordinating a divalent metal ion (Lee et al., 1995). Full coordination of the metal ion is accomplished by a residue from the ligand that contributes a carboxylate-containing side chain of an aspartate or glutamate residue. In addition, integrin I domains adopt multiple conformations that differ in their ligand affinities, alternating between 'open' or high affinity conformations and 'closed' or low affinity conformations (Shimaoka et al., 2003). The threonine in the MIDAS makes direct contact with the metal ion in the open conformation but not the closed, and mutation of this threonine is predicted to lock the I domain in a closed conformation (Li et al., 1998).

Consistent with a role for the ANTXR MIDAS, PA-ANTXR interactions are abolished in the presence of EDTA, and mutation of MIDAS residues diminishes binding to PA (Bradley et al., 2003; Scobie and Young, 2006). Furthermore, mutation of the solvent-exposed carboxylate side chain of aspartic acid 683 (D683) in PA abolishes binding of PA to its receptor, suggesting that PA mimics a natural MIDAS ligand (Bradley et al., 2003). However, this D683 residue appears to be more important for PA-ANTXR1 interaction, since certain PA variants mutated at this amino acid bind ANTXR2 but not ANTXR1 (Bradley et al., 2003; Scobie et al., 2006). Despite this finding, the crystal structure of ANTXR2 bound to PA reveals that the carboxylate side chain of PA (D683) directly interacts with the ANTXR2 MIDAS-coordinated cation as predicted for ANTXR1 (Bradley et al., 2003; Lacy et al., 2004a; Santelli et al., 2004). Finally, mutation of the threonine in the MIDAS domain has a strong inhibitory effect in PA binding to ANTXR1, but results in minimal inhibition of ANTXR2–PA binding (Bradley et al., 2003; Scobie and Young, 2006). The differential MIDAS requirements for binding to PA highlight differences in toxin/receptor affinities that may be important for pathogenesis.

PA_{83} binds to the ANTXR2 at a 1:1 ratio, and the same is predicted for ANTXR1 (Wigelsworth et al., 2004). Interestingly, the binding affinities (K_D) of PA to the isolated I

domains from ANTXR1 and ANTXR2 are strikingly different from each other. The binding affinity to the ANTXR1 I domain is 1.1 mM and 130 nM in the presence of Ca^{2+} and Mg^{2+} respectively (Scobie et al., 2005). Binding affinity to ANTXR2, however, is approximately 1000-fold higher, with K_D values or 780 pM or 170 pM in the presence of Ca^{2+} and Mg^{2+} respectively (Wigelsworth et al., 2004). The high binding affinity observed for PA-ANTXR2 interactions is due primarily to a very slow dissociation rate: 9.2×10^{-6} s^{-1} in Mg^{2+} and 8.4×10^{-5} s^{-1} in Ca^{2+} (Wigelsworth et al., 2004). Additional differences between the receptors come with the observation that PA-ANTXR2 binding is less dependent on interactions mediated by the metal ion than PA-ANTXR1 binding. The reduced requirement of PA-ANTXR2 for metal ions may be partly explained by the larger contact surface produced upon binding of PA to the ANTXR2 I-domain (Lacy et al., 2004b; Santelli et al., 2004). The higher affinity of ANTXR2 has been exploited for use as a soluble decoy, and has been compared both in vitro and in vivo using soluble versions of each receptor (Bradley et al., 2001; Scobie et al., 2005).

Receptors play a crucial role in PA prepore to pore formation. First, receptors are suggested to maintain the prepore in close proximity to the host membrane and in the correct orientation for proper pore formation (Sun et al., 2007). Second, the identity of the receptor (ANTXR1 versus ANTXR2) may dictate the manner in which EF and LF are delivered to the cytosol. Indeed, PA has the ability to form pores at physiologic pH in the absence of receptor, while a low pH is required when PA is bound to its receptor (Blaustein et al., 1989; Koehler and Collier, 1991; Milne and Collier, 1993; Miller et al., 1999). It has been proposed that the receptor acts as a molecular clamp that restricts pore formation until an acidic environment is reached. Interestingly, the receptor to which PA is bound dictates the pH at which pore formation occurs (Rainey et al., 2005). Pore formation occurs at pH ~ 6.0 when PA is bound to ANTXR1, whereas a lower pH (~5.0) is needed for pore formation when PA is bound to ANTXR2. Mutagenesis studies of both ANTXR1 and ANTXR2 reveal that residues contacting PA domain 2 or those on the neighbouring edge of PA domain 4 are involved in determining pH threshold for pore formation (Scobie *et al.*, 2007). These results suggest that ANTXR2 is the stronger molecular clamp since pore formation occurs at a much lower pH. This would also suggest that pore formation occurs within distinct endosomal compartments, depending on which receptor PA is associated with.

Co-receptor for anthrax toxin

Low-density lipoprotein receptor 6 (LRP6) was proposed to serve as a coreceptor for anthrax toxin by binding to ANTXR1 and ANTXR2 and promoting endocytosis (Wei *et al.*, 2006). Wei and coworkers found that knockdown of LRP6 using an antisense EST or siRNA in human M2182 prostate carcinoma cells leads to reduced PA binding, internalization and pore formation. LRP6 binds to both ANTXR1 and ANTXR2 *in vitro* and overexpression of a dominant negative form of LRP6 lacking the cytoplasmic domain renders M2182 cells resistant to toxin. Finally, siRNA-mediated knockdown of LRP6 in murine RAW 264.7 macrophage-like cells results in a modest decrease in sensitivity to LT.

However, other groups have failed to recapitulate the necessity of LRP6 in toxin internalization. Young and coworkers demonstrated that neither LRP6 nor the highly related protein LRP5 are necessary for PA-mediated internalization of anthrax toxin in rodent systems (Young *et al.*, 2007). Mice deficient in LRP5 or LRP6, and their macrophages *ex vivo*, are equally sensitive to PA + FP59 (a fusion of LF_N and the catalytic domain of *Pseudomonas* exotoxin A) compared with wild-type littermate control animals (Young *et al.*, 2007). Additionally, when siRNA against LRP6 is introduced into LRP5 knockout mouse embryonic fibroblasts (MEFs) or LRP5 siRNA introduced into LRP6 knockout MEFs, no change in LF internalization is seen (Young *et al.*, 2007). Furthermore, ANTXR deficient Chinese hamster ovary cells stably expressing either ANTXR1 or ANTXR2 together with siRNAs specific for either LRP5 or LRP6 are capable of internalizing toxin. A third study addressed the possibility that the LRP6 requirement might be human-specific, which they hypothesized could explain the discrepancy between the first two studies (Ryan and Young, 2008). Using siRNA against human LRP5 and LRP6 the authors show that knockdown of either of these proteins or knockdown of both simultaneously has no effect on the kinetics of toxin uptake in human HeLa cells. Thus, species-specificity is not a likely explanation to account for differences in LRP6 requirements. The significance of LRP6 in anthrax disease and its putative role as a coreceptor for anthrax toxin still remains to be determined.

Toxin internalization

The first step in cellular intoxication by anthrax toxin is binding of PA to either ANTXR1 or ANTXR2 on the host cell surface (Fig. 3.1). Binding of PA to the host cell surface allows for processing of PA by furin-like proteases (Klimpel *et al.*, 1992; Gordon *et al.*, 1995). Furin cleavage removes a 20-kDa fragment (PA_{20}) from the N-terminus, leaving a 63-kDa fragment (PA_{63}) bound to the receptor. Dissociation of PA_{20} exposes the binding sites for EF and LF. Removal of PA_{20} also eliminates a steric block allowing for PA oligomerization (Petosa *et al.*, 1997). Upon cleavage, PA_{20} and PA_{63} remain associated through non-covalent forces, though dissociation occurs within a minute of proteolytic activation (Christensen *et al.*, 2005). Given the fast dissociation of PA_{20}, it is unlikely that this is a rate-limiting step in intoxication.

The fast dissociation of PA_{20} is followed by a slow isomerization step of PA_{63} that is hypothesized to potentiate oligomerization (Christensen *et al.*, 2005). Oligomerization of PA_{63} is a dynamic process involving several PA_{63} species (Christensen *et al.*, 2006). LF has been proposed to both promote the assembly of the prepore at low concentrations, and paradoxically to inhibit prepore assembly at high LF concentrations. This can be explained if LF binds a monomeric form of PA_{63} followed by interaction with a second PA_{63} monomer to promote assembly of the LF-bridged PA dimer complex [$(PA_{63})_2 \cdot LF$] (Mogridge *et al.*, 2002b; Christensen *et al.*, 2006; Chvyrkova *et al.*, 2007). In this model, formation of the heterotrimeric [$(PA_{63})_2 \cdot LF$] complex is blocked at high LF concentrations because excess LF promotes heterodimeric [$(PA_{63}) \cdot LF$] formation, thereby depleting free PA_{63}. The [$(PA_{63}) \cdot LF$]

complex cannot contribute to prepore formation due to steric constraints that limit the number of LF molecules per prepore (Mogridge et al., 2002a). In contrast, low LF concentrations favour $[(PA_{63})_2 \cdot LF]$ formation, which then combine to assemble the $[(PA_{63})_7 \cdot LF_3]$ prepore complex, a species with a very low dissociation rate (Christensen et al., 2006). Oligomerization may be further promoted through dimerization of ANTXRs, which would stabilize intermediate oligomerization states of receptor–PA complexes (Go et al., 2006). Dimerization of receptor is mediated by interactions within the receptor transmembrane domains, and mutations that abrogate self-association of the transmembrane domain reduce the rate of PA heptamer formation.

Proteolytic processing of PA is a crucial step, as uncleaved PA does not oligomerize and is not endocytosed (Beauregard et al., 2000). Clustering of ANTXR driven by PA_{63} oligomerization causes receptors to associate with lipid rafts, which are specialized cholesterol and glycosphingolipid-rich microdomains of the plasma membrane (Abrami et al., 2003). Receptor clustering is also facilitated by dimerization through interactions within the receptor transmembrane domains (Go et al., 2006). Furthermore, while the cytoplasmic tail of receptor is not absolutely necessary for intoxication (Liu and Leppla, 2003; Liu et al., 2007), post-translational modifications in this domain contribute to efficient toxin uptake. Specifically, palmitoylation and ubiquitination play counteracting roles in receptor mediated anthrax toxin entry (Abrami et al., 2006). Palmitoylation appears to segregate ANTXR away from lipid rafts, thereby slowing internalization. Exclusion from rafts also separates receptors from the E3 ubiquitin (Ub) ligase Cbl that is responsible for ubiquitination of ANTXR. Ubiquitination triggers rapid endocytosis of receptor. Interestingly, while palmitoylation is typically thought to promote the association of proteins with lipid rafts, Abrami and coworkers found that a mutant ANTXR unable to be palmitoylated is entirely associated with lipid rafts (Abrami et al., 2006). In contrast, an inhibitor of palmitoylation blocks association of both PA_{63} and mutant ANTXR with lipid rafts, suggesting that while palmitoylation of receptor acts as a negative regulator for

lipid raft association, a second palmitoylated protein is likely involved in trafficking receptor to these membrane microdomains.

Ubiquitination of ANTXR at lipid rafts is predicted to promote interactions with proteins of the endocytic pathway, leading to the rapid uptake of toxin. Dynamin and EPS15 have been proposed to play a role in this process, thereby implicating clathrin (Abrami et al., 2003; Lu et al., 2004). However, a recent study suggests that while dynamin contributes to toxin entry, a complete block of the clathrin-mediated dynamin-dependent mechanism of internalization is not sufficient to block toxicity (Boll et al., 2004). It is possible that an increase in endosomal pH due to an interference with vesicle trafficking, and not a reduction in endocytosis, is responsible for the reduced toxin activity following dynamin inhibition.

Endosomal trafficking

Internalized toxin is trafficked to an acidic endosomal compartment where the low pH triggers a conformational change in PA, causing a shift from the prepore to the pore form with concomitant insertion into the endosomal membrane (Friedlander, 1986; Gordon et al., 1988; Milne and Collier, 1993; Milne et al., 1994). Trafficking involves ARAP3, a cellular protein that functions as a GTPase-activating protein (GAP) for RhoA and Arf6 (Lu et al., 2004). RNAi targeting ARAP3 results in a 10-fold resistance to anthrax toxin, but not diphtheria toxin or pseudomonas exotoxin A (Lu et al., 2004).

Recent work suggests that the catalytic subunits may not reach the cytosol directly through the PA pore, but are instead initially translocated into intraluminal vesicles (Abrami et al., 2004). In this model, toxin is internalized and trafficked to early endosomes where they are sorted to 'vesicular regions', which are areas of accumulated membrane invaginations and the site of endosomal carrier vesicles/multivesicular body (ECV/MVB) formation. Traffic into vesicular regions is dependent on the ε subunit of the COPI coatomer (ε -COP) which is involved in the generation of ECV/MVBs and transport to late endosomes. Further, the ability of LF to gain access to the cytosol depends on the cellular protein ALIX (apoptosis-linked gene-2 interacting protein X), a homologue of yeast class E vacuolar protein sorting vps31 involved in MVB sorting and biogenesis (Abrami *et al.*, 2004). Whereas early work suggested that the prepore toxin complex must reach late endosomes for membrane insertion of PA and translocation to occur, Abrami and coworkers found that pore associated solely with early endosomes (Abrami *et al.*, 2004). In fact, PA is internalized, trafficked to early endosomes and then rapidly degraded (Abrami *et al.*, 2003). These observations suggest that translocation of the catalytic subunits occurs before reaching late endosomes. In this mode of translocation, the catalytic subunits enter the lumen of intraluminal vesicles instead of the cytosol (Fig. 3.1). Transport of toxin to late endosomes occurs via ECV/MVBs that detach from early endosomes via microtubules and are trafficked to late endosomes. Once in late endosomes PA heptamer is rapidly degraded but the catalytic subunits remain intact, presumably because localization to the lumen of ECV/MVBs protects them from degradative enzymes. At this point, back fusion of intraluminal vesicle membranes with the limiting late-endosomal membrane results in release of catalytic subunits into the cytosol. Importantly, the finding that PA pore formation occurs at different pH thresholds depending on which receptor is bound may influence interpretation of these results. It is expected that PA bound to ANTXR1 will form pores at pH values associated with early endosomes, whereas ANTXR2-bound PA will require more acidic pH and thus likely not form a pore until the late endosome is reached. This suggests that EF and LF can translocate into the cytosol at different stages along the endocytic pathway. Indeed, whereas Abrami and coworkers suggested late endosomes to be involved in LF entry into the cytosol, another study proposed that LF reached the cytosol solely from early endosomes (Guidi-Rontani *et al.*, 2000). Based on these findings, the contributions that each of the endosomal compartments play in translocation of the catalytic subunit is not entirely clear. The cell lines used in the various studies addressing endosomal trafficking differ in their expression of the two anthrax toxin receptors, which is likely to influence the findings. Future experiments will need to account for expression of ANTXR1 and/or ANTXR2 as an important variable influencing toxin entry.

Translocation of the toxin catalytic subunits

PA binding and processing on the cell surface is crucial for binding the catalytic subunits, promoting endocytosis, and translocation of EF and LF into the cytosol. Endosomal acidification causes a large structural change in the PA prepore subunits in which amino acids 285–340 of each PA monomer peel away from the core of domain 2 to form part of a 14-stranded transmembrane extended β-barrel that inserts into the host membrane (Blaustein *et al.*, 1989; Benson *et al.*, 1998; Miller *et al.*, 1999; Nassi *et al.*, 2002; Katayama *et al.*, 2008). This prepore to pore conversion is thought to be facilitated by protonation of PA at glutamate or aspartate side chains. Protonation of histidine does not contribute directly to PA conformational changes, although it may be important in pore formation by altering the receptor-mediated pH clamp (Blaustein *et al.*, 1989; Wimalasena *et al.*, 2007).

The PA pore has a mushroom-shaped cap with a diameter of ~125 Å and a stem ~100 Å long (Katayama *et al.*, 2008). The pore spans the endosomal membrane and is a cation-selective and ion-conductive channel with a lumen size estimated to be ~15 Å wide (Blaustein *et al.*, 1989; Krantz *et al.*, 2004). Given the size of the pore, it is predicted that it would accommodate the passage of secondary structures only. This would suggest a requirement for unfolding of the protein to be translocated, which in fact appears to be the case since inhibition of protein unfolding blocks translocation (Wesche *et al.*, 1998). Unfolding of LF and EF for translocation is likely mediated by the same acidic endosomal conditions that drive PA pore formation, and decreased pH is sufficient to destabilize the native structures of EF and LF (Krantz *et al.*, 2004). The acid-destabilized protein is then translocated across the channel from N- to C-terminus, and movement is driven by the positive transmembrane potential (Krantz *et al.*, 2004). Seven phenylalanine residues, one residue contributed by each of the seven PA monomers at position 427 (Phe427), form a phenylalanine clamp (φ-clamp) that seals the channel and promotes translocation of unfolded proteins (Krantz et al., 2006). Upon conversion of the prepore to the pore, the seven Phe427 residues converge within the lumen of the pore, generating a radially symmetric heptad, with these residues remaining luminal and solvent exposed (Lacy et al., 2004a). A structural scaffold created from intersubunit interactions mediates positioning of the Phe427 residues within the pore, providing optimal interaction with the substrate to be translocated (Melnyk and Collier, 2006). Mutation of Phe427 to Ala greatly abolishes the ability of PA to translocate enzymatic A-chains into cells while increasing ion conductance (Sellman et al., 2001; Krantz et al., 2005). Indeed, a role for Phe427 in protein translocation has been demonstrated in vitro using planar lipid bilayers (Krantz et al., 2005). For translocation to occur, LF and EF must bind the prepore with the N-terminal portion placed at the entrance of the channel. The seven Phe427 residues in PA are crucial for interaction with the N-terminal domains of the catalytic subunits. Furthermore, placement of the φ-clamp at a constriction point in the translocation pathway allows it to mediate substrate translocation by means of a ΔpH-driven Brownian ratchet mechanism (Krantz et al., 2006). The φ-clamp likely binds hydrophobic segments of the translocating substrate, thereby reducing the energy barrier for unfolding of the protein. Interestingly, mutational analysis reveals that the Phe427 residue plays a dual role, functioning in both heptameric prepore to pore formation and catalytic subunit translocation (Sun et al., 2008). These findings suggest a central role for the Phe427 residue in transporting the enzymatic moieties across the membrane.

Our understanding of translocation through the PA-pore derives mainly from in vitro studies measuring electrical conductance across planar lipid bilayers. In this system, translocation of the catalytic subunits is initiated upon entry of LF_N (or EF_N) into the pore, which promotes translocation in an N- to C- terminal direction (Zhang et al., 2004a). Entry of the N-terminus is pH and/or voltage dependent; at small positive voltages LF_N (or EF_N) will enter and block ion conductance of the pore whereas large positive voltages mediate complete translocation (Zhang et al., 2004a,b, 2006; Neumeyer et al., 2006a,b).

Acidic conditions are predicted to promote protonation of histidine and acidic residues, destabilizing LF_N and promoting entry into the PA pore. In the context of a cell, the acidic conditions encountered in the endosome may promote protonation of certain residues of the unfolding catalytic moiety further promoting translocation through the cation-selective pore (Blaustein et al., 1989; Krantz et al., 2006). Upon transport of the unfolded chain across the membrane, the less acidic environment drives deprotonation, ensuring that transport of the unfolded polypeptide is a unidirectional process (Zhang et al., 2004b).

While translocation in vitro requires only PA and the N-terminus of either LF or EF, this process may be facilitated by cytosolic factors in vivo (Tamayo et al., 2008). Indeed, the N-terminus of LF interacts with ζ-COP and β-COP of the COPI coatomer complex, the complex also involved in toxin trafficking through MVB/ECVs (Abrami et al., 2003; Tamayo et al., 2008). Furthermore, addition of ζ-COP and β-COP to toxin preloaded endosomal vesicles greatly enhances translocation of LF (Tamayo et al., 2008). Interestingly, a basal level of translocation is observed in the absence of coatomer proteins, consistent with translocation in artificial planar lipid bilayers (Krantz et al., 2004). The basal level of translocation observed in the partially purified endosomal vesicles may be partly due to the established proton gradient as well as gating by the φ-clamp.

Given these observations, a translocation model is proposed where acidic pH of the endosome drives unfolding of the catalytic moieties and insertion of PA into the membrane. The transmembrane potential then drives translocation, facilitated by the φ-clamp. COPI interacts with the emerging N-terminal peptide and further promotes translocation by acting as a cytosolic chaperone. Alternatively, LF and EF may be translocated into the lumen in MVBs and gain access to the cytosol following backfusion of the MVBs with the limiting membrane of the late endosome (Abrami et al., 2004). In either case, LF and EF are delivered to the perinuclear region rather than the cell periphery (Abrami et al., 2004; Dal Molin et al., 2006). Interestingly, following translocation, LF is found free within the cytosol whereas EF remains associated with the membrane vesicles (Guidi-Rontani et al., 2000). This mode of delivery of EF results in an unevenly distributed rise in cAMP with decreasing concentration from the perinuclear area to the periphery (Dal Molin et al., 2006). The resulting intracellular distribution of cAMP may be a specific feature of the mechanism of intoxication by EF and may highlight the importance of compartmentalization of signalling events. LF, however, diffuses away into the cytosol upon translocation. The manner of delivery of LF leads to differential cleavage rates for different MKK species depending on their localization within the cytosol, which may also have important consequences for toxin action (Abrami et al., 2004).

Cellular effects of anthrax toxin

Much of what we know about the molecular mechanisms of ET and LT and their downstream effects derives from studies employing cultured cells. Murine macrophages were the first cell type for which LT-induced cell death was observed, and many of the subsequent studies on this toxin focused on the effects of LT on macrophages. The adenylate cyclase activity of EF was first identified using Chinese hamster ovary (CHO-K1) cells. Current research efforts are expanding our knowledge regarding both the cell types and cellular pathways affected by these two toxins.

Interestingly, while the AB toxin model implies that the binding 'B' moiety has no toxic activity on its own, recent studies indicate that PA induces death in cells overexpressing ANTXR1 (Salles et al., 2006; Taft and Weiss, 2008). In cell culture, PA-induced cytotoxicity is likely a result of pore formation in endosomal membranes, though toxicity is not seen with ANTXR2 overexpression (Salles et al., 2006; Taft and Weiss, 2008) (F.J. Maldonado-Arocho, K.A. Averette, and K.A. Bradley, unpublished). In addition, PA induces cytolysis of red blood cells, which is enhanced in the presence of LF or EF (Wu et al., 2003). Future studies will need to address whether these PA-mediated effects contribute to pathology in animal models before their significance is understood. Alternatively, PA-mediated cytotoxicity of ANTXR1 overexpressing cells may have important implications for the use

of anthrax toxin to treat cancer as described in 'Future perspectives' at the end of this chapter.

As the number of cell types and cellular processes affected by anthrax toxin expands, it is clear that both EF and LF affect multiple cellular functions, and that these functions often depend on the cell type under study. In addition, there is overlap in the pathways affected by LF and EF, with some functions being synergistic, while others appear antagonistic. Both toxins affect cell viability, proliferation, adhesion, migration, immune cell activation and cytokine production as described below.

Cellular effects of lethal toxin

Lethal factor cleaves MKKs, disrupts MAPK signalling and causes a variety of phenotypes depending on the cell type. The cellular response to LT is determined by the dependence of a target cell on MAPK signalling for cellular processes such as proliferation, differentiation, mobility, viability, activation and/or cytokine production. In addition, murine macrophages and DCs expressing specific 'lethal toxin sensitivity' (LT^S) alleles of the gene Nalp1b undergo a rapid caspase-1 dependent inflammatory cell death, termed pyroptosis. Disruption of MKK signalling does not appear to be sufficient for pyroptosis and it is currently unknown what signal is ultimately responsible for the rapid cytolysis. Since the host response to LT varies between cell types, understanding these differences will aid in interpreting the contribution of this toxin to the disease anthrax (Fukao, 2004; Mourez, 2004; Banks et al., 2006). The following sections summarize the cellular responses and highlight trends in current research.

Cytokine production and immune regulation
Cytokines and chemokines are cellular communication molecules important for innate and adaptive immune responses. These molecules bridge innate and adaptive immunity, recruit and activate immune cells and modulate the host response to pathogens. LT both suppresses and induces cytokine production in numerous cell types, including macrophages, DCs, T-cells, ECs and neutrophils (Erwin et al., 2001; Agrawal et al., 2003; Comer et al., 2005a; Fang et al., 2005;

Paccani *et al.*, 2005; Batty *et al.*, 2006; Depeille *et al.*, 2007; Barson *et al.*, 2008).

Macrophages that die through pyroptosis in the presence of LT release inflammatory molecules IL-1β and IL-18. These cytokines are produced through cleavage of pro-IL-1β and pro-IL-18 by interleukin-1 converting enzyme (ICE), also known as caspase-1. LT activates caspase-1 in LT^S macrophages through an unknown mechanism leading to maturation of the two cytokines and release following cellular lysis. The role of these cytokines in *B. anthracis* infection are unknown; however, the rapid release of these cytokines may be associated with LT toxicity in whole animals (Hanna *et al.*, 1993; Moayeri *et al.*, 2003; Moayeri *et al.*, 2004). Alternatively, pyroptosis and release of IL-1β and IL-18 may represent a host defence against anthrax infection. Indeed, there is an inverse relationship between LT sensitivity and sensitivity to spore challenge in mice (Welkos *et al.*, 1986; Moayeri *et al.*, 2004). Interestingly, other than the cytokines described above, the immune response to anthrax is remarkably non-inflammatory indicating that anthrax toxin actually represses cytokines (Moayeri *et al.*, 2003).

LT inhibits cytokine production in multiple cell types through disruption of MKK signalling. Specifically, LT inhibits IL-8 production and secretion in ECs through destabilization of the 3′ untranslated region of IL-8 mRNA (Batty *et al.*, 2006). LT also suppresses chemokine production in human neutrophil NB-4 cells and macrophages (Erwin *et al.*, 2001; Barson *et al.*, 2008). Messenger RNA levels of the immune cell chemoattractants IL-8, CCL20, CCL3 and CCL4 decrease in LT-treated NB-4 cells (Barson *et al.*, 2008). LT also blocks LPS-mediated induction of mRNAs encoding the proinflammatory cytokines tumour necrosis factor α (TNF-α), IL-6, IL-1α, IL-1β, and IL-1 receptor antagonist (IL-1RA) in macrophages (Erwin *et al.*, 2001).

Other molecules of the innate and adaptive immune system are differentially regulated in the presence of LT. Superoxide is a toxic compound released by phagocytes and neutrophils to kill invading microorganisms. There are conflicting reports as to whether LT increases or decreases superoxide production in murine neutrophils

(Crawford *et al.*, 2006; Xu *et al.*, 2008). LT may be disrupting neutrophil bactericidal activity through deregulation of superoxide production. LT is also a non-competitive inhibitor of glucocorticoid and progesterone receptors (Webster *et al.*, 2003). Specifically, LT disrupts glucocorticoid receptor (GR)-DNA binding by an unknown mechanism that may be mediated through disruption of p38 signalling. It is unclear what the role of nuclear hormone receptor inhibition has on infection, but the hypothalamus–pituitary–adrenal (HPA) axis and hormone receptors influence inflammatory responses and are a common target of numerous bacterial and viral products. The function of the HPA axis and GR repression in LT-treated mice is discussed in the section 'Animal models.'

Proliferation, differentiation and activation

Macrophages and DCs are key components of the innate immune system required for controlling *B. anthracis* infection (Cote *et al.*, 2004; Kang *et al.*, 2005; Cote *et al.*, 2006; Hu *et al.*, 2006; Ribot *et al.*, 2006). Phagocytosis of spores leads to rapid activation of DCs, an activation that is silenced soon after, presumably by production of anthrax toxin from the germinating spore (Agrawal *et al.*, 2003; Ruthel *et al.*, 2004; Brittingham *et al.*, 2005; Cleret *et al.*, 2006). Bacilli express toxin inside the host cell, contributing to escape and establishment of a systemic infection (Guidi-Rontani *et al.*, 1999; Guidi-Rontani *et al.*, 2001; Banks *et al.*, 2005). Following escape, bacilli continue to secrete LT, which inhibits cytokine secretion, impairs phagocyte bactericidal activity and promotes cell death (Erwin *et al.*, 2001, Tournier, 2005; Agrawal, 2003, Reig, 2008; Park *et al.*, 2002; Alileche *et al.*, 2005; Ribot *et al.*, 2006).

LT inhibits proliferation of undifferentiated monocytic cells, precursors of macrophages and DCs (Kassam *et al.*, 2005). LT also inhibits differentiation of human monocytic cells in response to phorbol myristate acetate (PMA) (Kassam *et al.*, 2005). Immature DC populations are maintained by circulating DC precursors derived from monocytes. By blocking monocyte proliferation and differentiation, LT is likely interfering with the ability of monocytes to replenish DC pools in tissues. In a monocytic cell line, THP-1, cell cycle arrest is corrected through activation of the phosphatidylinositol 3-kinase/Akt signalling pathway (Ha *et al.*, 2007b). Counteracting the effects of LT on monocytes may represent a path to future therapeutic approaches.

LT also inhibits proliferation of melanocytes, activated T-cells and B-cells (Koo *et al.*, 2002; Comer *et al.*, 2005a; Fang *et al.*, 2006) and inhibits T-cell activation and B-cell antibody production (Comer *et al.*, 2005a; Fang *et al.*, 2005; Paccani *et al.*, 2005, Fang, 2006). While no overall cytotoxic effect is observed in T-cells and B-cells in response to toxin, inhibiting proper T-cell and B-cell function may block the development of a protective immune response. However, the benefits of targeting adaptive immunity for a pathogen that induces rapid and high rates of mortality are currently unclear. One possibility is that targeting adaptive immune system may play a role in cutaneous anthrax, which presents with a more prolonged sequela.

Mobility and adhesion
LT enables *B. anthracis* to evade host immunity and directly induces pathology through inhibition of cellular motility and changes in cell–cell and cell–extracellular matrix adhesion. Neutrophils are scavengers for invading microorganisms and require proper mobility to phagocytose bacteria. LT inhibits actin-based mobility in neutrophils by blocking phosphorylation of heat shock protein 27 (Hsp27) (During *et al.*, 2005; During *et al.*, 2007). Hsp27 is specifically phosphorylated by p38 leading to reorganization of the actin cytoskeleton in response to stress and chemokines (Guay *et al.*, 1997; Jog *et al.*, 2007). Disruption of p38 signalling, through LT or p38 inhibitors, severely inhibits neutrophil chemotaxis. By targeting actin-based motility, LT debilitates the innate immune system thus promoting *B. anthracis* survival.

Many pathogens have evolved the ability to alter host cell adhesion. LT treated primary human lung microvascular ECs display decreased cell-cell adhesion leading to endothelial barrier dysfunction (Warfel *et al.*, 2005). ECs treated with LT also show increased central actin stress fibres and altered VE-cadherin distribution, another sign of EC barrier dysfunction (Warfel

et al., 2005). In the presence of TNF-α, LT induces increased vascular cell adhesion molecule-1 (VCAM-1) expression and monocyte adhesion to human lung microvascular ECs via an IFN regulatory factor-1-dependent mechanism (Steele *et al.*, 2005; Warfel and D'Agnillo, 2008). Increased expression of EC adhesion molecules can lead to an accumulation of leukocytes at the vessel cell wall that may contribute to inflammation and vasculitis. Vascular collapse is associated with anthrax and is induced in LT-animal models. Thus, LT-induced vasculitis and endothelial barrier function may contribute to LT-induced pathology, with ECs as critical targets of this toxin.

Cell death: apoptosis
In vitro, LT causes apoptotic death of differentiated and immortalized human monocytic cell lines, many immortalized cancer cell lines, ECs and activated macrophages (Koo *et al.*, 2002; Park *et al.*, 2002; Kirby, 2004; Abi-Habib *et al.*, 2005; Kassam *et al.*, 2005; Ding *et al.*, 2008; Rouleau *et al.*, 2008). Macrophages that do not undergo pyroptosis in response to LT, referred to as 'resistant' macrophages (LT^R), can be sensitized to LT-mediated apoptosis through stimulation with bacterial components, such as lipoteichoic acid or lipopolysaccharide (LPS), or tumour necrosis factor-α (Park *et al.*, 2002; Kim *et al.*, 2003). LT-induced apoptosis of activated macrophages requires mammalian target of rapamycin (mTOR) and p38 signalling (Kim *et al.*, 2003; Hsu *et al.*, 2004). Cell death is consistent with caspase-3 mediated apoptosis, through activation of caspase-3, -6, -7, and -9 and release of cytochrome c from mitochondria. Additionally, the dsRNA-responsive kinase PKR, as well as its downstream effectors eukaryotic initiation factor 2α (eIF-2α) and interferon response factor 3 (IRF3), are required for LT-induced apoptosis of activated macrophages (Hsu *et al.*, 2004). PKR is similarly required for apoptosis of macrophages treated with *Salmonella typhimurium* or *Yersinia pseudotuberculosis*, suggesting that this is a general pathway critical for pathogen-induced apoptosis.

LT also induces caspase-dependent apoptosis in ECs (Kirby, 2004). In addition to direct LT-induced apoptosis, ECs are sensitive to the release of toxic compounds during LT-mediated lysis of nearby macrophages (Kirby, 2004; Pandey and Warburton, 2004). Endothelial barrier dysfunction occurs in LT-treated human endothelial monolayers and in a zebrafish model of LT-intoxication, though barrier dysfunction occurs prior to measurable cell death (Warfel et al., 2005; Bolcome et al., 2008). In contrast, Warfel et al. did not detect apoptosis or necrosis in LT-treated ECs, but found reduced metabolic activity in these cells (Warfel et al., 2005). Thus, while LT can induce apoptosis in ECs, endothelial barrier dysfunction does not appear to require cell death. The significance of LT-mediated apoptosis is currently not known.

Pyroptosis

Different strains of mice have varying levels of sensitivity to LT, both at the whole animal level and with respect to LT-sensitivity of macrophages ex vivo (Welkos et al., 1986; Singh et al., 1989; Roberts et al., 1998; Boyden and Dietrich, 2006). While some exceptions exist, whole animal sensitivity generally correlates with macrophage sensitivity to LT in mice (Moayeri et al., 2004). LT induces pyroptosis, a rapid, caspase-1-dependent, pro-inflammatory death in macrophages and DCs harbouring LT^S alleles of Nalp1b. Thus, studies of the rapid macrophage cytolysis induced by LT may provide insight into effects of LT toxaemia.

Most of what we know about LT-induced pyroptosis has been elucidated in LT^S macrophages, but immature DCs also undergo pyroptosis and may have similar requirements for cell death (Reig et al., 2008). While LT does not induce pyroptosis in human cells, this form of cell death does occur in human macrophages in response to various other bacterial stimuli (Fink and Cookson, 2005; Labbe and Saleh, 2008). Therefore, LT-induced pyroptosis in murine macrophages is an interesting model to study pyroptosis in response to a bacterial toxin (Fink and Cookson, 2005).

The gene responsible for pyroptosis sensitivity to LT is NACHT-, LRR-, and PYD-containing protein 1 paralogue b, or Nalp1b. Nalp1b is a member of intracellular toll-like receptors, also called NOD-like receptors (NLRs). NLRs recognize danger signals or foreign molecules and signal to a cell that there is a foreign invader (Martinon et al., 2007). In inbred mice, there are multiple alleles of Nalp1b that are strain-specific (Boyden and Dietrich, 2006). It is now clear that LT^S macrophages contain alleles of Nalp1b that lead to caspase-1 activation in response to the enzymatic activity of LF (Boyden and Dietrich, 2006). Introducing the sensitivity allele of Nalp1b into resistant macrophages shifts these cells to a sensitive phenotype. Conversely, knocking down expression of the sensitive allele in macrophages by RNAi partially protects them from LT. Therefore, LT^S alleles function in a dominant manner and macrophages containing one or more copies of a sensitivity allele are sensitive to LT. It is unknown what cell types and tissues express Nalp1b, though expression may be restricted as macrophages and DCs are the only cell types reported to undergo pyroptosis in response to LT.

Some NLRs, including human NALP1, bind to caspase-1, directly or through an adaptor molecule, and activate caspase-1 inflammasomes (Faustin et al., 2007; Martinon et al., 2007; Martinon and Tschopp, 2007). NLRs are activated in response to at least eleven known bacteria or bacterial products, forming active inflammasome complexes in certain cell types (Martinon and Tschopp, 2007). In LT^S macrophages, LF activates caspase-1 and induces IL-1β and IL-18 release dependent on the presence of a sensitive Nalp1b allele (Cordoba-Rodriguez et al., 2004; Boyden and Dietrich, 2006).

The nature of the signal that Nalp1b recognizes is not currently known. LF catalytic activity is required (Klimpel et al., 1994), however MKK inactivation is not sufficient to induce pyroptosis (Kim et al., 2003). In the presence of the LT signal, Nalp1b binds to caspase-1, directly or indirectly, and induces a conformational change within the inflammasome. This triggers the auto-cleavage of caspase-1 and the production of IL-1β and IL-18. Caspase-1 is required for rapid lysis of sensitive macrophages (Boyden and Dietrich, 2006; Muehlbauer et al., 2007; Squires et al., 2007; Wickliffe et al., 2008a). Whether IL-1β or IL-18 release is required is under debate (Wickliffe et al., 2008a), but evidence suggests that IL-1β and IL-18 release is dependent on cell death, rather than required for cell death (Wickliffe et al., 2008a). Interestingly, activation of the NALP3-inflammasome in resistant macrophages does not sensitize these cells to LT, suggesting that the Nalp1b-inflammasome is specifically required, and caspase-1 activation, though required, is not sufficient for LT-induced pyroptosis (Wickliffe *et al.*, 2008a).

Another important molecular complex required for LT-induced pyroptosis is the proteasome. Proteasome inhibitors protect from LT (Tang and Leppla, 1999) and can protect cells very late in intoxication, right before membrane perturbation and cell lysis (Alileche *et al.*, 2006; Wickliffe *et al.*, 2008a). Proteasome inhibitors also prevent caspase-1 activation when added early in intoxication with LT, and abrogate caspase-1 activation when added late (Squires *et al.*, 2007; Wickliffe *et al.*, 2008a). Again, this appears to be Nalp1b-inflammasome specific, as caspase-1 is still activated in a NALP3-inflammasome in the presence of proteasome inhibitors (Wickliffe *et al.*, 2008a). Interestingly, *Shigella flexneri* and *Chlamydia trachomatis* also induce caspase-1 activation dependent on proteasome activity (Hilbi *et al.*, 2000; Lu *et al.*, 2000). Therefore, there may be a common protein, or different proteins depending on the stimulus and/or NLR, that inhibits inflammasome formation and must be degraded prior to the formation of functional inflammasomes (Willingham *et al.*, 2007).

Additional support for the role of proteasome-mediated protein degradation in LT-cytolysis is the requirement of N-end rule degradation in this process. The N-end rule determines protein stability based on the N-terminal residue. Inhibitors of N-end rule proteasomal degradation protect macrophages from caspase-1 activation and cell death (Wickliffe *et al.*, 2008b). Additionally, bestatin methyl ester, an aminopeptidase inhibitor that blocks N-end rule degradation, protects LT^S Balb/cJ macrophages from LT and also moderately protects Balb/cJ mice from LT-intoxication (Wickliffe *et al.*, 2008b).

LT increases ion fluxes in LT^S macrophages, which is essential for membrane permeability and cell lysis (Hanna *et al.*, 1992; Squires *et al.*, 2007; Wickliffe *et al.*, 2008a). Potassium chloride, potassium channel blockers, and sucrose are potent inhibitors of LT-induced cell lysis and caspase-1 activation (Alileche *et al.*, 2006; Wickliffe *et al.*,

2008a). This places ion flux perturbation as an early event that controls caspase-1 inflammasome activation. Potassium efflux is likely a secondary signal required for inflammasome activation since it also controls NALP3-inflammasome formation (Petrilli *et al.*, 2007b; Wickliffe *et al.*, 2008a). The PA pore may play a role in potassium efflux since other pore-forming toxins, such as staphylococcal alpha-toxin and gramicidin, also lead to potassium efflux and caspase-1 activation (Walev *et al.*, 1995). Both proteasome activity and potassium chloride efflux trigger cell lysis upstream of caspase-1 activation, but it is not clear how these two events relate to each other. However, only proteasome inhibitors protect cells from mitochondrial impairment, a downstream event in LT-induced cell death.

Mitochondrial function is pivotal to cellular homeostasis and disruption of the mitochondrial membrane leads to both apoptosis and necrosis (Pinton *et al.*, 2001; Zhivotovsky and Orrenius, 2001). Mitochondrial membrane potential (MMP) and succinate dehydrogenase activity (SDH), two measures of mitochondrial function, both decrease prior to LT-induced cell death (Alileche *et al.*, 2006). Addition of potassium chloride to LT-treated cells protects against plasma membrane disruption, even when added just prior to cell lysis. However, potassium chloride does not protect against mitochondrial disruption, as MMP and SDH activity still decrease in these cells. These cells ultimately succumb to toxin; potassium chloride appears to only delay cell death by a few hours if added after a drop in MMP. Therefore, it appears that mitochondrial dysfunction is a key step in LT-induced cytolysis.

Toxin-induced resistance (TIR), a phenomenon by which LT^S macrophages become resistant to pyroptosis induced by LT, is also linked to mitochondrial homeostasis. TIR is induced when macrophages are pre-treated with a sublethal dose of LT: the cells that survive become resistant to subsequent challenge with LT even though they internalize toxin (Salles *et al.*, 2003). Following LT challenge, MAPK signalling is disrupted, and particularly, p38 disruption also induces a TIR-like resistance (Ha *et al.*, 2007a). Inhibiting p38 destabilizes mRNA transcripts of two mitochondrial proteins: Bcl-2/adenovirus

E1B 19-kDa interacting protein 3 (Bnip3) and Bnip3-like (Bnip3L) (Ha et al., 2007a). Following sublethal LT-treatment of LT^S macrophages, Bnip3 and Bnip3L transcript levels and protein levels decrease. Knocking down Bnip3 or Bnip3L mRNA levels using siRNA also protects cells from LT, and overexpression of Bnip3 and Bnip3L in TIR cells converts them back to LT-sensitive (Ha et al., 2007a). Bnip3 and Bnip3L are members of the Bcl2-family and are pro-apoptotic proteins present at the mitochondrial membrane (Yasuda et al., 1998). It is likely that Bnip3 and Bnip3L are required for LT-mediated cell death, through disruption of the mitochondrial membrane or sequestering anti-apoptotic proteins. It is unclear whether Bnip3 and/or Bnip3L play a direct role in LT-induced cell death or if the absence of these proteins makes cells less susceptible to LT via indirect mechanisms.

Activation of pyroptosis is also dependent on signalling through other key proteins. MAPK inhibition by LT prevents ribosomal-S6 kinase-2 (RSK) activation and phosphorylation of the RSK target, CCAAT/enhancer binding protein (C/EBPβ) (Buck and Chojkier, 2007). In primary macrophages of the LT^S mouse strain FVB, expression of a dominant active form of C/EBPβ protects macrophages from LT-induced apoptosis (Buck and Chojkier, 2007). In this system, caspase-8 appears to be required for LT-induced cell death and constitutive activation of C/EBPβ prevents caspase-8 activation. Previously, Popov et al. found that a caspase-8 inhibitor could protect sensitive macrophages from low-dose LT-induced apoptosis, as could inhibitors of caspase-1, -2, -3, -4, and -6 (Popov et al., 2002). Further investigation into the role C/EBPβ and caspase-8 in LT-induced cytolysis is warranted.

Finally, multiple research groups have discovered proteins that may be involved in pyroptosis through inhibitor studies and siRNA knockdown. Pharmacological inhibitors of phospholipase A2, phospholipase C, serine/threonine phosphatase, tyrosine-specific protein kinase, or metallopeptidase/leukotriene A4 hydrolase protect macrophages from LT (Bhatnagar et al., 1989; Lin et al., 1996; Shin et al., 1999; Kau et al., 2002). Removing calcium from culture media

or adding a calcium chelator also protects macrophages from LT-induced cytolysis (Bhatnagar et al., 1989). Lithium chloride sensitizes cells to LT, suggesting a role for GSK-3β in toxicity (Tucker et al., 2003). Inhibiting protein translation with puromycin or cycloheximide protects from LT, suggesting that synthesis of new proteins is required for LT sensitivity (Bhatnagar and Friedlander, 1994). However, a conflicting report shows that blocking new protein synthesis with inhibitors of transcription or translation does not prevent LT-mediated cell death (Levin et al., 2008). Using siRNA knockdown in LT^S macrophages, the following genes were found to potentially be involved in LT sensitivity: J κ recombination signal binding protein (Rbp-jκ), cytohesin 4 (Pscd4), lysosomal-associated protein transmembrane 5 (Laptm5), hypoxia-inducible factor 1-α (HIF1-α), and subunit c of vacuolar H$^+$-ATPases (Atp6v0c) (Kim et al., 2007). The mechanisms by which these genes and proteins are involved in LT-mediated pyroptosis are still under investigation.

In summary, the events that occur following LF translocation into sensitive cells include: (1) cleavage of LF substrates/MKKs, (2) ion flux and potassium efflux, (3) proteasome cleavage of unknown target(s), (4) mitochondrial dysfunction, (5) caspase-1 inflammasome activation and production of mature IL-1β and IL-18, (6) plasma membrane permeability, and (7) cellular lysis and release of IL-1β and IL-18. In addition to these steps, multiple host proteins are implicated in promoting or counteracting LT-induced pyroptosis and future work will likely seek to validate these contributions. It must be emphasized that the role of rapid cell lysis of macrophages and immature DCs is still unclear in an in vivo anthrax infection. However, the type of cell death induced by LT appears to be similar to cytolysis induced by multiple pathogens via NLR/caspase-1 activation and necrosis (Kanneganti et al., 2007; Lamkanfi et al., 2007; Petrilli et al., 2007a; Ting et al., 2008; Wilmanski et al., 2008).

Genomics and proteomics of LT-treated macrophages
Comparative genomics and proteomics have been employed to elucidate changes in gene expression and protein levels following LT intoxication,

providing insight into the host response to LT. These studies highlight the abundant changes that occur in response to LT, as well as the diversity of signalling pathways and organelles affected by LT.

Treatment of LT^S macrophages with a lethal dose of LT leads to transcriptional changes in genes related to MAPK signalling, Wnt signalling, inflammatory response, cytoskeletal regulation, energy production, protein metabolism, signalling, stress, and transcription (Tucker *et al.*, 2003; Comer *et al.*, 2005b). Each gene group portrays trends in transcriptional changes: nearly all genes relating to cytoskeletal regulation, energy production or protein metabolism are down-regulated, while most inflammatory genes and transcription factors are up-regulated (Comer *et al.*, 2005b).

Sublethal challenge of LT^S macrophages with LT reveals that very few significant changes occur in gene expression at 3 and 6 hours post-treatment (Bergman *et al.*, 2005). When cells are co-treated with LPS and a sublethal dose of LT, numerous genes are differentially expressed compared to LPS-only treated cells (Bergman *et al.*, 2005). This is not surprising since LPS induces expression of numerous MAPK signalling-dependent genes that are blocked by LF. This finding confirms previous observations that LT blocks pro-inflammatory signals and supports the hypothesis that this toxin functions to enable immune evasion.

Gene expression data thus far has failed to explain LT-mediated, *Nalp1b*-dependent toxicity in macrophages. This is not surprising since LT, at high concentrations, is capable of lysing macrophages in as little as 45 minutes. Therefore, it is unlikely that changes in gene expression contribute significantly to toxicity in such a short time interval. Events that occur at the protein level such as degradation, phosphorylation or ubiquitination cannot be assessed through comparative genomics. Therefore, analysing the macrophage proteome in response to LT may unveil novel responses and important players in cytolysis.

Multiple groups have profiled proteomic changes in LT^S macrophages following toxin challenge. Consistent findings include changes in glycolysis, ATP production (ATP synthase β subunit), cytoskeletal arrangement (β-actin, vimentin), and the stress response (GRP78, Enolase-1, HSP70) (Chandra *et al.*, 2005; Kuhn *et al.*, 2006; Sapra *et al.*, 2006; Kim *et al.*, 2008b).

All proteomic studies report a change in HSP70, a heat shock protein that is up-regulated in response to many toxic compounds and thermal or oxidative stress. Some discrepancy exists with one group reporting a decrease in HSP70 (Chandra *et al.*, 2005), and other groups finding an increase in HSP70 following LT treatment (Kuhn *et al.*, 2006; Sapra *et al.*, 2006). The role of HSP70 in LT-toxicity is unknown. However, LT-treated cells may recognize a danger or stress signal that leads to an increase in HSP70 as a protective factor.

A recent proteomic study suggests that there may be other targets of LF (Kim *et al.*, 2008b). One protein, neuronal nitric oxide synthase (nNOS), was identified as a direct target of LF cleavage. In cell lysates treated with recombinant LF or in cells treated with LT, nNOS is fractionated into two or three bands, suggesting that it may be cleaved twice by LT. Interestingly, the cleavage site within nNOS does not match the consensus sequence for MKK cleavage, exemplifying the idea that the LF active site is promiscuous and may have other targets. While additional substrates of LF have been hypothesized, nNOS is the first non-MKK target of LF enzymatic activity to be reported. Determining the role of nNOS in LT-toxicity will be an important area of future research.

Cellular effects of oedema toxin

Historically, ET has received less attention than LT, as it was believed that ET contributed less significantly to disease (Beall *et al.*, 1962; Pezard *et al.*, 1991). Recent work, however, has prompted an increased interest in studying ET. It is now recognized that ET not only plays an immunomodulatory role during the initial stages of disease development, but it also induces cellular and histopathological changes that contribute to the circulatory shock associated with the disease. Furthermore, in contrast to earlier findings, recent evidence indicates that ET induces cell death in certain cell types and is lethal to mice (Firoved *et al.*, 2005; Hong *et al.*, 2005; Voth *et al.*, 2005). Many of the ET-induced effects on the host can be linked to the ability of EF to alter intracellular cAMP levels, therefore leading to changes in cell signalling pathways and cellular functions.

ET alters host cell gene expression

ET promotes gene transcriptional changes in different cell types, including macrophages, DCs and ECs (Comer *et al.*, 2006; Maldonado-Arocho *et al.*, 2006; Hong *et al.*, 2007; Raymond *et al.*, 2007; Xu *et al.*, 2007). ET-induced transcriptional changes are linked to the ability of EF to raise cAMP levels in cells. Cyclic AMP is a second messenger involved in signal transduction of several signalling cascades. Signalling via cAMP is mediated through several downstream effectors, including protein kinase A (PKA) and Epac. As expected, ET-induced transcriptional changes are linked to activation of these two downstream effectors, though other pathways may also exist (Hong *et al.*, 2005; Maldonado-Arocho *et al.*, 2006; Hong *et al.*, 2007; Raymond *et al.*, 2007).

Epac is a guanine nucleotide-exchange factor (GEF) for Rap1, a host protein regulating cellular adhesion and spreading (de Rooij *et al.*, 1998; Bos, 2005). ET induces expression of Epac-related activators of Rap1. Interestingly, ET also induces direct activation of Rap1 through Epac (Hong *et al.*, 2007). It is the activation of Rap1 that is responsible for the inhibition of chemotaxis observed in ECs, an inhibition that is independent of PKA.

Upon activation by cAMP, the catalytic subunit of PKA translocates to the nucleus where it phosphorylates various transcription factors leading to transcriptional changes of target genes (Daniel *et al.*, 1998). ET-induced activation of PKA promotes Epac-independent cell rounding, aggregation and detachment of cells, specifically in Y-1 adrenal cells, 293T kidney cells and MEFs (Hong *et al.*, 2005). In macrophages, ET-induced PKA activation inhibits expression of several genes including phospholipase A2 (sPLA-IIA) (Comer *et al.*, 2006; Raymond *et al.*, 2007). PKA activation also leads to induction of various transcription factors including the cAMP-responsive element binding (CREB) protein. CREB recognizes the CRE sequence present in cAMP responsive genes, allowing binding and formation of a transcription complex (Mayr and Montminy,

2001). ET induces activation of CREB via PKA (Park *et al.*, 2005; Maldonado-Arocho *et al.*, 2006), and several genes differentially regulated following ET treatment are dependent on CREB activation, including genes responsible for macrophage migration and survival (Park *et al.*, 2005; Kim *et al.*, 2008a). Yet other ET-affected genes, including sPLA2-IIA, are independent of CREB, indicating that other PKA-activated transcription factors play a role in ET-induced transcriptional changes (Raymond *et al.*, 2007).

ET-induced activation of PKA also leads to up-regulation of the ANTXR genes in monocyte-derived cells (macrophages and DCs) (Comer *et al.*, 2006; Maldonado-Arocho *et al.*, 2006; Xu *et al.*, 2007). Increased surface expression of ANTXRs leads to increased PA binding and more rapid accumulation of LF and EF in the host cytosol (Maldonado-Arocho *et al.*, 2006). Receptor levels may be limiting for toxin internalization *in vivo*, and high levels of PA promote receptor endocytosis, thereby further limiting receptor availability (Molnar and Altenbern, 1963). Indeed, exposure to high doses of PA *in vitro* leads to a reduced availability of cell surface receptors (Maldonado-Arocho *et al.*, 2006). Therefore, the ability of ET to induce surface receptor expression may be a mechanism that ensures cells are continually intoxicated as toxin concentrations increase. Finally, the observation that the ET-induced increase in ANTXR expression is monocyte-lineage specific may be indicative of the importance of targeting these immune cells by anthrax toxin.

Activation of PKA and Epac by ET highlights how this toxin may alter cellular signalling networks through cAMP production. The transcriptional changes mediated through these, as well as other, cell signalling networks may be unique to ET due to the manner in which this toxin induces cAMP production. ET is delivered to the cytosol from late endosomes, generating an uneven distribution of cAMP levels with decreasing concentration from the perinuclear area to the periphery (Dal Molin *et al.*, 2006). The manner in which ET raises cAMP levels distinguishes it from other cAMP elevating toxins, leading to gene transcription changes that are unique from other cAMP elevating agents.

Effects of ET on cell viability

Previously, it was believed that ET did not affect cellular viability or induce cytotoxicity. Multiple studies indicate that ET does not kill ECs, neutrophils, macrophages, T-cells, B-cells, fibroblasts, hepatoma cells or melanoma cells (O'Brien *et al.*, 1985; Comer *et al.*, 2005a; Park *et al.*, 2005; Maldonado-Arocho *et al.*, 2006; Hong *et al.*, 2007; Kim *et al.*, 2008a). In fact, ET treatment actually protects macrophages from LPS-induced apoptosis, a response dependent on ET-induced CREB activation (Park *et al.*, 2005). However, this model is now challenged by the findings of Voth and colleagues, who report ET-mediated cytotoxicity in macrophages (Voth *et al.*, 2005). ET-induced cell death is cell type specific and depends on the system under study (Hong *et al.*, 2005; Voth *et al.*, 2005). Cells undergoing EF-induced death show morphological changes leading to cell lysis, including cell rounding, aggregation and cell detachment (Hong *et al.*, 2005; Voth *et al.*, 2005). Interestingly, the cell death observed after ET intoxication does not show any of the hallmarks of apoptosis or oncosis, suggesting a unique mechanism of cell death (Voth *et al.*, 2005). Given these new results, future experiments will be needed to identify cell types that are sensitive to ET-mediated killing, to determine if activation status or differences between distinct functional subsets of macrophages are important for ET-sensitivity, and to define the mechanism by which cell death occurs.

ET effects on cell function

ET has long been recognized to play a role in promoting immune evasion by influencing phagocytes (O'Brien et al., 1985). It is now clear that ET affects both innate and adaptive arms of the immune system, having multiple effects on many immune cell types. The cellular effects of ET span beyond those of immune cells suggesting that this toxin may have a broader impact in the host. Given the pleotropic effects of cAMP signalling, and the ability of PA to mediate EF entry into nearly all cell types, future research should focus not only on what cellular events occur in response to ET, but which of the many responses induced by ET are functionally important.

Effects of ET on immune cell function According to the current model, a major role of ET is to dampen the host immune response. The immune response to an invading pathogen is modulated through the release of cytokines and chemokines. ET affects cytokine secretion profiles of monocytes and T-cells, and selectively inhibits production of IL-12p70 and TNF-α by DCs (Hoover *et al.*, 1994; Paccani *et al.*, 2005; Tournier *et al.*, 2005).

Release of cytokines and chemokines is important for proper recruitment of various immune cells to the site of infection. In addition to altering cytokine production, ET differentially modulates the chemotactic response of different cell types to these signals. Early work demonstrated that ET promotes neutrophil chemotaxis towards fMLP, although the significance of this during a *B. anthracis* infection is not clear (Wade *et al.*, 1985). In addition, ET promotes migration of macrophages in a cAMP and CREB dependent manner (Kim *et al.*, 2008a). Expression of EF by internalized vegetative bacilli is proposed to promote macrophage migration, enhancing the ability of bacilli to progress to a systemic infection in a Trojan horse model. Contrasting work, however, reports that ET inhibits macrophage and T-cell chemotaxis through perturbation of chemokine receptor signalling (Rossi Paccani *et al.*, 2007). Discrepancies between the two studies may be due to differences in the assay setup and/or the manner in which macrophages are generated. Distinct functional subsets of macrophages have been described, and their functional phenotype may change in response to alterations in their microenvironment (Stout *et al.*, 2005). It is conceivable that ET may have differential effects on macrophages depending on their phenotypic programming.

Recruitment of immune cells is crucial especially at the initiation of disease to limit the spread of infection. Both neutrophils and macrophages arrive at sites of infection and mediate clearance of infection through engulfment and killing of the invading pathogen. ET, however, limits the ability of neutrophils to clear infection by inhibiting both the oxidative burst as well as phagocytosis of *B. anthracis* (O'Brien *et al.*, 1985). In addition, ET inhibits macrophage production of sPLA2-IIA, which is bactericidal to

Gram-positive bacteria (Raymond *et al.*, 2007). By inhibiting the bactericidal activity of these cells, ET promotes the survival of germinated bacilli.

Effects of ET on vascular cell function In addition to the immunomodulatory role so far discussed, ET also plays a role in vascular damage, a characteristic of anthrax infection (Moayeri *et al.*, 2003; Firoved *et al.*, 2005; Kuo *et al.*, 2007). ET inhibits VEGF-induced EC chemotaxis (Hong *et al.*, 2007), suppressing angiogenesis and inhibiting the ability of ECs to repair damage to blood vessels. ET also inhibits thrombin and ADP-induced rabbit platelet aggregation (Alam *et al.*, 2006). Platelet aggregation diminishes the ability of blood to coagulate, an outcome that may be important for colonization and invasion.

Whole-animal effects of anthrax toxins

Whole-animal response to LT

LT causes death of animals, and injection of purified LT mimics pathology of anthrax. *B. anthracis* strains lacking LF are dramatically attenuated in most animal models, suggesting that LF plays a critical role in anthrax infection. Multiple cell types express anthrax toxin receptor and may be targets for LT during *in vivo* intoxication, including cells of the innate and adaptive immune system. In this section, whole animal response to LT will be discussed, highlighting the similarities and differences between models, and the plethora of pathological features found in LT-treated animals. LT-mediated pathology is associated with TNF-α-independent shock, and is characterized by alterations in cardiac, vascular, pulmonary and immune system function.

Animals subjected to LT intoxication show signs of a shock-like death (Shafazand *et al.*, 1999; Lent *et al.*, 2001; Prince, 2003; Sherer *et al.*, 2007a). Multiple studies have confirmed that most animals undergo what appears to be hypoxic shock in response to LT (Moayeri *et al.*, 2003; Prince, 2003; Cui *et al.*, 2004; Sherer *et al.*, 2007a). Shock results from inadequate delivery of oxygen and substrates to tissues, resulting in metabolic deficiencies. Cells starved for oxygen shift from the efficient aerobic catabolism of glu-cose to an anaerobic pathway leading to production and accumulation of lactic acid (Levy, 2006). Under hypoxic conditions, cells are unable to support cellular homeostasis and ultimately die. Without treatment, widespread cell death ensues, resulting in multiple organ failure, vascular leakage and death.

LT-treated animals show signs of cell death and organ failure (Moayeri *et al.*, 2003; Cui *et al.*, 2004; Firoved *et al.*, 2005; Kuo *et al.*, 2007, 2008). Severe cell death is seen in bone marrow, spleen and liver of LT-treated mice (Moayeri *et al.*, 2003). In mice, there is a decrease in platelet count (Culley *et al.*, 2005); however, there is no significant difference in white blood cell count or platelet count in rats (Kuo *et al.*, 2008). LT injection results in decreased heart rate, hypotension associated with left ventricular systolic function, acidosis and increased haemoglobin levels, all consistent with septic shock (Cui *et al.*, 2004; Watson *et al.*, 2007a,b; Kuo *et al.*, 2008).

LT treated mice show hepatic dysfunction and hypoalbuminaemia, which can cause vascular leakage (Moayeri *et al.*, 2003). Intradermal administration of LT results in vascular leakage within 15 to 25 minutes in some mouse strains, which can be blocked with an inhibitor of mast cell degranulation (Gozes *et al.*, 2006). LT challenge of animals results in extravascular fluid collection and an increase in the concentration of red blood cells in the blood (Culley *et al.*, 2005; Cui *et al.*, 2007). Normal saline treatment, however, actually worsens outcome of LT treated rats (Sherer *et al.*, 2007b). Rats treated with LT have higher haemoglobin and haematocrit levels that may be due to hypoxia or vascular leakage (Kuo *et al.*, 2008). Importantly, the vascular leakage and haemorrhages in LT-treated animals is similar to the pathology observed in *B. anthracis* infection suggesting that the pathology may be LT-dependent (Moayeri *et al.*, 2003; Cui *et al.*, 2004; Culley *et al.*, 2005; Gozes *et al.*, 2006; Sherer *et al.*, 2007a).

One of the hallmarks of inhalation anthrax, and a pathology seen in LT-treated animals, is pulmonary oedema and pleural effusion (Beall and Dalldorf, 1966; Gray and Archer, 1967; Barakat *et al.*, 2002; Kuo *et al.*, 2008). In LT-treated mice, pulmonary damage appears to play a role in toxicity through decreased blood oxygenation, potentially leading to cardiac dysfunction and organ failure (Moayeri *et al.*, 2003). In rats, however, respiratory failure does not appear to be a contributor to death: histopathological studies of LT-treated rats do not support pulmonary injury as a significant pathology and rats show signs of shock prior to pulmonary effects (Cui *et al.*, 2004). Although pulmonary dysfunction may not be the cause of death, it is present in all animal models and probably exacerbates LT toxicity (Sherer *et al.*, 2007a).

LT efficiently decreases cytokine production in most animal models, both in LT intoxication models and in *B. anthracis* infection when comparing wild-type with LF-deficient bacteria (Cui *et al.*, 2004; Drysdale *et al.*, 2007; Watson *et al.*, 2007a). In the rat model, LT inhibits the inflammatory response elicited by intravascular LPS or intratracheal challenge with *E. coli* (Cui *et al.*, 2004). LT treatment to suppress cytokine production during bacterial infection or sepsis does not seem feasible, however, since animals actually die earlier in the presence of LT and *E. coli*, despite showing signs of decreased inflammation and cytokine response. Therefore, unlike the cytokine response to LPS, animals injected with LT do not produce a cytokine storm required for lethality, and in most cases LT decreases the cytokine response (Moayeri *et al.*, 2003; Cui *et al.*, 2004). Through inhibition of cytokine production LT likely contributes to shock by helping *B. anthracis* evade the host immune system resulting in increased toxin production.

In contrast to cytokine repression described above, there is a rapid and transitory cytokine response in mice harbouring LT^S alleles of *Nalp1b*. This response is characterized by an increase in serum KC, MCP-1/JE, MIP-2, eotaxin and IL-1β (Moayeri *et al.*, 2003). Of note, mice with LT^S macrophages display earlier signs of toxicity when injected with LT, but mice containing either LT^S or LT^R macrophages show similar pathology and ultimately succumb to liver necrosis and pleural oedema, likely through hypoxic tissue injury (Moayeri *et al.*, 2003). However, exceptions to this correlation exist, implying that macrophage sensitivity and resultant cytokine burst are not the only genetic factors controlling mouse susceptibility to toxin (Moayeri *et al.*, 2004). This is supported by the finding that at

least three quantitative trait loci are associated with LT-sensitivity in mice (McAllister *et al.*, 2003). In contrast, rat sensitivity to LT seems to correlate with macrophage susceptibility to LT, though the data here is more limited (Nye *et al.*, 2008). Therefore, macrophages likely play a role in the susceptibility of animals to LT, but other factors may dominate in certain genetic backgrounds.

In many animal models of bacterial and viral infections, disruption of the HPA axis leads to increased mortality and/or inflammation. Conversely, addition of exogenous glucocorticoids to these animals during infection leads to decreased mortality and/or decreased inflammation. Both LT^S and LT^R adrenalectomized (ADX) mice (removal of the adrenal glands that produce glucocorticoids) are more susceptible to LT (Moayeri *et al.*, 2005). Interestingly, LT inhibits GR activation by dexamethasone (DEX), a synthetic glucocorticoid (Moayeri *et al.*, 2005). DEX treatment, however, does not protect LT-intoxicated ADX mice. In contrast, DEX sensitizes non-ADX LT^R mice to LT, suggesting that GR activation is detrimental *in vivo* when combined with LT (Moayeri *et al.*, 2005). This is an unexpected result, since previous work demonstrates that DEX improves outcome in mice challenged with various toxins or bacterial products, such as endotoxin, *E. coli*, shiga toxin, *Clostridium difficile* toxin A, and *Staphylococcus aureus* enterotoxin B (Webster and Sternberg, 2004). It is unclear why mice treated with DEX are more sensitive LT and whether this result translates to humans.

Whole-animal response to oedema toxin

Early experiments showed that injection of ET into experimental animals did not lead to death (Beall *et al.*, 1962), suggesting a more supportive role for ET during infection (Pezard *et al.*, 1991, 1993). Recent work, however, demonstrates that ET is cytotoxic to certain cell types and is able to cause death in mice in the absence of LT (Firoved *et al.*, 2005; Voth *et al.*, 2005). ET induces death in both LT^S and LT^R mouse strains (Firoved *et al.*, 2005; Firoved *et al.*, 2007). In mice, ET injection results in widespread histopathological lesions in disparate tissues and organs, increased expression of several cytokines, lymphocytolysis, increased monocytes and neutrophils in circulation and development of hypotension and bradycardia. Lymphocyte depletion is predicted to impair the ability of the host to clear infection, further supporting the immunomodulatory role of ET during infection.

Among the tissues affected by ET, the cortex of the adrenal glands shows the most extensive and consistent damage, characterized by adrenal necrosis (Firoved *et al.*, 2005). Damage to the adrenal glands is observed even at sublethal doses of ET. As mentioned in the previous section, alteration of the HPA axis in mice through adrenalectomy or treatment with DEX increases sensitivity to LT (Moayeri *et al.*, 2005). Consistent with this, ET-induced damage to the adrenal glands also sensitizes mice to LT (Firoved *et al.*, 2007). Thus, the synergy observed between ET and LT in animal models may be due, at least in part, to adrenal damage induced by ET.

ET also mediates significant cardiac pathology in mice and rats. In mice, ET-induced pathology includes cardiomyocyte necrosis accompanied by bradycardia and hypotension with no tachycardia (Firoved *et al.*, 2007). The absence of tachycardia is a surprising finding because this is a characteristic response to increased cAMP levels. In rats, however, ET induces tachycardia when administered as an infusion over time (Watson *et al.*, 2007a). In addition, ET leads to decreased blood pressure and pulmonary haemorrhage, consistent with damage to the cardiovascular system observed during an anthrax infection (Cui *et al.*, 2007; Kuo *et al.*, 2007; Watson *et al.*, 2007a). The effects of ET on the cardiovascular system are additive to those induced by LT leading to anthrax toxin-induced shock. Interestingly, in contrast to mice, mortality in rats requires a higher dose of ET than LT and the time to death is more rapid in LT-treated rats. This suggests that the animal species, toxin dose and method of administration influence the host response to anthrax toxins.

In addition to rodent models, zebrafish (*Danio rerio*) and fruit flies (*Drosophila melanogaster*) have been used to dissect the mechanism and effects of ET (Voth *et al.*, 2005; Guichard *et al.*, 2006). Given that anthrax toxin acts on conserved components of essential host signalling

pathways, these animal models are useful tools in understanding the effects of toxin. Although *Drosophila* does not express anthrax toxin receptors, the catalytic subunits can be expressed in cells. Expression of EF in the developing wing imaginal discs of *Drosophila* causes patterning defects similar to those caused by mutations that affect hedgehog (Hh) signalling (Guichard *et al.*, 2006). The Hh-like phenotype is consistent with the inhibitory role cAMP-dependent PKA signalling has on suppression of Hh response, further validating the ability of EF to activate PKA through increases in cAMP levels.

The developing zebrafish embryo is a useful model to study anthrax toxin (Tucker *et al.*, 2003; Voth *et al.*, 2005; Bolcome *et al.*, 2008). The zebrafish presents several advantages for studying effects of bacterial toxins on tissues including a short developmental period and optically transparency, enabling convenient visualization of organ systems. Furthermore, unlike *Drosophila*, zebrafish encode homologues of ANTXR1 and ANTXR2 genes. Intoxication of the zebrafish Pac2 embryonic fibroblasts with ET decreases cell viability by approximately 50% and induces morphological changes including cell elongation and eventual rounding (Voth *et al.*, 2005). Intoxicating zebrafish embryos with ET induces pleotropic effects including tissue and organ damage, swim bladder defects and death (Voth *et al.*, 2005).

Animal models of anthrax – role of toxin during infection

While it is generally believed that LT is the major virulence factor contributing to death in several, but not all, animal models, the manner by which it causes disease is still being elucidated. Furthermore, ET, once considered a secondary player, is now thought to play a more significant role in disease progression. Toxin deficient strains of *B. anthracis* have been employed to test the importance of individual toxin subunits in pathogenesis (Welkos *et al.*, 1986; Pezard *et al.*, 1991; Welkos *et al.*, 1993; Heninger *et al.*, 2006; Glomski *et al.*, 2007a). Results from these studies indicate that the role of toxin depends heavily on the specific animal model employed, as well as the presence or absence of capsule. In the absence of capsule, toxins are required for virulence with LT contributing more than ET in mouse models (Welkos *et al.*, 1986; Pezard *et al.*, 1991). Anthrax toxin also contributes to dissemination when spores are injected subcutaneously (Glomski *et al.*, 2007a). In contrast, the absence of toxin expression does not affect virulence of encapsulated strains in certain mouse models of infection (Welkos *et al.*, 1993; Heninger *et al.*, 2006). Interestingly, histopathological differences are observed between mice infected with the encapsulated and toxigenic parent strain compared with toxin deficient mutants, suggesting that toxins are important for affecting host responses even in this model (Heninger *et al.*, 2006). In particular, infection with EF deficient strains results in neutrophil infiltration, decreased levels of necrosis/apoptosis, and postnecrotic depletion in spleens. Based on their observations, the authors suggest that ET might play a more significant role than LT in producing pathological changes in this model. Finally, these models must be interpreted with care, as it has been recognized for some time that mice are difficult to protect using vaccines targeted at anthrax toxin components. Thus, the mouse model may not be as toxin dependent as other models such as guinea pig, rabbit, and non-human primates (Smith and Keppie, 1954; Ivins *et al.*, 1986; Ivins *et al.*, 1998; Moayeri and Leppla, 2004; Shoop *et al.*, 2005).

Despite limitations of the mouse model, this system has led to important discoveries as to the role of toxins in virulence. The contribution of LT and ET to pathogenesis is more complex than originally believed and it is now clear that each plays multiple roles in virulence. There is growing evidence that anthrax toxin promotes evasion from the host innate immune system early in infection, thereby contributing to colonization and subsequent bacteraemia and shock characteristic of anthrax. Given their strategic location in many of the portals of entry, DCs and macrophages are among the first immune cells to encounter *B. anthracis* spores. Upon phagocytosis, spores germinate and rapidly produce toxin, thereby blocking DC maturation and promoting escape from macrophages (Guidi-Rontani *et al.*, 1999; Banks *et al.*, 2005; Brittingham *et al.*, 2005; Cleret *et al.*, 2006, 2007, 2008).

Late in infection, systemic toxin levels rapidly increase, achieving concentrations up to 100 μg/ml (Mabry *et al.*, 2006). At this stage, anthrax toxin is more than just immunomodulatory and directly induces vascular collapse and multi-organ failure, though specific cellular targets have yet to be identified. It is this late-stage toxaemia that accounts for the point of no return first described by Smith and Keppie. One prediction of this model is that inhibitors of LT and/or ET should provide therapeutic value beyond the time at which antibiotics are no longer beneficial. Indeed, studies in rats and rabbits have confirmed this prediction, indicating that toxins do in fact contribute directly to the mortality associated with anthrax (Smith and Keppie, 1954; Ivins *et al.*, 1986; Shoop *et al.*, 2005).

Future perspectives

Despite the fact that anthrax toxin has been studied for over 50 years, we still do not fully understand the molecular and cellular events triggered in the host by this toxin that lead to immune evasion and induction of pathology. Future advances in the field are likely to come from several avenues, including, but not limited to, the identification and characterization of toxin inhibitors and the use of new animal models. Finally, as our understanding of anthrax toxin function increases, it may be possible to harness the activities of this toxin for beneficial purposes such as treatment of cancers.

Toxin inhibitors are important in that they serve both as therapeutics and as tools to dissect toxin mechanism of action. The first known toxin inhibitor was antisera directed against PA, which remains a potent means to block toxin action (Gladstone, 1946; Baillie, 2006). Indeed, humanized monoclonal antibodies and antibody fragments have progressed to pharmaceutical development. In addition, inhibitors have now been described that target each of the three subunits, as well as host response to toxin (Rainey and Young, 2004). These inhibitors are drug-like small molecules, peptides, variants of ANTXRs, or dominant negative versions of the toxin molecules themselves. Blocking toxin action can provide insight into the mechanism as well as the host response to anthrax toxin. For instance, isolation of dominant negative PA led to the discovery of the φ-clamp mechanism of toxin translocation (Sellman *et al.*, 2001; Krantz *et al.*, 2005). PA variants that cannot bind receptor protect animals from challenge, indicating that processing of PA can occur prior to cell surface binding (Moayeri *et al.*, 2007). Antibodies against PA domain 1 protect animals from intoxication indicating that this fragment of PA, previously ignored, may serve a functional role that has yet to be discovered (Rivera *et al.*, 2006). Finally, toxin inhibitors can teach us about host defence mechanisms, as exemplified by the finding that host-encoded peptides called defensins block LT at multiple steps (Kim *et al.*, 2005; Wang *et al.*, 2006).

Advances in our understanding of anthrax toxin will probably result from further studies of molecular events induced by LT and ET as well as increased use of whole-animal models of intoxication and infection. While mice, rats, rabbits and guinea pigs have traditionally been used to study anthrax toxin, the zebrafish system is emerging as a powerful tool to study both LT and ET. Zebrafish studies support a role for GSK-3β in LT sensitivity, LT targeting of the host vasculature, and ET-mediated toxicity (Tucker *et al.*, 2003; Voth *et al.*, 2005; Bolcome *et al.*, 2008). Based on the genetic tractability, ease of use and transparent nature of zebrafish, this system is an attractive model for studying anthrax toxin.

Finally, one very exciting and important aspect of anthrax toxin research is the potential application of LT as a therapy to treat human cancers. MKK proteins control activity of three major signal transduction pathways, including p38, ERK, and JNK. Thus, it is not surprising that MKK signalling is important for a wide variety of processes associated with cancers. For example, the majority of metastatic melanomas possess the V599E B-RAF mutation and are dependent on MKK signalling for survival (Abi-Habib *et al.*, 2005). Following the discovery that LF inactivates MKK proteins (Duesbery *et al.*, 1998), several groups began development of LT as an anticancer agent (Frankel *et al.*, 2002). Cells from prostate, colon, renal, and breast cancers, as well as leukaemia-, neuroblastoma-, melanoma-, and fibrosarcoma-derived cells are all sensitive to LT-mediated killing *in vitro* in a manner that correlates with MKK inactivation (Koo *et al.*, 2002; Abi-Habib *et al.*, 2005; Ding *et al.*, 2008; Rouleau *et al.*, 2008). In addition,

in vivo models have been used to demonstrate that LT induces regression of xenograft tumours using melanoma, fibrosarcoma, neuroblastoma, and non-small cell lung and renal cell carcinoma models in mice (Frankel *et al.*, 2002; Abi-Habib *et al.*, 2006; Su *et al.*, 2007; Huang *et al.*, 2008; Rouleau *et al.*, 2008).

In contrast to the tumour cell lines described above, many cancer cell types do not die in response to LT treatment *in vitro* despite expressing functional toxin receptors. In the case of V12 H-*ras*-transformed 3T3 cells, LT reverts the transformed phenotype *in vitro* without inducing direct cytotoxicity (Duesbery *et al.*, 2001). Surprisingly, xenograft tumours using these cells are highly sensitive to LT, with significant tumour size reduction and tumour cell necrosis (Duesbery *et al.*, 2001). Further analysis reveals that LT-treated tumours have greatly reduced vasculature (Duesbery *et al.*, 2001). These results indicate that LT promotes tumour destruction through mechanisms beyond direct cytotoxicity to tumour cells, which is supported by more recent studies indicating that tumour vasculature is targeted by LT in melanoma, lung, and colon carcinoma xenograft models (Liu *et al.*, 2008). The identification that ANTXR1/TEM8, a protein up-regulated on tumour neovasculature, serves as a receptor for LT may help explain this. Indeed, endothelial cells are highly dependent on MKK signalling during angiogenesis (Depeille *et al.*, 2007), and the combination of higher ANTXR1/TEM8 expression and increased dependence on MKK signalling make these cells especially sensitive to LT (Alfano *et al.*, 2008). Interestingly, a recent report demonstrates that PA alone (without EF or LF) blocks vascular endothelial growth factor (VEGF)-induced and basic fibroblast growth factor (bFGF)-induced angiogenesis in a corneal neovascularization assay (Rogers *et al.*, 2007). The mechanism of this inhibition is not known, but may be related to blocking the natural function of ANTXR1/TEM8 and/or ANTXR2.

One concern associated with LT as a therapeutic is that PA binds to both ANTXR1 and ANTXR2, and ANTXR2 is expressed on a wide variety of normal cell types (Scobie *et al.*, 2003). Therefore, tumours are not specifically targeted using wild type PA. In order to increase tumour targeting and reduce side-effects, Leppla and colleagues have engineered versions of PA in which the furin cleavage site required for oligomerization and LF binding is replaced with consensus cleavage sites for proteases up-regulated on the surface of tumours (i.e. matrix metalloproteinases MMP-2 and MMP-9, and urokinase plasminogen activator) (Liu *et al.*, 2000; Liu *et al.*, 2001; Liu *et al.*, 2003; Su *et al.*, 2007). As expected, these PA variants induce tumour regression with reduced toxicity to the animals. The re-engineering of anthrax toxin for beneficial purposes is a result of decades of research into the basic and cellular mechanisms of this toxin and point the way to future uses for this family of proteins.

References

Abi-Habib, R.J., Singh, R., Liu, S., Bugge, T.H., Leppla, S.H., and Frankel, A.E. (2006). A urokinase-activated recombinant anthrax toxin is selectively cytotoxic to many human tumor cell types. Mol. Cancer Ther. 5, 2556–2562.

Abi-Habib, R.J., Urieto, J.O., Liu, S., Leppla, S.H., Duesbery, N.S., and Frankel, A.E. (2005). BRAF status and mitogen-activated protein/extracellular signal-regulated kinase kinase 1/2 activity indicate sensitivity of melanoma cells to anthrax lethal toxin. Mol. Cancer Ther. 4, 1303–1310.

Abrami, L., Leppla, S.H., and van der Goot, F.G. (2006). Receptor palmitoylation and ubiquitination regulate anthrax toxin endocytosis. J. Cell Biol. 172, 309–320.

Abrami, L., Lindsay, M., Parton, R.G., Leppla, S.H., and van der Goot, F.G. (2004). Membrane insertion of anthrax protective antigen and cytoplasmic delivery of lethal factor occur at different stages of the endocytic pathway. J. Cell Biol. 166, 645–651.

Abrami, L., Liu, S., Cosson, P., Leppla, S.H., and van der Goot, F.G. (2003). Anthrax toxin triggers endocytosis of its receptor via a lipid raft-mediated clathrin-dependent process. J. Cell Biol. 160, 321–328.

Agrawal, A., Lingappa, J., Leppla, S.H., Agrawal, S., Jabbar, A., Quinn, C., and Pulendran, B. (2003). Impairment of dendritic cells and adaptive immunity by anthrax lethal toxin. Nature 424, 329–334.

Ahuja, N., Kumar, P., and Bhatnagar, R. (2001). Hydrophobic residues Phe552, Phe554, Ile562, Leu566, and Ile574 are required for oligomerization of anthrax protective antigen. Biochem. Biophys. Res. Commun. 287, 542–549.

Alam, S., Gupta, M., and Bhatnagar, R. (2006). Inhibition of platelet aggregation by anthrax edema toxin. Biochem. Biophys. Res. Commun. 339, 107–114.

Alfano, R.W., Leppla, S.H., Liu, S., Bugge, T.H., Duesbery, N.S., and Frankel, A.E. (2008). Potent inhibition of tumor angiogenesis by the matrix metalloproteinase-activated anthrax lethal toxin: implications for broad anti-tumor efficacy. Cell Cycle 7, 745–749.

Alileche, A., Serfass, E.R., Muehlbauer, S.M., Porcelli, S.A., and Brojatsch, J. (2005). Anthrax lethal toxin-mediated killing of human and murine dendritic cells impairs the adaptive immune response. PLoS Pathogen. 1, e19.

Alileche, A., Squires, R.C., Muehlbauer, S.M., Lisanti, M.P., and Brojatsch, J. (2006). Mitochondrial impairment is a critical event in anthrax lethal toxin-induced cytolysis of murine macrophages. Cell Cycle 5, 100–106.

Arora, N., Klimpel, K.R., Singh, Y., and Leppla, S.H. (1992). Fusions of anthrax toxin lethal factor to the ADP-ribosylation domain of Pseudomonas exotoxin A are potent cytotoxins which are translocated to the cytosol of mammalian cells. J. Biol. Chem. 267, 15542–15548.

Arora, N., and Leppla, S.H. (1993). Residues 1–254 of anthrax toxin lethal factor are sufficient to cause cellular uptake of fused polypeptides. J. Biol. Chem. 268, 3334–3341.

Baillie, L.W. (2006). Past, imminent and future human medical countermeasures for anthrax. J. Appl. Microbiol. 101, 594–606.

Banks, D.J., Barnajian, M., Maldonado-Arocho, F.J., Sanchez, A.M., and Bradley, K.A. (2005). Anthrax toxin receptor 2 mediates *Bacillus anthracis* killing of macrophages following spore challenge. Cell. Microbiol. 7, 1173–1185.

Banks, D.J., Ward, S.C., and Bradley, K.A. (2006). New insights into the functions of anthrax toxin. Expert Rev. Mol. Med. 8, 1–18.

Barakat, L.A., Quentzel, H.L., Jernigan, J.A., Kirschke, D.L., Griffith, K., Spear, S.M., Kelley, K., Barden, D., Mayo, D., Stephens, D.S., *et al.* (2002). Fatal inhalational anthrax in a 94-year-old Connecticut woman. JAMA 287, 863–868.

Bardwell, A.J., Abdollahi, M., and Bardwell, L. (2004). Anthrax lethal factor-cleavage products of MAPK (mitogen-activated protein kinase) kinases exhibit reduced binding to their cognate MAPKs. Biochem. J. 378, 569–577.

Barson, H.V., Mollenkopf, H., Kaufmann, S.H., and Rijpkema, S. (2008). Anthrax lethal toxin suppresses chemokine production in human neutrophil NB-4 cells. Biochem. Biophys. Res. Commun. 374, 288–293.

Bartkus, J.M., and Leppla, S.H. (1989). Transcriptional regulation of the protective antigen gene of *Bacillus anthracis*. Infect. Immun. 57, 2295–2300.

Batty, S., Chow, E.M., Kassam, A., Der, S.D., and Mogridge, J. (2006). Inhibition of mitogen-activated protein kinase signalling by *Bacillus anthracis* lethal toxin causes destabilization of interleukin-8 mRNA. Cell. Microbiol. 8, 130–138.

Beall, F.A., and Dalldorf, F.G. (1966). The pathogenesis of the lethal effect of anthrax toxin in the rat. J. Infect. Dis. 116, 377–389.

Beall, F.A., Taylor, M.J., and Thorne, C.B. (1962). Rapid lethal effect in rats of a third component found upon fractionating the toxin of *Bacillus anthracis*. J. Bacteriol. 83, 1274–1280.

Beatty, M.E., Ashford, D.A., Griffin, P.M., Tauxe, R.V., and Sobel, J. (2003). Gastrointestinal anthrax: review of the literature. Arch Intern Med *163*, 2527–2531.

Beauregard, K.E., Collier, R.J., and Swanson, J.A. (2000). Proteolytic activation of receptor-bound anthrax protective antigen on macrophages promotes its internalization. Cell. Microbiol. *2*, 251–258.

Bell, S.E., Mavila, A., Salazar, R., Bayless, K.J., Kanagala, S., Maxwell, S.A., and Davis, G.E. (2001). Differential gene expression during capillary morphogenesis in 3D collagen matrices: regulated expression of genes involved in basement membrane matrix assembly, cell cycle progression, cellular differentiation and G-protein signaling. J Cell Sci *114*, 2755–2773.

Benson, E.L., Huynh, P.D., Finkelstein, A., and Collier, R.J. (1998). Identification of residues lining the anthrax protective antigen channel. Biochemistry *37*, 3941–3948.

Bergman, N.H., Passalacqua, K.D., Gaspard, R., Shetron-Rama, L.M., Quackenbush, J., and Hanna, P.C. (2005). Murine macrophage transcriptional responses to *Bacillus anthracis* infection and intoxication. Infect. Immun. *73*, 1069–1080.

Bhatnagar, R., and Friedlander, A.M. (1994). Protein synthesis is required for expression of anthrax lethal toxin cytotoxicity. Infect. Immun. *62*, 2958–2962.

Bhatnagar, R., Singh, Y., Leppla, S.H., and Friedlander, A.M. (1989). Calcium is required for the expression of anthrax lethal toxin activity in the macrophagelike cell line J774A.1. Infect. Immun. *57*, 2107–2114.

Bischof, T.S., Hahn, B.L., and Sohnle, P.G. (2007). Characteristics of spore germination in a mouse model of cutaneous anthrax. J. Infect. Dis. *195*, 888–894.

Blanke, S.R., Milne, J.C., Benson, E.L., and Collier, R.J. (1996). Fused polycationic peptide mediates delivery of diphtheria toxin A chain to the cytosol in the presence of anthrax protective antigen. Proc. Natl. Acad. Sci. USA *93*, 8437–8442.

Blaustein, R.O., Koehler, T.M., Collier, R.J., and Finkelstein, A. (1989). Anthrax toxin: channel-forming activity of protective antigen in planar phospholipid bilayers. Proc. Natl. Acad. Sci. USA *86*, 2209–2213.

Bolcome, R.E., 3rd, Sullivan, S.E., Zeller, R., Barker, A.P., Collier, R.J., and Chan, J. (2008). Anthrax lethal toxin induces cell death-independent permeability in zebrafish vasculature. Proc. Natl. Acad. Sci. USA *105*, 2439–2444.

Boll, W., Ehrlich, M., Collier, R.J., and Kirchhausen, T. (2004). Effects of dynamin inactivation on pathways of anthrax toxin uptake. Eur. J. Cell Biol. *83*, 281–288.

Bonuccelli, G., Sotgia, F., Frank, P.G., Williams, T.M., de Almeida, C.J., Tanowitz, H.B., Scherer, P.E., Hotchkiss, K.A., Terman, B.I., Rollman, B., et al. (2005). ATR/TEM8 is highly expressed in epithelial cells lining *Bacillus anthracis'* three sites of entry: implications for the pathogenesis of anthrax infection. Am. J. Physiol. Cell Physiol. *288*, C1402–1410.

Bos, J.L. (2005). Linking Rap to cell adhesion. Curr. Opin. Cell Biol. *17*, 123–128.

Boyden, E.D., and Dietrich, W.F. (2006). Nalp1b controls mouse macrophage susceptibility to anthrax lethal toxin. Nat Genet *38*, 240–244.

Bradley, K.A., Mogridge, J., Jonah, G., Rainey, A., Batty, S., and Young, J.A. (2003). Binding of anthrax toxin to its receptor is similar to alpha integrin-ligand interactions. J. Biol. Chem. *278*, 49342–49347.

Bradley, K.A., Mogridge, J., Mourez, M., Collier, R.J., and Young, J.A. (2001). Identification of the cellular receptor for anthrax toxin. Nature *414*, 225–229.

Bragg, T.S., and Robertson, D.L. (1989). Nucleotide sequence and analysis of the lethal factor gene (lef) from *Bacillus anthracis*. Gene *81*, 45–54.

Brittingham, K.C., Ruthel, G., Panchal, R.G., Fuller, C.L., Ribot, W.J., Hoover, T.A., Young, H.A., Anderson, A.O., and Bavari, S. (2005). Dendritic cells endocytose *Bacillus anthracis* spores: implications for anthrax pathogenesis. J. Immunol. *174*, 5545–5552.

Brossier, F., Sirard, J.C., Guidi-Rontani, C., Duflot, E., and Mock, M. (1999). Functional analysis of the carboxy-terminal domain of *Bacillus anthracis* protective antigen. Infect. Immun. *67*, 964–967.

Buck, M., and Chojkier, M. (2007). C/EBPbeta phosphorylation rescues macrophage dysfunction and apoptosis induced by anthrax lethal toxin. Am. J. Physiol. Cell Physiol. *293*, C1788–1796.

Carson-Walter, E.B., Watkins, D.N., Nanda, A., Vogelstein, B., Kinzler, K.W., and St Croix, B. (2001). Cell surface tumor endothelial markers are conserved in mice and humans. Cancer Res. *61*, 6649–6655.

Chandra, H., Gupta, P.K., Sharma, K., Mattoo, A.R., Garg, S.K., Gade, W.N., Sirdeshmukh, R., Maithal, K., and Singh, Y. (2005). Proteome analysis of mouse macrophages treated with anthrax lethal toxin. Biochim Biophys Acta *1747*, 151–159.

Chopra, A.P., Boone, S.A., Liang, X., and Duesbery, N.S. (2003). Anthrax lethal factor proteolysis and inactivation of MAPK kinase. J. Biol. Chem. *278*, 9402–9406.

Christensen, K.A., Krantz, B.A., and Collier, R.J. (2006). Assembly and disassembly kinetics of anthrax toxin complexes. Biochemistry *45*, 2380–2386.

Christensen, K.A., Krantz, B.A., Melnyk, R.A., and Collier, R.J. (2005). Interaction of the 20 kDa and 63 kDa fragments of anthrax protective antigen: kinetics and thermodynamics. Biochemistry *44*, 1047–1053.

Chvyrkova, I., Zhang, X.C., and Terzyan, S. (2007). Lethal factor of anthrax toxin binds monomeric form of protective antigen. Biochem. Biophys. Res. Commun. *360*, 690–695.

Cleret, A., Quesnel-Hellmann, A., Mathieu, J., Vidal, D., and Tournier, J.N. (2006). Resident CD11c+ lung cells are impaired by anthrax toxins after spore infection. J. Infect. Dis. *194*, 86–94.

Cleret, A., Quesnel-Hellmann, A., Vallon-Eberhard, A., Verrier, B., Jung, S., Vidal, D., Mathieu, J., and Tournier, J.N. (2007). Lung dendritic cells rapidly mediate anthrax spore entry through the pulmonary route. J. Immunol. *178*, 7994–8001.

Comer, J.E., Chopra, A.K., Peterson, J.W., and Konig, R. (2005a). Direct inhibition of T-lymphocyte activation by anthrax toxins *in vivo*. Infect. Immun. *73*, 8275–8281.

Comer, J.E., Galindo, C.L., Chopra, A.K., and Peterson, J.W. (2005b). GeneChip analyses of global transcriptional responses of murine macrophages to the lethal toxin of *Bacillus anthracis*. Infect. Immun. *73*, 1879–1885.

Comer, J.E., Galindo, C.L., Zhang, F., Wenglikowski, A.M., Bush, K.L., Garner, H.R., Peterson, J.W., and Chopra, A.K. (2006). Murine macrophage transcriptional and functional responses to *Bacillus anthracis* edema toxin. Microb. Pathog. *41*, 96–110.

Cordoba-Rodriguez, R., Fang, H., Lankford, C.S., and Frucht, D.M. (2004). Anthrax lethal toxin rapidly activates caspase-1/ICE and induces extracellular release of interleukin (IL)-1beta and IL-18. J. Biol. Chem. *279*, 20563–20566.

Cote, C.K., DiMezzo, T.L., Banks, D.J., France, B., Bradley, K.A., and Welkos, S.L. (2008). Early interactions between fully virulent *Bacillus anthracis* and macrophages that influence the balance between spore clearance and development of a lethal infection. Microbes Infect. *10*, 613–619.

Cote, C.K., Rea, K.M., Norris, S.L., van Rooijen, N., and Welkos, S.L. (2004). The use of a model of *in vivo* macrophage depletion to study the role of macrophages during infection with *Bacillus anthracis* spores. Microb. Pathog. *37*, 169–175.

Cote, C.K., Van Rooijen, N., and Welkos, S.L. (2006). Roles of macrophages and neutrophils in the early host response to *Bacillus anthracis* spores in a mouse model of infection. Infect. Immun. *74*, 469–480.

Craig, J.P. (1965). A permeability factor (toxin) found in cholera stools and culture filtrates and its neutralization by convalescent cholera sera. Nature *207*, 614–616.

Crawford, M.A., Aylott, C.V., Bourdeau, R.W., and Bokoch, G.M. (2006). *Bacillus anthracis* toxins inhibit human neutrophil NADPH oxidase activity. J. Immunol. *176*, 7557–7565.

Cui, X., Li, Y., Li, X., Laird, M.W., Subramanian, M., Moayeri, M., Leppla, S.H., Fitz, Y., Su, J., Sherer, K., and Eichacker, P.Q. (2007). *Bacillus anthracis* edema and lethal toxin have different hemodynamic effects but function together to worsen shock and outcome in a rat model. J. Infect. Dis. *195*, 572–580.

Cui, X., Moayeri, M., Li, Y., Li, X., Haley, M., Fitz, Y., Correa-Araujo, R., Banks, S.M., Leppla, S.H., and Eichacker, P.Q. (2004). Lethality during continuous anthrax lethal toxin infusion is associated with circulatory shock but not inflammatory cytokine or nitric oxide release in rats. Am. J. Physiol. Regul. Integr. Comp. Physiol. *286*, R699–709.

Culley, N.C., Pinson, D.M., Chakrabarty, A., Mayo, M.S., and Levine, S.M. (2005). Pathophysiological manifestations in mice exposed to anthrax lethal toxin. Infect. Immun. *73*, 7006–7010.

Cunningham, K., Lacy, D.B., Mogridge, J., and Collier, R.J. (2002). Mapping the lethal factor and edema factor binding sites on oligomeric anthrax protective antigen. Proc. Natl. Acad. Sci. USA *99*, 7049–7053.

Dai, Z., and Koehler, T.M. (1997). Regulation of anthrax toxin activator gene (atxA) expression in *Bacillus anthracis*: temperature, not CO2/bicarbonate, affects AtxA synthesis. Infect. Immun. *65*, 2576–2582.

Dai, Z., Sirard, J.C., Mock, M., and Koehler, T.M. (1995). The atxA gene product activates transcription of the anthrax toxin genes and is essential for virulence. Mol. Microbiol. *16*, 1171–1181.

Dal Molin, F., Tonello, F., Ladant, D., Zornetta, I., Zamparo, I., Di Benedetto, G., Zaccolo, M., and Montecucco, C. (2006). Cell entry and cAMP imaging of anthrax edema toxin. Embo J *25*, 5405–5413.

Daniel, P.B., Walker, W.H., and Habener, J.F. (1998). Cyclic AMP signaling and gene regulation. Annu Rev Nutr *18*, 353–383.

de Rooij, J., Zwartkruis, F.J., Verheijen, M.H., Cool, R.H., Nijman, S.M., Wittinghofer, A., and Bos, J.L. (1998). Epac is a Rap1 guanine-nucleotide-exchange factor directly activated by cyclic AMP. Nature *396*, 474–477.

Depeille, P., Young, J.J., Boguslawski, E.A., Berghuis, B.D., Kort, E.J., Resau, J.H., Frankel, A.E., and Duesbery, N.S. (2007). Anthrax lethal toxin inhibits growth of and vascular endothelial growth factor release from endothelial cells expressing the human herpes virus 8 viral G protein coupled receptor. Clin Cancer Res. *13*, 5926–5934.

Ding, Y., Boguslawski, E.A., Berghuis, B.D., Young, J.J., Zhang, Z., Hardy, K., Furge, K., Kort, E., Frankel, A.E., Hay, R.V., *et al.* (2008). Mitogen-activated protein kinase kinase signaling promotes growth and vascularization of fibrosarcoma. Mol Cancer Ther *7*, 648–658.

Dixon, T.C., Meselson, M., Guillemin, J., and Hanna, P.C. (1999). Anthrax. N Engl J Med *341*, 815–826.

Dowling, O., Difeo, A., Ramirez, M.C., Tukel, T., Narla, G., Bonafe, L., Kayserili, H., Yuksel-Apak, M., Paller, A.S., Norton, K., *et al.* (2003). Mutations in capillary morphogenesis gene-2 result in the allelic disorders juvenile hyaline fibromatosis and infantile systemic hyalinosis. Am. J. Hum. Genet. *73*, 957–966.

Drum, C.L., Yan, S.Z., Bard, J., Shen, Y.Q., Lu, D., Soelaiman, S., Grabarek, Z., Bohm, A., and Tang, W.J. (2002). Structural basis for the activation of anthrax adenylyl cyclase exotoxin by calmodulin. Nature *415*, 396–402.

Drysdale, M., Olson, G., Koehler, T.M., Lipscomb, M.F., and Lyons, C.R. (2007). Murine innate immune response to virulent toxigenic and nontoxigenic *Bacillus anthracis* strains. Infect. Immun. *75*, 1757–1764.

Duesbery, N.S., Resau, J., Webb, C.P., Koochekpour, S., Koo, H.M., Leppla, S.H., and Vande Woude, G.F. (2001). Suppression of ras-mediated transformation and inhibition of tumor growth and angiogenesis by anthrax lethal factor, a proteolytic inhibitor of multiple MEK pathways. Proc. Natl. Acad. Sci. USA *98*, 4089–4094.

Duesbery, N.S., Webb, C.P., Leppla, S.H., Gordon, V.M., Klimpel, K.R., Copeland, T.D., Ahn, N.G., Oskarsson, M.K., Fukasawa, K., Paull, K.D., and Vande Woude, G.F. (1998). Proteolytic inactivation of MAP-kinase-kinase by anthrax lethal factor. Science *280*, 734–737.

During, R.L., Gibson, B.G., Li, W., Bishai, E.A., Sidhu, G.S., Landry, J., and Southwick, F.S. (2007). Anthrax lethal toxin paralyzes actin-based motility by blocking Hsp27 phosphorylation. EMBO J. 26, 2240–2250.

During, R.L., Li, W., Hao, B., Koenig, J.M., Stephens, D.S., Quinn, C.P., and Southwick, F.S. (2005). Anthrax lethal toxin paralyzes neutrophil actin-based motility. J. Infect. Dis. 192, 837–845.

Erwin, J.L., DaSilva, L.M., Bavari, S., Little, S.F., Friedlander, A.M., and Chanh, T.C. (2001). Macrophage-derived cell lines do not express proinflammatory cytokines after exposure to Bacillus anthracis lethal toxin. Infect. Immun. 69, 1175–1177.

Escuyer, V., and Collier, R.J. (1991). Anthrax protective antigen interacts with a specific receptor on the surface of CHO-K1 cells. Infect. Immun. 59, 3381–3386.

Ezzell, J.W., Jr., and Abshire, T.G. (1992). Serum protease cleavage of Bacillus anthracis protective antigen. J. Gen. Microbiol. 138, 543–549.

Fang, H., Cordoba-Rodriguez, R., Lankford, C.S., and Frucht, D.M. (2005). Anthrax lethal toxin blocks MAPK kinase-dependent IL-2 production in CD4+ T-cells. J. Immunol. 174, 4966–4971.

Fang, H., Xu, L., Chen, T.Y., Cyr, J.M., and Frucht, D.M. (2006). Anthrax lethal toxin has direct and potent inhibitory effects on B cell proliferation and immunoglobulin production. J. Immunol. 176, 6155–6161.

Faustin, B., Lartigue, L., Bruey, J.M., Luciano, F., Sergienko, E., Bailly-Maitre, B., Volkmann, N., Hanein, D., Rouiller, I., and Reed, J.C. (2007). Reconstituted NALP1 inflammasome reveals two-step mechanism of caspase-1 activation. Mol. Cell 25, 713–724.

Fink, S.L., and Cookson, B.T. (2005). Apoptosis, pyroptosis, and necrosis: mechanistic description of dead and dying eukaryotic cells. Infect. Immun. 73, 1907–1916.

Firoved, A.M., Miller, G.F., Moayeri, M., Kakkar, R., Shen, Y., Wiggins, J.F., McNally, E.M., Tang, W.J., and Leppla, S.H. (2005). Bacillus anthracis edema toxin causes extensive tissue lesions and rapid lethality in mice. Am. J. Pathol. 167, 1309–1320.

Firoved, A.M., Moayeri, M., Wiggins, J.F., Shen, Y., Tang, W.J., and Leppla, S.H. (2007). Anthrax edema toxin sensitizes DBA/2J mice to lethal toxin. Infect. Immun. 75, 2120–2125.

Frankel, A.E., Bugge, T.H., Liu, S., Vallera, D.A., and Leppla, S.H. (2002). Peptide toxins directed at the matrix dissolution systems of cancer cells. Protein Pept. Lett. 9, 1–14.

Friedlander, A.M. (1986). Macrophages are sensitive to anthrax lethal toxin through an acid-dependent process. J. Biol. Chem. 261, 7123–7126.

Fukao, T. (2004). Immune system paralysis by anthrax lethal toxin: the roles of innate and adaptive immunity. Lancet Infect. Dis. 4, 166–170.

Gaur, R., Gupta, P.K., Goyal, A., Wels, W., and Singh, Y. (2002). Delivery of nucleic acid into mammalian cells by anthrax toxin. Biochem. Biophys. Res. Commun. 297, 1121–1127.

Gladstone, G. (1946). Immunity of anthrax: protective antigen present in cell-free culture filtrates. Br. J. Exp. Pathol. 27, 393–410.

Glomski, I.J., Corre, J.P., Mock, M., and Goossens, P.L. (2007a). Noncapsulated toxinogenic Bacillus anthracis

presents a specific growth and dissemination pattern in naive and protective antigen-immune mice. Infect. Immun. 75, 4754–4761.

Glomski, I.J., Piris-Gimenez, A., Huerre, M., Mock, M., and Goossens, P.L. (2007b). Primary involvement of pharynx and peyer's patch in inhalational and intestinal anthrax. PLoS Pathog. 3, e76.

Go, M.Y., Kim, S., Partridge, A.W., Melnyk, R.A., Rath, A., Deber, C.M., and Mogridge, J. (2006). Self-association of the transmembrane domain of an anthrax toxin receptor. J. Mol. Biol. 360, 145–156.

Gordon, V.M., Klimpel, K.R., Arora, N., Henderson, M.A., and Leppla, S.H. (1995). Proteolytic activation of bacterial toxins by eukaryotic cells is performed by furin and by additional cellular proteases. Infect. Immun. 63, 82–87.

Gordon, V.M., Leppla, S.H., and Hewlett, E.L. (1988). Inhibitors of receptor-mediated endocytosis block the entry of Bacillus anthracis adenylate cyclase toxin but not that of Bordetella pertussis adenylate cyclase toxin. Infect. Immun. 56, 1066–1069.

Gordon, V.M., Rehemtulla, A., and Leppla, S.H. (1997). A role for PACE4 in the proteolytic activation of anthrax toxin protective antigen. Infect. Immun. 65, 3370–3375.

Gozes, Y., Moayeri, M., Wiggins, J.F., and Leppla, S.H. (2006). Anthrax lethal toxin induces ketotifen-sensitive intradermal vascular leakage in certain inbred mice. Infect. Immun. 74, 1266–1272.

Gray, I., and Archer, L.J. (1967). Metabolic changes in nicotinamide adenine dinucleotide in response to anthrax toxin. J. Bacteriol. 93, 36–39.

Green, B.D., Battisti, L., Koehler, T.M., Thorne, C.B., and Ivins, B.E. (1985). Demonstration of a capsule plasmid in Bacillus anthracis. Infect. Immun. 49, 291–297.

Guay, J., Lambert, H., Gingras-Breton, G., Lavoie, J.N., Huot, J., and Landry, J. (1997). Regulation of actin filament dynamics by p38 map kinase-mediated phosphorylation of heat shock protein 27. J Cell Sci 110 (Pt 3), 357–368.

Guichard, A., Park, J.M., Cruz-Moreno, B., Karin, M., and Bier, E. (2006). Anthrax lethal factor and edema factor act on conserved targets in Drosophila. Proc. Natl. Acad. Sci. USA 103, 3244–3249.

Guidi-Rontani, C. (2002). The alveolar macrophage: the Trojan horse of Bacillus anthracis. Trends Microbiol 10, 405–409.

Guidi-Rontani, C., Levy, M., Ohayon, H., and Mock, M. (2001). Fate of germinated Bacillus anthracis spores in primary murine macrophages. Mol. Microbiol.42, 931–938.

Guidi-Rontani, C., Weber-Levy, M., Labruyere, E., and Mock, M. (1999). Germination of Bacillus anthracis spores within alveolar macrophages. Mol. Microbiol.31, 9–17.

Guidi-Rontani, C., Weber-Levy, M., Mock, M., and Cabiaux, V. (2000). Translocation of Bacillus anthracis lethal and oedema factors across endosome membranes. Cell. Microbiol. 2, 259–264.

Ha, S.D., Ng, D., Lamothe, J., Valvano, M.A., Han, J., and Kim, S.O. (2007a). Mitochondrial proteins Bnip3 and Bnip3L are involved in anthrax lethal toxin-induced macrophage cell death. J. Biol. Chem. *282*, 26275–26283.

Ha, S.D., Ng, D., Pelech, S.L., and Kim, S.O. (2007b). Critical role of the phosphatidylinositol 3-kinase/Akt/glycogen synthase kinase-3 signaling pathway in recovery from anthrax lethal toxin-induced cell cycle arrest and MEK cleavage in macrophages. J. Biol. Chem. *282*, 36230–36239.

Hadjifrangiskou, M., Chen, Y., and Koehler, T.M. (2007). The alternative sigma factor sigmaH is required for toxin gene expression by *Bacillus anthracis*. J. Bacteriol. *189*, 1874–1883.

Hadjifrangiskou, M., and Koehler, T.M. (2008). Intrinsic curvature associated with the coordinately regulated anthrax toxin gene promoters. Microbiology *154*, 2501–2512.

Hammond, S.E., and Hanna, P.C. (1998). Lethal factor active-site mutations affect catalytic activity *in vitro*. Infect. Immun. *66*, 2374–2378.

Hanks, S., Adams, S., Douglas, J., Arbour, L., Atherton, D.J., Balci, S., Bode, H., Campbell, M.E., Feingold, M., Keser, G., *et al.* (2003). Mutations in the gene encoding capillary morphogenesis protein 2 cause juvenile hyaline fibromatosis and infantile systemic hyalinosis. Am. J. Hum. Genet. *73*, 791–800.

Hanna, P.C., Acosta, D., and Collier, R.J. (1993). On the role of macrophages in anthrax. Proc. Natl. Acad. Sci. USA *90*, 10198–10201.

Hanna, P.C., Kochi, S., and Collier, R.J. (1992). Biochemical and physiological changes induced by anthrax lethal toxin in J774 macrophage-like cells. Mol Biol Cell 3, 1269–1277.

Heninger, S., Drysdale, M., Lovchik, J., Hutt, J., Lipscomb, M.F., Koehler, T.M., and Lyons, C.R. (2006). Toxin-deficient mutants of *Bacillus anthracis* are lethal in a murine model for pulmonary anthrax. Infect. Immun. *74*, 6067–6074.

Hilbi, H., Puro, R.J., and Zychlinsky, A. (2000). Tripeptidyl peptidase II promotes maturation of caspase-1 in Shigella flexneri-induced macrophage apoptosis. Infect. Immun. *68*, 5502–5508.

Hong, J., Beeler, J., Zhukovskaya, N.L., He, W., Tang, W.J., and Rosner, M.R. (2005). Anthrax edema factor potency depends on mode of cell entry. Biochem. Biophys. Res. Commun. *335*, 850–857.

Hong, J., Doebele, R.C., Lingen, M.W., Quilliam, L.A., Tang, W.J., and Rosner, M.R. (2007). Anthrax edema toxin inhibits endothelial cell chemotaxis via Epac and Rap1. J. Biol. Chem. *282*, 19781–19787.

Hoover, D.L., Friedlander, A.M., Rogers, L.C., Yoon, I.K., Warren, R.L., and Cross, A.S. (1994). Anthrax edema toxin differentially regulates lipopolysaccharide-induced monocyte production of tumor necrosis factor alpha and interleukin-6 by increasing intracellular cyclic AMP. Infect. Immun. *62*, 4432–4439.

Hotchkiss, K.A., Basile, C.M., Spring, S.C., Bonuccelli, G., Lisanti, M.P., and Terman, B.I. (2005). TEM8 expression stimulates endothelial cell adhesion and migration by regulating cell-matrix interactions on collagen. Exp Cell Res *305*, 133–144.

Hsu, L.C., Park, J.M., Zhang, K., Luo, J.L., Maeda, S., Kaufman, R.J., Eckmann, L., Guiney, D.G., and Karin, M. (2004). The protein kinase PKR is required

for macrophage apoptosis after activation of Toll-like receptor 4. Nature *428*, 341–345.

Hu, H., Sa, Q., Koehler, T.M., Aronson, A.I., and Zhou, D. (2006). Inactivation of *Bacillus anthracis* spores in murine primary macrophages. Cell. Microbiol. *8*, 1634–1642.

Huang, D., Ding, Y., Luo, W.M., Bender, S., Qian, C.N., Kort, E., Zhang, Z.F., VandenBeldt, K., Duesbery, N.S., Resau, J.H., and Teh, B.T. (2008). Inhibition of MAPK kinase signaling pathways suppressed renal cell carcinoma growth and angiogenesis *in vivo*. Cancer Res. *68*, 81–88.

Ivins, B.E., Ezzell, J.W., Jr., Jemski, J., Hedlund, K.W., Ristroph, J.D., and Leppla, S.H. (1986). Immunization studies with attenuated strains of *Bacillus anthracis*. Infect. Immun. *52*, 454–458.

Ivins, B.E., Pitt, M.L., Fellows, P.F., Farchaus, J.W., Benner, G.E., Waag, D.M., Little, S.F., Anderson, G.W., Jr., Gibbs, P.H., and Friedlander, A.M. (1998). Comparative efficacy of experimental anthrax vaccine candidates against inhalation anthrax in rhesus macaques. Vaccine *16*, 1141–1148.

Jog, N.R., Jala, V.R., Ward, R.A., Rane, M.J., Haribabu, B., and McLeish, K.R. (2007). Heat shock protein 27 regulates neutrophil chemotaxis and exocytosis through two independent mechanisms. J. Immunol. *178*, 2421–2428.

Kang, T.J., Fenton, M.J., Weiner, M.A., Hibbs, S., Basu, S., Baillie, L., and Cross, A.S. (2005). Murine macrophages kill the vegetative form of *Bacillus anthracis*. Infect. Immun. *73*, 7495–7501.

Kanneganti, T.D., Lamkanfi, M., and Nunez, G. (2007). Intracellular NOD-like receptors in host defense and disease. Immunity *27*, 549–559.

Kassam, A., Der, S.D., and Mogridge, J. (2005). Differentiation of human monocytic cell lines confers susceptibility to *Bacillus anthracis* lethal toxin. Cell. Microbiol. *7*, 281–292.

Katayama, H., Janowiak, B.E., Brzozowski, M., Juryck, J., Falke, S., Gogol, E.P., Collier, R.J., and Fisher, M.T. (2008). GroEL as a molecular scaffold for structural analysis of the anthrax toxin pore. Nat. Struct. Mol. Biol. *15*, 754–760.

Kau, J.H., Lin, C.G., Huang, H.H., Hsu, H.L., Chen, K.C., Wu, Y.P., and Lin, H.C. (2002). Calyculin A sensitive protein phosphatase is required for *Bacillus anthracis* lethal toxin induced cytotoxicity. Curr. Microbiol. *44*, 106–111.

Kim, C., Gajendran, N., Mittrucker, H.W., Weiwad, M., Song, Y.H., Hurwitz, R., Wilmanns, M., Fischer, G., and Kaufmann, S.H. (2005). Human alpha-defensins neutralize anthrax lethal toxin and protect against its fatal consequences. Proc. Natl. Acad. Sci. USA *102*, 4830–4835.

Kim, C., Wilcox-Adelman, S., Sano, Y., Tang, W.J., Collier, R.J., and Park, J.M. (2008a). Antiinflammatory cAMP signaling and cell migration genes co-opted by the anthrax bacillus. Proc. Natl. Acad. Sci. USA *105*, 6150–6155.

Kim, J., Park, H., Myung-Hyun, J., Han, S.H., Chung, H., Lee, J.S., Park, J.S., and Yoon, M.Y. (2008b). The effects of anthrax lethal factor on the macrophage proteome: potential activity on nitric oxide synthases. Arch. Biochem. Biophy.s *472*, 58–64.

Kim, S.O., Ha, S.D., Lee, S., Stanton, S., Beutler, B., and Han, J. (2007). Mutagenesis by retroviral insertion in chemical mutagen-generated quasi-haploid mammalian cells. Biotechniques *42*, 493–501.

Kim, S.O., Jing, Q., Hoebe, K., Beutler, B., Duesbery, N.S., and Han, J. (2003). Sensitizing anthrax lethal toxin-resistant macrophages to lethal toxin-induced killing by tumor necrosis factor-alpha. J. Biol. Chem. *278*, 7413–7421.

Kim, T.G., Galloway, D.R., and Langridge, W.H. (2004). Synthesis and assembly of anthrax lethal factor-cholera toxin B-subunit fusion protein in transgenic potato. Mol. Biotechnol. *28*, 175–183.

King, H.K., and Stein, J.H. (1950). The non-toxicity of *Bacillus anthracis* cell material. J. Gen. Microbiol. *4*, 48–52.

Kirby, J.E. (2004). Anthrax lethal toxin induces human endothelial cell apoptosis. Infect. Immun. *72*, 430–439.

Klimpel, K.R., Arora, N., and Leppla, S.H. (1994). Anthrax toxin lethal factor contains a zinc metalloprotease consensus sequence which is required for lethal toxin activity. Mol. Microbiol.*13*, 1093–1100.

Klimpel, K.R., Molloy, S.S., Thomas, G., and Leppla, S.H. (1992). Anthrax toxin protective antigen is activated by a cell surface protease with the sequence specificity and catalytic properties of furin. Proc. Natl. Acad. Sci. USA *89*, 10277–10281.

Koehler, T.M., and Collier, R.J. (1991). Anthrax toxin protective antigen: low-pH-induced hydrophobicity and channel formation in liposomes. Mol. Microbiol. *5*, 1501–1506.

Koo, H.M., VanBrocklin, M., McWilliams, M.J., Leppla, S.H., Duesbery, N.S., and Woude, G.F. (2002). Apoptosis and melanogenesis in human melanoma cells induced by anthrax lethal factor inactivation of mitogen-activated protein kinase kinase. Proc. Natl. Acad. Sci. USA *99*, 3052–3057.

Krantz, B.A., Finkelstein, A., and Collier, R.J. (2006). Protein translocation through the anthrax toxin transmembrane pore is driven by a proton gradient. J. Mol. Biol. *355*, 968–979.

Krantz, B.A., Melnyk, R.A., Zhang, S., Juris, S.J., Lacy, D.B., Wu, Z., Finkelstein, A., and Collier, R.J. (2005). A phenylalanine clamp catalyzes protein translocation through the anthrax toxin pore. Science *309*, 777–781.

Krantz, B.A., Trivedi, A.D., Cunningham, K., Christensen, K.A., and Collier, R.J. (2004). Acid-induced unfolding of the amino-terminal domains of the lethal and edema factors of anthrax toxin. J. Mol. Biol. *344*, 739–756.

Kuhn, J.F., Hoerth, P., Hoehn, S.T., Preckel, T., and Tomer, K.B. (2006). Proteomics study of anthrax lethal toxin-treated murine macrophages. Electrophoresis *27*, 1584–1597.

Kumar, P., Ahuja, N., and Bhatnagar, R. (2002). Anthrax edema toxin requires influx of calcium for inducing cyclic AMP toxicity in targeT-cells. Infect. Immun. *70*, 4997–5007.

Kuo, S.R., Willingham, M.C., Bour, S.H., Andreas, E.A., Park, S.K., Jackson, C., Duesbery, N.S., Leppla, S.H., Tang, W.J., and Frankel, A.E. (2007). Anthrax toxin-induced shock in rats is associated with pulmonary edema and hemorrhage. Microb. Pathog. *44*, 467–472

Kuo, S.R., Willingham, M.C., Bour, S.H., Andreas, E.A., Park, S.K., Jackson, C., Duesbery, N.S., Leppla, S.H., Tang, W.J., and Frankel, A.E. (2008). Anthrax toxin-induced shock in rats is associated with pulmonary edema and hemorrhage. Microb. Pathog. *44*, 467–472.

Labbe, K., and Saleh, M. (2008). Cell death in the host response to infection. Cell Death Differ. *15*, 1339–1349.

Lacy, D.B., Wigelsworth, D.J., Melnyk, R.A., Harrison, S.C., and Collier, R.J. (2004a). Structure of heptameric protective antigen bound to an anthrax toxin receptor: a role for receptor in pH-dependent pore formation. Proc. Natl. Acad. Sci. USA *101*, 13147–13151.

Lacy, D.B., Wigelsworth, D.J., Scobie, H.M., Young, J.A., and Collier, R.J. (2004b). Crystal structure of the von Willebrand factor A domain of human capillary morphogenesis protein 2: an anthrax toxin receptor. Proc. Natl. Acad. Sci. USA *101*, 6367–6372.

Lamkanfi, M., Kanneganti, T.D., Franchi, L., and Nunez, G. (2007). Caspase-1 inflammasomes in infection and inflammation. J. Leukoc. Biol. *82*, 220–225.

Lee, J.O., Rieu, P., Arnaout, M.A., and Liddington, R. (1995). Crystal structure of the A domain from the alpha subunit of integrin CR3 (CD11b/CD18). Cell *80*, 631–638.

Lent, M., Hirshberg, A., and Margolis, G. (2001). Systemic toxins. Signs, symptoms & management of patients in septic shock. Jems *26*, 54–65; quiz 66–67.

Leppla, S.H. (1982). Anthrax toxin edema factor: a bacterial adenylate cyclase that increases cyclic AMP concentrations of eukaryotic cells. Proc. Natl. Acad. Sci. USA *79*, 3162–3166.

Leppla, S.H. (1984). *Bacillus anthracis* calmodulin-dependent adenylate cyclase: chemical and enzymatic properties and interactions with eucaryotic cells. Adv. Cyclic Nucleotide Protein Phosphorylation Res. *17*, 189–198.

Levin, T.C., Wickliffe, K.E., Leppla, S.H., and Moayeri, M. (2008). Heat shock inhibits caspase-1 activity while also preventing its inflammasome-mediated activation by anthrax lethal toxin. Cell. Microbiol. *10*, 2434–2446.

Levy, B. (2006). Lactate and shock state: the metabolic view. Curr. Opin. Crit. Care *12*, 315–321.

Li, R., Rieu, P., Griffith, D.L., Scott, D., and Arnaout, M.A. (1998). Two functional states of the CD11b A-domain: correlations with key features of two Mn2+-complexed crystal structures. J. Cell. Biol. *143*, 1523–1534.

Lin, C.G., Kao, Y.T., Liu, W.T., Huang, H.H., Chen, K.C., Wang, T.M., and Lin, H.C. (1996). Cytotoxic effects of anthrax lethal toxin on macrophage-like cell line J774A.1. Curr. Microbiol. *33*, 224–227.

Lincoln, R.E., Hodges, D.R., Klein, F., Mahlandt, B.G., Jones, W.I., Jr., Haines, B.W., Rhian, M.A., and Walker, J.S. (1965). Role of the lymphatics in the pathogenesis of anthrax. J. Infect. Dis. *115*, 481–494.

Little, S.F., and Lowe, J.R. (1991). Location of receptor-binding region of protective antigen from *Bacillus anthracis*. Biochem. Biophys. Res. Commun. *180*, 531–537.

Liu, S., Aaronson, H., Mitola, D.J., Leppla, S.H., and Bugge, T.H. (2003). Potent antitumor activity of a urokinase-activated engineered anthrax toxin. Proc. Natl. Acad. Sci. USA *100*, 657–662.

Liu, S., Bugge, T.H., and Leppla, S.H. (2001). Targeting of tumor cells by cell surface urokinase plasminogen activator-dependent anthrax toxin. J. Biol. Chem. *276*, 17976–17984.

Liu, S., and Leppla, S.H. (2003). Cell surface tumor endothelium marker 8 cytoplasmic tail-independent anthrax toxin binding, proteolytic processing, oligomer formation, and internalization. J. Biol. Chem. *278*, 5227–5234.

Liu, S., Leung, H.J., and Leppla, S.H. (2007). Characterization of the interaction between anthrax toxin and its cellular receptors. Cell. Microbiol. *9*, 977–987.

Liu, S., Netzel-Arnett, S., Birkedal-Hansen, H., and Leppla, S.H. (2000). Tumor cell-selective cytotoxicity of matrix metalloproteinase-activated anthrax toxin. Cancer Res. *60*, 6061–6067.

Liu, S., Wang, H., Currie, B.M., Molinolo, A., Leung, H.J., Moayeri, M., Basile, J.R., Alfano, R.W., Gutkind, J.S., Frankel, A.E., *et al.* (2008). Matrix metalloproteinase-activated anthrax lethal toxin demonstrates high potency in targeting tumor vasculature. J. Biol. Chem. *283*, 529–540.

Lu, H., Shen, C., and Brunham, R.C. (2000). Chlamydia trachomatis infection of epithelial cells induces the activation of caspase-1 and release of mature IL-18. J. Immunol. *165*, 1463–1469.

Lu, Q., Wei, W., Kowalski, P.E., Chang, A.C., and Cohen, S.N. (2004). EST-based genome-wide gene inactivation identifies ARAP3 as a host protein affecting cellular susceptibility to anthrax toxin. Proc. Natl. Acad. Sci. USA *101*, 17246–17251.

Mabry, R., Brasky, K., Geiger, R., Carrion, R., Jr., Hubbard, G.B., Leppla, S., Patterson, J.L., Georgiou, G., and Iverson, B.L. (2006). Detection of anthrax toxin in the serum of animals infected with *Bacillus anthracis* by using engineered immunoassays. Clin Vaccine Immunol *13*, 671–677.

Maldonado-Arocho, F.J., Fulcher, J.A., Lee, B., and Bradley, K.A. (2006). Anthrax oedema toxin induces anthrax toxin receptor expression in monocyte-derived cells. Mol. Microbiol.*61*, 324–337.

Martinon, F., Gaide, O., Petrilli, V., Mayor, A., and Tschopp, J. (2007). NALP inflammasomes: a central role in innate immunity. Semin. Immunopathol. *29*, 213–229.

Martinon, F., and Tschopp, J. (2007). Inflammatory caspases and inflammasomes: master switches of inflammation. Cell Death Differ. *14*, 10–22.

Matsumoto, G. (2003). Bioterrorism. Anthrax powder: state of the art? Science *302*, 1492–1497.

Mayr, B., and Montminy, M. (2001). Transcriptional regulation by the phosphorylation-dependent factor CREB. Nat Rev. Mol. Cell. Biol. *2*, 599–609.

McAllister, R.D., Singh, Y., du Bois, W.D., Potter, M., Boehm, T., Meeker, N.D., Fillmore, P.D., Anderson, L.M., Poynter, M.E., and Teuscher, C. (2003). Susceptibility to anthrax lethal toxin is controlled by three linked quantitative trait loci. Am. J. Pathol *163*, 1735–1741.

Melnyk, R.A., and Collier, R.J. (2006). A loop network within the anthrax toxin pore positions the phenylalanine clamp in an active conformation. Proc. Natl. Acad. Sci. USA *103*, 9802–9807.

Mikesell, P., Ivins, B.E., Ristroph, J.D., and Dreier, T.M. (1983). Evidence for plasmid-mediated toxin production in *Bacillus anthracis*. Infect. Immun. *39*, 371–376.

Miller, C.J., Elliott, J.L., and Collier, R.J. (1999). Anthrax protective antigen: prepore-to-pore conversion. Biochemistry *38*, 10432–10441.

Milne, J.C., Blanke, S.R., Hanna, P.C., and Collier, R.J. (1995). Protective antigen-binding domain of anthrax lethal factor mediates translocation of a heterologous protein fused to its amino- or carboxy-terminus. Mol. Microbiol. *15*, 661–666.

Milne, J.C., and Collier, R.J. (1993). pH-dependent permeabilization of the plasma membrane of mammalian cells by anthrax protective antigen. Mol. Microbiol. *10*, 647–653.

Milne, J.C., Furlong, D., Hanna, P.C., Wall, J.S., and Collier, R.J. (1994). Anthrax protective antigen forms oligomers during intoxication of mammalian cells. J. Biol. Chem. *269*, 20607–20612.

Moayeri, M., Haines, D., Young, H.A., and Leppla, S.H. (2003). *Bacillus anthracis* lethal toxin induces TNF-alpha-independent hypoxia-mediated toxicity in mice. J. Clin. Invest. *112*, 670–682.

Moayeri, M., and Leppla, S.H. (2004). The roles of anthrax toxin in pathogenesis. Curr. Opin. Microbiol. *7*, 19–24.

Moayeri, M., Martinez, N.W., Wiggins, J., Young, H.A., and Leppla, S.H. (2004). Mouse susceptibility to anthrax lethal toxin is influenced by genetic factors in addition to those controlling macrophage sensitivity. Infect. Immun. *72*, 4439–4447.

Moayeri, M., Webster, J.I., Wiggins, J.F., Leppla, S.H., and Sternberg, E.M. (2005). Endocrine perturbation increases susceptibility of mice to anthrax lethal toxin. Infect. Immun. *73*, 4238–4244.

Moayeri, M., Wiggins, J.F., and Leppla, S.H. (2007). Anthrax protective antigen cleavage and clearance from the blood of mice and rats. Infect. Immun. *75*, 5175–5184.

Mock, M., and Fouet, A. (2001). Anthrax. Annu. Rev. Microbiol. *55*, 647–671.

Mogridge, J., Cunningham, K., and Collier, R.J. (2002a). Stoichiometry of anthrax toxin complexes. Biochemistry *41*, 1079–1082.

Mogridge, J., Cunningham, K., Lacy, D.B., Mourez, M., and Collier, R.J. (2002b). The lethal and edema factors of anthrax toxin bind only to oligomeric forms of the protective antigen. Proc. Natl. Acad. Sci. USA *99*, 7045–7048.

Mogridge, J., Mourez, M., and Collier, R.J. (2001). Involvement of domain 3 in oligomerization by the protective antigen moiety of anthrax toxin. J. Bacteriol. *183*, 2111–2116.

Molnar, D.M., and Altenbern, R.A. (1963). Alterations in the biological activity of protective antigen of *Bacillus anthracis* toxin. Proc. Soc. Exp. Biol. Med. *114*, 294–297.

Mourez, M. (2004). Anthrax toxins. Rev. Physiol. Biochem. Pharmacol. *152*, 135–164.

Mourez, M., Yan, M., Lacy, D.B., Dillon, L., Bentsen, L., Marpoe, A., Maurin, C., Hotze, E., Wigelsworth, D., Pimental, R.A., et al. (2003). Mapping dominant-negative mutations of anthrax protective antigen by scanning mutagenesis. Proc. Natl. Acad. Sci. USA *100*, 13803–13808.

Muehlbauer, S.M., Evering, T.H., Bonuccelli, G., Squires, R.C., Ashton, A.W., Porcelli, S.A., Lisanti, M.P., and Brojatsch, J. (2007). Anthrax lethal toxin kills macrophages in a strain-specific manner by apoptosis or caspase-1-mediated necrosis. Cell Cycle *6*, 758–766.

Nanda, A., Carson-Walter, E.B., Seaman, S., Barber, T.D., Stampfl, J., Singh, S., Vogelstein, B., Kinzler, K.W., and St Croix, B. (2004). TEM8 interacts with the cleaved C5 domain of collagen alpha 3(VI). Cancer Res. *64*, 817–820.

Nassi, S., Collier, R.J., and Finkelstein, A. (2002). PA63 channel of anthrax toxin: an extended beta-barrel. Biochemistry *41*, 1445–1450.

Neumeyer, T., Tonello, F., Dal Molin, F., Schiffler, B., and Benz, R. (2006a). Anthrax edema factor, voltage-dependent binding to the protective antigen ion channel and comparison to LF binding. J. Biol. Chem. *281*, 32335–32343.

Neumeyer, T., Tonello, F., Dal Molin, F., Schiffler, B., Orlik, F., and Benz, R. (2006b). Anthrax lethal factor (LF) mediated block of the anthrax protective antigen (PA) ion channel: effect of ionic strength and voltage. Biochemistry *45*, 3060–3068.

Novak, J.M., Stein, M.P., Little, S.F., Leppla, S.H., and Friedlander, A.M. (1992). Functional characterization of protease-treated *Bacillus anthracis* protective antigen. J. Biol. Chem. *267*, 17186–17193.

Nye, S.H., Wittenburg, A.L., Evans, D.L., O'Connor, J.A., Roman, R.J., and Jacob, H.J. (2008). Rat survival to anthrax lethal toxin is likely controlled by a single gene. Pharmacogenomics J *8*, 16–22.

O'Brien, J., Friedlander, A., Dreier, T., Ezzell, J., and Leppla, S. (1985). Effects of anthrax toxin components on human neutrophils. Infect. Immun. *47*, 306–310.

Okinaka, R., Cloud, K., Hampton, O., Hoffmaster, A., Hill, K., Keim, P., Koehler, T., Lamke, G., Kumano, S., Manter, D., et al. (1999a). Sequence, assembly and analysis of pX01 and pX02. J. Appl. Microbiol. *87*, 261–262.

Okinaka, R.T., Cloud, K., Hampton, O., Hoffmaster, A.R., Hill, K.K., Keim, P., Koehler, T.M., Lamke, G., Kumano, S., Mahillon, J., et al. (1999b). Sequence and organization of pXO1, the large *Bacillus anthracis* plasmid harboring the anthrax toxin genes. J. Bacteriol. *181*, 6509–6515.

Paccani, S.R., Tonello, F., Ghittoni, R., Natale, M., Muraro, L., D'Elios, M.M., Tang, W.J., Montecucco, C., and Baldari, C.T. (2005). Anthrax toxins suppress T lymphocyte activation by disrupting antigen receptor signaling. J. Exp. Med. *201*, 325–331.

Pandey, J., and Warburton, D. (2004). Knock-on effect of anthrax lethal toxin on macrophages potentiates cytotoxicity to endothelial cells. Microbes Infect. *6*, 835–843.

Pannifer, A.D., Wong, T.Y., Schwarzenbacher, R., Renatus, M., Petosa, C., Bienkowska, J., Lacy, D.B., Collier, R.J., Park, S., Leppla, S.H., et al. (2001). Crystal structure of the anthrax lethal factor. Nature *414*, 229–233.

Park, J.M., Greten, F.R., Li, Z.W., and Karin, M. (2002). Macrophage apoptosis by anthrax lethal factor through p38 MAP kinase inhibition. Science *297*, 2048–2051.

Park, J.M., Greten, F.R., Wong, A., Westrick, R.J., Arthur, J.S., Otsu, K., Hoffmann, A., Montminy, M., and Karin, M. (2005). Signaling pathways and genes that inhibit pathogen-induced macrophage apoptosis – CREB and NF-kappaB as key regulators. Immunity *23*, 319–329.

Petosa, C., Collier, R.J., Klimpel, K.R., Leppla, S.H., and Liddington, R.C. (1997). Crystal structure of the anthrax toxin protective antigen. Nature *385*, 833–838.

Petrilli, V., Dostert, C., Muruve, D.A., and Tschopp, J. (2007a). The inflammasome: a danger sensing complex triggering innate immunity. Curr. Opin. Immunol. *19*, 615–622.

Petrilli, V., Papin, S., Dostert, C., Mayor, A., Martinon, F., and Tschopp, J. (2007b). Activation of the NALP3 inflammasome is triggered by low intracellular potassium concentration. Cell Death Differ. *14*, 1583–1589.

Pezard, C., Berche, P., and Mock, M. (1991). Contribution of individual toxin components to virulence of *Bacillus anthracis*. Infect. Immun. *59*, 3472–3477.

Pezard, C., Duflot, E., and Mock, M. (1993). Construction of *Bacillus anthracis* mutant strains producing a single toxin component. J. Gen. Microbiol *139*, 2459–2463.

Phillips, Z.E., and Strauch, M.A. (2002). Bacillus subtilis sporulation and stationary phase gene expression. Cell Mol. Life Sci. *59*, 392–402.

Pinton, P., Ferrari, D., Rapizzi, E., Di Virgilio, F., Pozzan, T., and Rizzuto, R. (2001). The Ca2+ concentration of the endoplasmic reticulum is a key determinant of ceramide-induced apoptosis: significance for the molecular mechanism of Bcl-2 action. EMBO J. *20*, 2690–2701.

Popov, S.G., Villasmil, R., Bernardi, J., Grene, E., Cardwell, J., Wu, A., Alibek, D., Bailey, C., and Alibek, K. (2002). Lethal toxin of *Bacillus anthracis* causes apoptosis of macrophages. Biochem. Biophys. Res. Commun. *293*, 349–355.

Prince, A.S. (2003). The host response to anthrax lethal toxin: unexpected observations. J. Clin. Invest. *112*, 656–658.

Rainey, G.J., Wigelsworth, D.J., Ryan, P.L., Scobie, H.M., Collier, R.J., and Young, J.A. (2005). Receptor-specific requirements for anthrax toxin delivery into cells. Proc. Natl. Acad. Sci. USA *102*, 13278–13283.

Rainey, G.J., and Young, J.A. (2004). Antitoxins: novel strategies to target agents of bioterrorism. Nat. Rev. Microbiol. *2*, 721–726.

Raymond, B., Leduc, D., Ravaux, L., Le Goffic, R., Candela, T., Raymondjean, M., Goossens, P.L., and Touqui, L. (2007). Edema toxin impairs anthracidal phospholipase A2 expression by alveolar macrophages. PLoS Pathog. *3*, e187.

Reig, N., Jiang, A., Couture, R., Sutterwala, F.S., Ogura, Y., Flavell, R.A., Mellman, I., and van der Goot, F.G. (2008). Maturation modulates caspase-1-independent responses of dendritic cells to anthrax lethal toxin. Cell. Microbiol. *10*, 1190–1207.

Ren, G., Quispe, J., Leppla, S.H., and Mitra, A.K. (2004). Large-scale structural changes accompany binding of lethal factor to anthrax protective antigen: a cryo-electron microscopic study. Structure *12*, 2059–2066.

Ribot, W.J., Panchal, R.G., Brittingham, K.C., Ruthel, G., Kenny, T.A., Lane, D., Curry, B., Hoover, T.A., Friedlander, A.M., and Bavari, S. (2006). Anthrax lethal toxin impairs innate immune functions of alveolar macrophages and facilitates *Bacillus anthracis* survival. Infect. Immun. *74*, 5029–5034.

Rivera, J., Nakouzi, A., Abboud, N., Revskaya, E., Goldman, D., Collier, R.J., Dadachova, E., and Casadevall, A. (2006). A monoclonal antibody to *Bacillus anthracis* protective antigen defines a neutralizing epitope in domain 1. Infect. Immun. *74*, 4149–4156.

Rmali, K.A., Al-Rawi, M.A., Parr, C., Puntis, M.C., and Jiang, W.G. (2004). Up-regulation of tumour endothelial marker-8 by interleukin-1beta and its impact in IL-1beta induced angiogenesis. Int. J. Mol. Med. *14*, 75–80.

Roberts, J.E., Watters, J.W., Ballard, J.D., and Dietrich, W.F. (1998). Ltx1, a mouse locus that influences the susceptibility of macrophages to cytolysis caused by intoxication with *Bacillus anthracis* lethal factor, maps to chromosome 11. Mol. Microbiol.*29*, 581–591.

Rogers, M.S., Christensen, K.A., Birsner, A.E., Short, S.M., Wigelsworth, D.J., Collier, R.J., and D'Amato, R.J. (2007). Mutant anthrax toxin B moiety (protective antigen) inhibits angiogenesis and tumor growth. Cancer Res. *67*, 9980–9985.

Rossi Paccani, S., Tonello, F., Patrussi, L., Capitani, N., Simonato, M., Montecucco, C., and Baldari, C.T. (2007). Anthrax toxins inhibit immune cell chemotaxis by perturbing chemokine receptor signalling. Cell. Microbiol. *9*, 924–929.

Rouleau, C., Menon, K., Boutin, P., Guyre, C., Yoshida, H., Kataoka, S., Perricone, M., Shankara, S., Frankel, A.E., Duesbery, N.S., *et al.* (2008). The systemic administration of lethal toxin achieves a growth delay of human melanoma and neuroblastoma xenografts: assessment of receptor contribution. Int. J. Oncol. *32*, 739–748.

Russell, B.H., Liu, Q., Jenkins, S.A., Tuvim, M.J., and Xu, Y. (2008). In vivo demonstration and quantification of intracellular *Bacillus anthracis* in lung epithelial cells. Infect. Immun. *76*, 3975–3983.

Russell, B.H., Vasan, R., Keene, D.R., Koehler, T.M., and Xu, Y. (2007). Potential dissemination of *Bacillus anthracis* utilizing human lung epithelial cells. Cell. Microbiol. *10*, 945–957.

Ruthel, G., Ribot, W.J., Bavari, S., and Hoover, T.A. (2004). Time-lapse confocal imaging of development of *Bacillus anthracis* in macrophages. J. Infect. Dis. *189*, 1313–1316.

Ryan, P.L., and Young, J.A. (2008). Evidence against a human cell-specific role for LRP6 in anthrax toxin entry. PLoS ONE *3*, e1817.

Saile, E., and Koehler, T.M. (2002). Control of anthrax toxin gene expression by the transition state regulator abrB. J. Bacteriol. *184*, 370–380.

Saile, E., and Koehler, T.M. (2006). *Bacillus anthracis* multiplication, persistence, and genetic exchange in the rhizosphere of grass plants. Appl. Environ. Microbiol. *72*, 3168–3174.

Salles, II, Tucker, A.E., Voth, D.E., and Ballard, J.D. (2003). Toxin-induced resistance in *Bacillus anthracis* lethal toxin-treated macrophages. Proc. Natl. Acad. Sci. USA *100*, 12426–12431.

Salles, II, Voth, D.E., Ward, S.C., Averette, K.M., Tweten, R.K., Bradley, K.A., and Ballard, J.D. (2006). Cytotoxic activity of *Bacillus anthracis* protective antigen observed in a macrophage cell line overexpressing ANTXR1. Cell. Microbiol. *8*, 1272–1281.

Santelli, E., Bankston, L.A., Leppla, S.H., and Liddington, R.C. (2004). Crystal structure of a complex between anthrax toxin and its host cell receptor. Nature *430*, 905–908.

Sanz, P., Teel, L.D., Alem, F., Carvalho, H.M., Darnell, S.C., and O'Brien, A.D. (2008). Detection of *Bacillus anthracis* spore germination *in vivo* by bioluminescence imaging. Infect. Immun. *76*, 1036–1047.

Sapra, R., Gaucher, S.P., Lachmann, J.S., Buffleben, G.M., Chirica, G.S., Comer, J.E., Peterson, J.W., Chopra, A.K., and Singh, A.K. (2006). Proteomic analyses of murine macrophages treated with *Bacillus anthracis* lethal toxin. Microb Pathog *41*, 157–167.

Scobie, H.M., Marlett, J.M., Rainey, G.J., Lacy, D.B., Collier, R.J., and Young, J.A. (2007). Anthrax toxin receptor 2 determinants that dictate the pH threshold of toxin pore formation. PLoS ONE *2*, e329.

Scobie, H.M., Rainey, G.J., Bradley, K.A., and Young, J.A. (2003). Human capillary morphogenesis protein 2 functions as an anthrax toxin receptor. Proc. Natl. Acad. Sci. USA *100*, 5170–5174.

Scobie, H.M., Thomas, D., Marlett, J.M., Destito, G., Wigelsworth, D.J., Collier, R.J., Young, J.A., and Manchester, M. (2005). A soluble receptor decoy protects rats against anthrax lethal toxin challenge. J. Infect. Dis. *192*, 1047–1051.

Scobie, H.M., Wigelsworth, D.J., Marlett, J.M., Thomas, D., Rainey, G.J., Lacy, D.B., Manchester, M., Collier, R.J., and Young, J.A. (2006). Anthrax toxin receptor 2-dependent lethal toxin killing *in vivo*. PLoS Pathog *2*, e111.

Scobie, H.M., and Young, J.A. (2006). Divalent metal ion coordination by residue T118 of anthrax toxin receptor 2 is not essential for protective antigen binding. PLoS ONE *1*, e99.

Sellman, B.R., Nassi, S., and Collier, R.J. (2001). Point mutations in anthrax protective antigen that block translocation. J. Biol. Chem. *276*, 8371–8376.

Shafa, F., Moberly, B.J., and Gerhardt, P. (1966). Cytological features of anthrax spores phagocytized *in vitro* by rabbit alveolar macrophages. J. Infect. Dis. *116*, 401–413.

Shafazand, S., Doyle, R., Ruoss, S., Weinacker, A., and Raffin, T.A. (1999). Inhalational anthrax: epidemiology, diagnosis, and management. Chest *116*, 1369–1376.

Shen, Y., Guo, Q., Zhukovskaya, N.L., Drum, C.L., Bohm, A., and Tang, W.J. (2004a). Structure of anthrax edema factor-calmodulin-adenosine 5'-(alpha,beta-methylene)-triphosphate complex reveals an alternative mode of ATP binding to the catalytic site. Biochem. Biophys. Res. Commun. *317*, 309–314.

Shen, Y., Zhukovskaya, N.L., Guo, Q., Florian, J., and Tang, W.J. (2005). Calcium-independent calmodulin binding and two-metal-ion catalytic mechanism of anthrax edema factor. EMBO J. *24*, 929–941.

Shen, Y., Zhukovskaya, N.L., Zimmer, M.I., Soelaiman, S., Bergson, P., Wang, C.R., Gibbs, C.S., and Tang, W.J. (2004b). Selective inhibition of anthrax edema factor by adefovir, a drug for chronic hepatitis B virus infection. Proc. Natl. Acad. Sci. USA *101*, 3242–3247.

Sherer, K., Li, Y., Cui, X., and Eichacker, P.Q. (2007a). Lethal and edema toxins in the pathogenesis of *Bacillus anthracis* septic shock: implications for therapy. Am. J. Respir. Crit. Care Med. *175*, 211–221.

Sherer, K., Li, Y., Cui, X., Li, X., Subramanian, M., Laird, M.W., Moayeri, M., Leppla, S.H., Fitz, Y., Su, J., and Eichacker, P.Q. (2007b). Fluid support worsens outcome and negates the benefit of protective antigen-directed monoclonal antibody in a lethal toxin-infused rat *Bacillus anthracis* shock model. Crit. Care Med. *35*, 1560–1567.

Shimaoka, M., Xiao, T., Liu, J.H., Yang, Y., Dong, Y., Jun, C.D., McCormack, A., Zhang, R., Joachimiak, A., Takagi, J., et al. (2003). Structures of the alpha L I domain and its complex with ICAM-1 reveal a shape-shifting pathway for integrin regulation. Cell *112*, 99–111.

Shin, S., Kim, Y.B., and Hur, G.H. (1999). Involvement of phospholipase A2 activation in anthrax lethal toxin-induced cytotoxicity. Cell. Biol. Toxicol. *15*, 19–29.

Shoop, W.L., Xiong, Y., Wiltsie, J., Woods, A., Guo, J., Pivnichny, J.V., Felcetto, T., Michael, B.F., Bansal, A., Cummings, R.T., et al. (2005). Anthrax lethal factor inhibition. Proc. Natl. Acad. Sci. USA *102*, 7958–7963.

Singh, Y., Klimpel, K.R., Goel, S., Swain, P.K., and Leppla, S.H. (1999). Oligomerization of anthrax toxin protective antigen and binding of lethal factor during endocytic uptake into mammalian cells. Infect. Immun. *67*, 1853–1859.

Singh, Y., Leppla, S.H., Bhatnagar, R., and Friedlander, A.M. (1989). Internalization and processing of *Bacillus anthracis* lethal toxin by toxin-sensitive and -resistant cells. J. Biol. Chem. *264*, 11099–11102.

Sirard, J.C., Mock, M., and Fouet, A. (1994). The three *Bacillus anthracis* toxin genes are coordinately regulated by bicarbonate and temperature. J. Bacteriol. *176*, 5188–5192.

Sirisanthana, T., and Brown, A.E. (2002). Anthrax of the gastrointestinal tract. Emerg. Infect. Dis. *8*, 649–651.

Smith, H., and Keppie, J. (1954). Observations on experimental anthrax; demonstration of a specific lethal factor produced *in vivo* by *Bacillus anthracis*. Nature *173*, 869–870.

Smith, H., and Stoner, H.B. (1967). Anthrax toxic complex. Fed. Proc. *26*, 1554–1557.

Squires, R.C., Muehlbauer, S.M., and Brojatsch, J. (2007). Proteasomes control caspase-1 activation in anthrax lethal toxin-mediated cell killing. J. Biol. Chem. *282*, 34260–34267.

St Croix, B., Rago, C., Velculescu, V., Traverso, G., Romans, K.E., Montgomery, E., Lal, A., Riggins, G.J., Lengauer, C., Vogelstein, B., and Kinzler, K.W. (2000). Genes expressed in human tumor endothelium. Science *289*, 1197–1202.

Steele, A.D., Warfel, J.M., and D'Agnillo, F. (2005). Anthrax lethal toxin enhances cytokine-induced VCAM-1 expression on human endothelial cells. Biochem. Biophys. Res. Commun. *337*, 1249–1256.

Sterne, M. (1937). Avirulent anthrax vaccine. Onderstepoort J. Vet. Sci. Animal Ind. *21*, 41–43.

Stout, R.D., Jiang, C., Matta, B., Tietzel, I., Watkins, S.K., and Suttles, J. (2005). Macrophages sequentially change their functional phenotype in response to changes in microenvironmental influences. J. Immunol. *175*, 342–349.

Strauch, M.A., Ballar, P., Rowshan, A.J., and Zoller, K.L. (2005). The DNA-binding specificity of the *Bacillus anthracis* AbrB protein. Microbiology *151*, 1751–1759.

Su, Y., Ortiz, J., Liu, S., Bugge, T.H., Singh, R., Leppla, S.H., and Frankel, A.E. (2007). Systematic urokinase-activated anthrax toxin therapy produces regressions of subcutaneous human non-small cell lung tumor in athymic nude mice. Cancer Res. *67*, 3329–3336.

Sun, J., Lang, A.E., Aktories, K., and Collier, R.J. (2008). Phenylalanine-427 of anthrax protective antigen functions in both pore formation and protein translocation. Proc. Natl. Acad. Sci. USA *105*, 4346–4351.

Sun, J., Vernier, G., Wigelsworth, D.J., and Collier, R.J. (2007). Insertion of anthrax protective antigen into liposomal membranes: effects of a receptor. J. Biol. Chem. *282*, 1059–1065.

Taft, S.C., and Weiss, A.A. (2008). Toxicity of anthrax toxin is influenced by receptor expression. Clin. Vaccine Immunol. *15*, 1330–1336.

Tamayo, A.G., Bharti, A., Trujillo, C., Harrison, R., and Murphy, J.R. (2008). COPI coatomer complex proteins facilitate the translocation of anthrax lethal factor across vesicular membranes *in vitro*. Proc. Natl. Acad. Sci. USA *105*, 5254–5259.

Tang, G., and Leppla, S.H. (1999). Proteasome activity is required for anthrax lethal toxin to kill macrophages. Infect. Immun. *67*, 3055–3060.

Ting, J.P., Willingham, S.B., and Bergstralh, D.T. (2008). NLRs at the intersection of cell death and immunity. Nat. Rev. Immunol. *8*, 372–379.

Tonello, F., Naletto, L., Romanello, V., Dal Molin, F., and Montecucco, C. (2004). Tyrosine-728 and glutamic acid-735 are essential for the metalloproteolytic activity of the lethal factor of *Bacillus anthracis*. Biochem. Biophys. Res. Commun. *313*, 496–502.

Tournier, J.N., Quesnel-Hellmann, A., Mathieu, J., Montecucco, C., Tang, W.J., Mock, M., Vidal, D.R., and Goossens, P.L. (2005). Anthrax edema toxin cooperates with lethal toxin to impair cytokine secretion during infection of dendritic cells. J. Immunol. *174*, 4934–4941.

Tucker, A.E., Salles, II, Voth, D.E., Ortiz-Leduc, W., Wang, H., Dozmorov, I., Centola, M., and Ballard, J.D. (2003). Decreased glycogen synthase kinase 3-beta levels and related physiological changes in *Bacillus anthracis* lethal toxin-treated macrophages. Cell. Microbiol. *5*, 523–532.

Turk, B.E., Wong, T.Y., Schwarzenbacher, R., Jarrell, E.T., Leppla, S.H., Collier, R.J., Liddington, R.C., and Cantley, L.C. (2004). The structural basis for substrate and inhibitor selectivity of the anthrax lethal factor. Nat. Struct. Mol. Biol. *11*, 60–66.

Turnbull, P.C. (2002). Introduction: anthrax history, disease and ecology. Curr. Top. Microbiol. Immunol. *271*, 1–19.

Varughese, M., Teixeira, A.V., Liu, S., and Leppla, S.H. (1999). Identification of a receptor-binding region within domain 4 of the protective antigen component of anthrax toxin. Infect. Immun. *67*, 1860–1865.

Vetter, S.M., and Schlievert, P.M. (2007). The two-component system Bacillus respiratory response A and B (BrrA-BrrB) is a virulence factor regulator in *Bacillus anthracis*. Biochemistry *46*, 7343–7352.

Vitale, G., Pellizzari, R., Recchi, C., Napolitani, G., Mock, M., and Montecucco, C. (1998). Anthrax lethal factor cleaves the N-terminus of MAPKKs and induces tyrosine/threonine phosphorylation of MAPKs in cultured macrophages. Biochem. Biophys. Res. Commun. *248*, 706–711.

Vitale, G., Pellizzari, R., Recchi, C., Napolitani, G., Mock, M., and Montecucco, C. (1999). Anthrax lethal factor cleaves the N-terminus of MAPKKS and induces tyrosine/threonine phosphorylation of MAPKS in cultured macrophages. J. Appl. Microbiol. *87*, 288.

Voth, D.E., Hamm, E.E., Nguyen, L.G., Tucker, A.E., Salles, II, Ortiz-Leduc, W., and Ballard, J.D. (2005). *Bacillus anthracis* oedema toxin as a cause of tissue necrosis and cell type-specific cytotoxicity. Cell. Microbiol. *7*, 1139–1149.

Wade, B.H., Wright, G.G., Hewlett, E.L., Leppla, S.H., and Mandell, G.L. (1985). Anthrax toxin components stimulate chemotaxis of human polymorphonuclear neutrophils. Proc. Soc. Exp. Biol. Med. *179*, 159–162.

Walev, I., Reske, K., Palmer, M., Valeva, A., and Bhakdi, S. (1995). Potassium-inhibited processing of IL-1 beta in human monocytes. Embo J *14*, 1607–1614.

Wang, W., Mulakala, C., Ward, S.C., Jung, G., Luong, H., Pham, D., Waring, A.J., Kaznessis, Y., Lu, W., Bradley, K.A., and Lehrer, R.I. (2006). Retrocyclins kill bacilli and germinating spores of *Bacillus anthracis* and inactivate anthrax lethal toxin. J. Biol. Chem. *281*, 32755–32764.

Warfel, J.M., and D'Agnillo, F. (2008). Anthrax lethal toxin enhances TNF-induced endothelial VCAM-1 expression via an IFN regulatory factor-1-dependent mechanism. J. Immunol. *180*, 7516–7524.

Warfel, J.M., Steele, A.D., and D'Agnillo, F. (2005). Anthrax lethal toxin induces endothelial barrier dysfunction. Am. J. Pathol. *166*, 1871–1881.

Watson, L.E., Kuo, S.R., Katki, K., Dang, T., Park, S.K., Dostal, D.E., Tang, W.J., Leppla, S.H., and Frankel, A.E. (2007a). Anthrax toxins induce shock in rats by depressed cardiac ventricular function. PLoS ONE 2, e466.

Watson, L.E., Mock, J., Lal, H., Lu, G., Bourdeau, R.W., Tang, W.J., Leppla, S.H., Dostal, D.E., and Frankel, A.E. (2007b). Lethal and edema toxins of anthrax induce distinct hemodynamic dysfunction. Front. Biosci. *12*, 4670–4675.

Webster, J.I., and Sternberg, E.M. (2004). Role of the hypothalamic-pituitary-adrenal axis, glucocorticoids and glucocorticoid receptors in toxic sequelae of exposure to bacterial and viral products. J. Endocrinol. *181*, 207–221.

Webster, J.I., Tonelli, L.H., Moayeri, M., Simons, S.S., Jr., Leppla, S.H., and Sternberg, E.M. (2003). Anthrax lethal factor represses glucocorticoid and progesterone receptor activity. Proc. Natl. Acad. Sci. USA *100*, 5706–5711.

Wei, W., Lu, Q., Chaudry, G.J., Leppla, S.H., and Cohen, S.N. (2006). The LDL receptor-related protein LRP6 mediates internalization and lethality of anthrax toxin. Cell *124*, 1141–1154.

Welkos, S.L., Keener, T.J., and Gibbs, P.H. (1986). Differences in susceptibility of inbred mice to *Bacillus anthracis*. Infect. Immun. *51*, 795–800.

Welkos, S.L., Vietri, N.J., and Gibbs, P.H. (1993). Non-toxigenic derivatives of the Ames strain of *Bacillus anthracis* are fully virulent for mice: role of plasmid pX02 and chromosome in strain-dependent virulence. Microb. Pathog. *14*, 381–388.

Werner, E., Kowalczyk, A.P., and Faundez, V. (2006). Anthrax toxin receptor 1/tumor endothelium marker 8 mediates cell spreading by coupling extracellular ligands to the actin cytoskeleton. J. Biol. Chem. *281*, 23227–23236.

Wesche, J., Elliott, J.L., Falnes, P.O., Olsnes, S., and Collier, R.J. (1998). Characterization of membrane translocation by anthrax protective antigen. Biochemistry 37, 15737–15746.

Wickliffe, K.E., Leppla, S.H., and Moayeri, M. (2008a). Anthrax lethal toxin-induced inflammasome formation and caspase-1 activation are late events dependent on ion fluxes and the proteasome. Cell. Microbiol. *10*, 332–343.

Wickliffe, K.E., Leppla, S.H., and Moayeri, M. (2008b). Killing of macrophages by anthrax lethal toxin: involvement of the N-end rule pathway. Cell. Microbiol. *10*, 1352–1362.

Wigelsworth, D.J., Krantz, B.A., Christensen, K.A., Lacy, D.B., Juris, S.J., and Collier, R.J. (2004). Binding stoichiometry and kinetics of the interaction of a human anthrax toxin receptor, CMG2, with protective antigen. J. Biol. Chem. *279*, 23349–23356.

Willingham, S.B., Bergstralh, D.T., O'Connor, W., Morrison, A.C., Taxman, D.J., Duncan, J.A., Barnoy, S., Venkatesan, M.M., Flavell, R.A., Deshmukh, M., et al. (2007). Microbial pathogen-induced necrotic cell death mediated by the inflammasome components CIAS1/cryopyrin/NLRP3 and ASC. Cell Host Microbe 2, 147–159.

Wilmanski, J.M., Petnicki-Ocwieja, T., and Kobayashi, K.S. (2008). NLR proteins: integral members of innate immunity and mediators of inflammatory diseases. J. Leukoc. Biol. 83, 13–30.

Wimalasena, D.S., Cramer, J.C., Janowiak, B.E., Juris, S.J., Melnyk, R.A., Anderson, D.E., Kirk, K.L., Collier, R.J., and Bann, J.G. (2007). Effect of 2-fluorohistidine labeling of the anthrax protective antigen on stability, pore formation, and translocation. Biochemistry 46, 14928–14936.

Wu, A.G., Alibek, D., Li, Y.L., Bradburne, C., Bailey, C.L., and Alibek, K. (2003). Anthrax toxin induces hemolysis: an indirect effect through polymorphonuclear cells. J. Infect. Dis. 188, 1138–1141.

Xia, Z.G., and Storm, D.R. (1990). A-type ATP binding consensus sequences are critical for the catalytic activity of the calmodulin-sensitive adenylyl cyclase from Bacillus anthracis. J. Biol. Chem. 265, 6517–6520.

Xu, L., Fang, H., and Frucht, D.M. (2008). Anthrax lethal toxin increases superoxide production in murine neutrophils via differential effects on MAPK signaling pathways. J. Immunol. 180, 4139–4147.

Xu, Q., Hesek, E.D., and Zeng, M. (2007). Transcriptional stimulation of anthrax toxin receptors by anthrax edema toxin and Bacillus anthracis Sterne spore. Microb. Pathog. 43, 37–45.

Yasuda, M., Theodorakis, P., Subramanian, T., and Chinnadurai, G. (1998). Adenovirus E1B-19K/BCL-2 interacting protein BNIP3 contains a BH3 domain and a mitochondrial targeting sequence. J. Biol. Chem. 273, 12415–12421.

Young, J.J., Bromberg-White, J.L., Zylstra, C., Church, J.T., Boguslawski, E., Resau, J.H., Williams, B.O., and Duesbery, N.S. (2007). LRP5 and LRP6 are not required for protective antigen-mediated internalization or lethality of anthrax lethal toxin. PLoS Pathog. 3, e27.

Zhang, S., Finkelstein, A., and Collier, R.J. (2004a). Evidence that translocation of anthrax toxin's lethal factor is initiated by entry of its N terminus into the protective antigen channel. Proc. Natl. Acad. Sci. USA 101, 16756–16761.

Zhang, S., Udho, E., Wu, Z., Collier, R.J., and Finkelstein, A. (2004b). Protein translocation through anthrax toxin channels formed in planar lipid bilayers. Biophys. J. 87, 3842–3849.

Zhivotovsky, B., and Orrenius, S. (2001). Current concepts in cell death. Curr. Protoc. Cell Biol. Chapter 18, Unit 18 11.

Subtilase Cytotoxin – a New Bacterial AB$_5$ Toxin Family

4

Adrienne W. Paton and James C. Paton

Abstract

Subtilase cytotoxin (SubAB) is the recently recognized prototype of a new AB$_5$ toxin family secreted by Shiga toxigenic *Escherichia coli* (STEC). Its A subunit is a subtilase-like serine protease and cytotoxicity for eukaryotic cells is due to a highly specific, single-site cleavage of BiP/GRP78, an essential Hsp70 family chaperone located in the ER. This cleavage triggers a severe ER stress response, ultimately resulting in apoptosis. The B subunit has specificity for glycans terminating in the sialic acid *N*-glycolylneuraminic acid. Although its actual role in human disease remains to be established, SubAB is lethal for mice and induces pathological features overlapping those seen in the haemolytic uraemic syndrome, a life-threatening complication of STEC infection.

Introduction

AB$_5$ toxins produced by pathogenic bacteria comprise an A subunit with enzymic activity and a B subunit pentamer responsible for interaction with glycan receptors on target eukaryotic cells (Fan *et al.*, 2000). Three long-established AB$_5$ toxin families are the Shiga toxins (Stx), Cholera toxin (Ctx) and the related *Escherichia coli* heat labile enterotoxins (LT), and pertussis toxin (Ptx). In each case, they are key virulence determinants of the bacteria that produce them (Shiga toxigenic *E. coli* [STEC] and *Shigella dysenteriae*, *Vibrio cholerae* and enterotoxigenic *E. coli*, and *Bordetella pertussis*, respectively). Collectively, these pathogens cause massive global morbidity and mortality, accounting for millions of deaths each year, particularly amongst children in developing countries. The AB$_5$ toxins exert their catastrophic effects by a two-step process involving B subunit-mediated entry of their respective target cells, followed by A subunit-dependent inhibition or corruption of essential host functions. The A subunits of Stx toxins have RNA-*N*-glycosidase activity, and cleave 28S rRNA, thereby inhibiting host protein synthesis. The A subunits of Ctx/LT and Ptx are ADP-ribosylases which modify distinct host G proteins, resulting in alteration of intracellular cAMP levels and dysregulation of ion transport mechanisms (Fan *et al.*, 2000). Subtilase cytotoxin is the recently recognized prototype of a fourth AB$_5$ toxin family, because of the distinct catalytic activity of its A subunit, which is a subtilase-like serine protease (Paton *et al.*, 2004). In this chapter, we will summarize the current state of knowledge regarding the molecular and cellular biology of this new toxin, its potential role in disease pathogenesis, and its applications as a tool in cell biology and perhaps as a therapeutic agent.

Initial discovery and preliminary characterization

Subtilase cytotoxin was discovered in a strain of STEC responsible for a small outbreak of haemolytic–uraemic syndrome (HUS) in South Australia. HUS is a life-threatening complication of STEC infection that has long been considered to be almost entirely due to the systemic effects of Stx (Paton and Paton, 1998). Interestingly, during this particular outbreak, the clinical presentation of the affected patients was atypical, with more marked neurological involvement than in previous cases of HUS seen at the same hospital. The causative STEC strain isolated from the patients belonged to serotype O113:H21 and produced Shiga toxin type 2 (Stx2) (Paton et al., 1999). O113:H21 is a prominent STEC serotype frequently associated with serious human disease. Indeed, it was among the first STEC serotypes to be causally associated with HUS (Karmali et al., 1985). However, unlike other prominent HUS-associated STEC serotypes such as O157:H7, O113:H21 strains lack the locus of enterocyte effacement (LEE), which encodes important accessory virulence traits that contribute to intestinal colonization and pathology. This led to the hypothesis that O113:H21 and perhaps some other virulent STEC strains might produce an additional cytotoxin capable of either augmenting the effects of Stx or causing additional pathology in its own right. A novel toxin operon was subsequently isolated by screening a cosmid gene bank of O113:H21 DNA constructed in E. coli K-12 for cytotoxicity on Vero cell monolayers. The cytopathic effect was distinct from that of Stx; it was maximal after three days incubation and was characterized by rounding of cells, detachment from the substratum, and loss of viability (Paton et al., 2004).

The operon is located on the O113:H21 megaplasmid pO113 (GenBank accession number AF399919.3). It consists of two closely linked genes, designated subA and subB. The subA gene encodes a 347 amino acid putative secreted protein with a modest degree of similarity to members of the Peptidase_S8 (subtilase) family of serine proteases (pfam00082.8). Its closest bacterial relative is the BA_2875 gene product of Bacillus anthracis (26% identity, 39% similarity over 246 amino acids). PROSITE analysis also indicated that SubA contains three conserved sequence domains, designated the catalytic triad, characteristic of members of the subtilase family (Siezen and Leunissen, 1997). The SubA domain sequences match the consensus sequences for the so-called Asp, His and Ser subtilase catalytic domains at 11/12, 10/11 and 10/11 positions, respectively, including the known critical active site residues Asp_{52}, His_{89} and Ser_{272} (Paton et al., 2004). The subB gene is 16 nucleotides downstream of subA and en-codes a 141 amino acid protein with significant similarities to putative exported proteins from Yersinia pestis (YPO0337; 56% identity, 79% similarity over 136 amino acids) and Salmonella Typhi (STY1891; 50% identity, 68% similarity over 117 amino acids). STY1891 has significant similarity (30% identity over 101 amino acids) to the S2 subunit of Ptx, but there is only 18% identity between SubB and the latter (Paton et al., 2004). Subcloning experiments demonstrated that both subA and subB genes were required for expression of cytotoxicity, and mutagenesis of any one of the critical A subunit residues essentially abolished cytotoxicity. This confirmed that serine protease activity was fundamental to the cytotoxic mechanism. Moreover, 1:5 stoichiometry for the SubA and SubB subunits was confirmed after purification of the SubAB holotoxin from recombinant E. coli expressing subAB, via a His_6 tag fused to the C-terminus of SubB. Purified SubAB was highly toxic for Vero cells, with a specific activity $>10^{10}$ 50% cytotoxic doses (CD_{50}) per mg. That is, < 0.1 pg of SubAB is sufficient to kill at least 50% of the ~3×10^4 Vero cells present in a microtitre plate well. For comparison, the specific Vero cell cytotoxicities reported for both major Stx toxin types are in the range of 1–10 pg/CD_{50} (Noda et al., 1987; Yutsudo et al., 1987; Richardson et al., 1992; Lindgren et al., 1994; Fujii et al., 2003). Thus, SubAB is 10–100 times more potent than Stx (Paton et al., 2004). Interestingly, Morinaga et al. (2007) have reported an additional vacuolating activity of SubAB, which appeared to be due to SubB alone. This was seen at early time points, before the SubA-dependent, protease-mediated cytotoxicity became apparent, but occurred only at high toxin doses (> 1 µg/ml). Lass et al. (2008) have recently reported a similar phenomenon in SubAB-treated HeLa cells, with formation of numerous vacuoles derived from elements of the ER, Golgi and probably also the mitochondria; these were located in the perinuclear region and at the cell periphery. Lipid droplets were also seen in the cytoplasm. Notably, however, these changes were not seen in cells treated with a non-toxic SubAB derivative with an A subunit Ser_{272}–Ala mutation (designated $SubA_{A272}B$), indicating that they are a consequence of the proteolytic activity of SubA.

Pathological features and role in human disease

Studies of the *in vivo* effects of SubAB have so far been limited to mice. Colonization of streptomycin-treated mice with *E. coli* K-12 carrying the *subAB* operon on a low copy-number plasmid resulted in dramatic weight loss (approximately 15%) over a 6-day period. In contrast, mice colonized with a clone producing the mutant toxin SubA$_{A272}$B continued to thrive. Interestingly, after six days, toxin-affected mice appeared to recover and gained weight, which correlated with production of toxin-neutralizing serum antibodies (Paton *et al.*, 2004). Prior immunization with purified SubA$_{A272}$B also protected mice from weight loss induced by colonization with *E. coli* expressing active SubAB (Talbot *et al.*, 2005). On the other hand, intraperitoneal injection of purified SubAB resulted in microangiopathic haemolytic anaemia, thrombocytopenia and renal impairment in mice, characteristics typical of Stx-induced HUS. SubAB caused extensive microvascular thrombosis and other histological damage in the brain, kidneys and liver, as well as dramatic splenic atrophy. Peripheral blood leukocytes were raised at 24 hours; there was also significant neutrophil infiltration in the liver, kidneys and spleen, and toxin-induced apoptosis at these sites (Wang *et al.*, 2007). Doses as low as 200 ng were invariably lethal (Paton *et al.*, 2004). These findings raise the possibility that SubAB directly contributes to pathology in humans infected with strains of STEC that produce both Stx and SubAB.

To date, only limited information is available on the distribution of SubAB-producing bacteria, based primarily on PCR screening of strain collections from specific geographic regions (Paton *et al.*, 2004; Paton and Paton, 2005; Izumiya *et al.*, 2006; Khaitan *et al.*, 2007; Cergole-Novella *et al.*, 2007). These preliminary studies indicate that the *subAB* operon is present in STEC isolates from Australia, Japan, USA and Brazil, belonging to a wide variety of serogroups, including O8, O23, O48, O77, O79, O82, O91, O96, O105, O113, O128, O146, O153, O163, O174, O178, OX18, OX25, and O-non-typable). Izumiya *et al.* (2006) have also reported the presence of sequence variants in *subAB* (about 90% identity to the published sequence) amongst the isolates

from Japan. It is not yet possible to determine whether there is any epidemiological association between production of SubAB and severity of STEC disease in humans or animals. Answering this question will require analysis of larger and more comprehensive strain collections than those examined to date. So far, *subAB* has been detected only in LEE-negative STEC, and there appears to be an association between presence of *subAB* and STEC carrying *stx$_2$*, rather than *stx$_1$* + *stx$_2$* or *stx$_1$* alone (Paton and Paton, 2005; Khaitan *et al.*, 2007). However, given that at least in O113:H21, the megaplasmid that carries the *subAB* operon is capable of conjugative transmission (Srimanote *et al.*, 2002), there is potential for wider dissemination amongst other *E. coli* pathotypes and possibly other Enterobacteriaceae. Thus, mass PCR screening of stool specimens from humans or animals, rather than testing specific isolates, will be needed to provide a more complete picture of the prevalence and distribution of SubAB-producing bacteria. Moreover, clinicians need to be made aware of the potential for a distinct disease spectrum in patients infected with such strains.

Intracellular target and cytotoxic mechanism

Knowledge that serine protease activity was pivotal to the cytotoxicity of SubAB led to proteomic identification of BiP (GRP78), a member of the Hsp70 family of chaperones located in the endoplasmic reticulum (ER), as its intracellular target (Paton *et al.*, 2006). The toxin cleaved a di-leucine motif in the hinge region connecting the N-terminal ATPase and C-terminal protein-binding domains of BiP. SubAB did not cleave any other proteins in the cell, including the most closely related Hsp70 chaperones. Structural analysis of SubA revealed an unusually deep active site cleft, accounting for this exquisite substrate specificity. The central role of BiP cleavage in the lethal mechanism was also confirmed by co-expressing a BiP derivative with a Leu$_{416}$Asp mutation that was resistant to the protease in transfected Vero cells, which protected the cells from SubAB cytotoxicity (Paton *et al.*, 2006). One of BiP's main functions is to mediate correct folding of nascent secretory proteins. BiP is also responsible for maintaining the permeability barrier of the ER membrane, as well as for targeting terminally mis-folded proteins via the Sec61 apparatus for degradation by the proteasome. BiP plays a crucial role in the unfolded protein response (UPR) as the ER stress-signalling master regulator. It also exhibits anti-apoptotic properties through interference with caspase activation (Gething, 1999; Hendershot, 2004). Thus, it is not surprising that disruption of BiP function has inevitably fatal consequences for the cell. However, no other cytotoxin has been shown to directly target chaperone proteins or components of the ER, and these findings reveal a previously un-described mechanism of inducing cell death.

There remains, however, a substantial lag between cleavage of BiP in SubAB-treated cells, which occurs within 15 minutes of addition of toxin to the culture medium, and death of the cells, which can take 24–48 hours to become apparent, even at high toxin doses. The precise molecular events that occur in between are poorly understood, although one of the likely consequences of BiP cleavage would be triggering of a severe ER stress response. ER stress can be induced by diverse perturbations in ER function and its purpose is to implement a series of changes in cellular activity, such that the ER stress is alleviated, enabling the cell to restore ER homeostasis and recover (Boyce and Yuan, 2006). ER stress responses include the UPR, which is characterized by: 1) transcriptional up-regulation of ER chaperones including BiP and other proteins, creating an ER milieu in which protein folding capacity is optimized; 2) activation of proteasome-dependent ER-associated degradation (ERAD) to remove unfolded proteins from the ER lumen; and 3) modulation of translation to slow down the traffic of nascent proteins into the ER compartment that require folding. However, if these responses fail to restore ER homeostasis, apoptosis may result (Boyce and Yuan, 2006).

In mammals, ER stress responses can be triggered by activation of three distinct ER membrane-spanning signalling molecules, namely PKR-like ER kinase (PERK), inositol-requiring enzyme 1 (IRE1), and activating transcription factor 6 (ATF6). The luminal domains of all three of these sentinel proteins are known to interact with BiP, and accumulation of unfolded proteins in the ER lumen is thought to recruit BiP away, allowing these sentinels to initiate signalling. By phosphorylation of its target eIF2α, activated PERK inhibits global protein synthesis, yet allows translation of mRNAs such as ATF4, a transcriptional activator for genes that ultimately assist in re-establishing ER homeostasis (Szegezdi et al., 2006). Release from BiP enables ATF6 to traffic to the Golgi, where it undergoes limited proteolysis, releasing a 50-kDa activated form, which translocates to the nucleus and binds to the ER stress response element, thus inducing genes encoding ER chaperones such as BiP, GRP94 and protein disulphide isomerase, as well as the transcription factors C/EBP homologous protein (CHOP) and X-box-binding protein 1 (XBP1). Activated IRE1 splices XBP1 mRNA into a form that can be translated into XBP1 protein, which then up-regulates genes encoding ER chaperones and the HSP40 family member P58^{IPK}. P58^{IPK} provides a feedback loop by binding and inhibiting PERK, thereby relieving the eIF2α-mediated translational block.

In Vero cells, treatment with SubAB rapidly activated all three ER stress signalling pathways (Wolfson et al., 2008). Activation of PERK was demonstrated by phosphorylation of eIF2α, which occurred within 30 minutes of toxin treatment, and correlated with inhibition of global protein synthesis. Activation of IRE1 was demonstrated by splicing of XBP1 mRNA, while ATF6 activation was demonstrated by depletion of the 90-kDa un-cleaved form, and appearance of the 50-kDa cleaved form. At least for PERK and IRE1, the rapidity with which ER stress signalling responses are triggered by exposure of cells to SubAB is consistent with the hypothesis that cleavage by the toxin causes BiP to dissociate from the signalling molecules, without the need for accumulation of unfolded proteins in the ER lumen. Downstream consequences that were detected during the following 24 hour period included up-regulation of GRP94, ATF4, EDEM, CHOP, and GADD34. BiP itself was also up-regulated at the mRNA level, but at the protein level, it continued to be degraded by SubAB in the ER lumen, and presumably this prevents restoration of ER homeostasis. Collectively, the above findings confirm induction of a severe and sustained ER stress response, and at 30 h, there was evidence of apoptosis. The relevance of these findings to the *in vivo* situation are supported by previous observations of CHOP induction in the liver (Paton *et al.*, 2006), as well as evidence of apoptosis in the kidneys, spleen and liver of SubAB-treated mice (Wang *et al.*, 2007). Morinaga *et al.* (2008) also reported transient inhibition of protein synthesis in SubAB-treated Vero cells due to PERK-mediated eIF2α phosphorylation, but interestingly, they also found that the toxin induced cell cycle arrest in G1 phase, possibly through down-regulation of cyclin D1 due to a combination of translational inhibition and proteasomal degradation.

Intracellular trafficking

The capacity of AB$_5$ toxins to damage mammalian cells is also dependent on B-subunit-mediated recognition of glycan receptors on the target cell surface, followed by internalization and intracellular trafficking, such that the catalytic A subunit has access to its substrate. In Vero cells, fluorescence co-localization with sub-cellular markers established that SubAB is trafficked from the cell surface to the ER via a retrograde pathway similar, but not identical, to those of Stx and Ctx, with their pathways converging at the Golgi. The clathrin inhibitor phenylarsine oxide prevented SubAB entry and BiP cleavage in SubAB-treated Vero, HeLa and N2A cells, while cholesterol depletion did not, demonstrating that, unlike either Stx or Ctx, which can also engage the lipid raft transport pathway, SubAB internalization and trafficking is exclusively clathrin dependent (Chong *et al.*, 2008). The trafficking of SubAB also differs from other AB$_5$ toxins in that its intracellular journey ends in the ER lumen, where its substrate is located. In contrast, the substrates for the other toxins are located in the cytosol, and this necessitates retro-translocation of their respective catalytic subunits across the ER membrane, by subversion of the Sec61 translocon (Lencer and Tsai, 2003; Yu and Haslam, 2005). Interestingly, at least for StxA, retro-translocation is believed to occur following interaction with BiP and another chaperone HEDJ/ERdj3 (Yu and Haslam, 2005). This raises the possibility that cleavage of BiP by SubAB may directly modulate entry of StxA into the cytosol, and

hence the *in vivo* consequences of Stx intoxication in patients infected with a bacterial strain producing both toxins. The critical role of BiP in Sec61-mediated retrotranslocation is also supported by the recent finding that SubAB inhibits ERAD, presumably through reduced trafficking of substrates (Lass *et al.*, 2008).

Receptor specificity

Specificity of toxin-receptor interactions is critical for pathogenesis, as it determines host susceptibility, tissue tropism, and the spectrum of pathology. Yahiro *et al.* (2006) reported that SubAB binds to α2β1integrin on the surface of Vero and HeLa cells, and that treatment of Vero cells with β1-integrin RNAi prevented the vacuolating activity of the toxin. It was not clear whether it was this, or some other toxin-receptor interaction, which is critical for the internalization and trafficking of SubAB that is required for BiP cleavage-mediated cytotoxicity. However, it raised the possibility that SubAB is more like Ptx, which utilizes a sialated glycoprotein receptor (Stein *et al.*, 1994), than Stx and Ctx, which recognize glycans displayed on glycolipids (Fan *et al.*, 2000).

Precise information on the binding specificity of SubAB to immobilized glycans has recently been obtained using fluorescent-labelled SubAB and a printed array of 320 glycan structures (Byres *et al.*, 2008). This showed a high degree of binding specificity for glycans terminating with α2–3-linked residues of the non-human sialic acid N-glycolylneuraminic acid (Neu5Gc). Much weaker binding was seen with those glycans that terminated in α2–3-linked N-acetylneuraminic acid (Neu5Ac), which differs by one hydroxyl group from Neu5Gc. Of all the glycans on the array, Neu5Gcα2–3Galβ1–4GlcNAcβ– bound SubAB best. The binding of SubAB to this glycan was reduced 20-fold if the Neu5Gc was changed to Neu5Ac; over 30-fold if the Neu5Gc linkage was changed from α2–3 to α2–6; and 100-fold if the sialic acid was removed. SubAB had a high affinity for terminal α2–3-linked Neu5Gc with little discrimination for the penultimate moiety. This high specificity of SubAB for Neu5Gc-terminating glycans is, to the best of our knowledge, unique amongst bacterial toxins.

The crystal structure of the *apo*-form of the SubB pentamer has also been solved to 2.0 Å resolution (Byres *et al.*, 2008). The SubB protomers form a homopentameric ring characteristic of the AB_5 toxin family, in which the 'top' of the molecule interacts with SubA, whilst the 'base' of the molecule potentially interacts with receptor. Despite very low sequence homology (3–18%) to the B subunits of other AB_5 toxins, the SubB protomer adopted the common OB (oligonucleotide/oligosaccharide-binding) fold (Murzin, 1993), typical of other AB_5 toxins, consistent with the fact that the toxin can bind glycans. The SubB protomer is of intermediate length (~ 120 residues) compared to other AB_5 B-protomers, which are either between 70 to 100 amino acids long (Stx, Ctx, and *E. coli* heat-labile enterotoxin [LT]), or about 200 amino acids (Ptx S2 and S3 subunits). Structural and sequence-based alignments revealed that SubB superposed well over the last C-terminal 100 residues of S2/3 subunits, which contain a shallow binding site for sialylated glycoproteins (Stein *et al.*, 1994; Byres *et al.*, 2008).

The crystal structure of the SubB pentamer was also determined in complex with Neu5Gc, as well as with one of the optimum binding trisaccharides Neu5Gcα2–3Galβ1–3GlcNAc. Neu5Gc bound to a shallow pocket halfway down the sides of the pentamer, whereas identical experiments using Neu5Ac failed to show any binding. The Neu5Gc was observed in all five binding sites of the pentamer, with the mode of binding being identical in each protomer. Comparisons of unliganded versus liganded SubB showed that no major structural rearrangements within SubB occurred upon Neu5Gc binding (Byres *et al.*, 2008). Interestingly, the Ptx S2/3 sialic acid binding site is also shallow and in the same location (Stein *et al.*, 1994). These findings differ from those for Stx, Ctx and LT, whose receptors are glycolipids rather than glycoproteins. The B subunits of these latter toxins have deep receptor binding pockets located on the membrane face of the toxin, at the opposite face to where subunit A interacts with the B pentamer (Merritt *et al.*, 1994; Merritt *et al.*, 1997, Ling *et al.*, 1998).

In the SubB–Neu5Gc complex, key interactions with the side chains of Asp_8, Ser_{12},

Glu_{36} and Tyr_{78} were observed. Neu5Gc differs from Neu5Ac by the addition of a hydroxyl on the methyl group of the N-Acetyl moiety of Neu5Ac. This extra hydroxyl present in Neu5Gc was observed to make additional crucial interactions with SubB; namely the extra hydroxyl points towards and interacts with Tyr_{78}^{OH} and also hydrogen bonds with the main chain of Met_{10}. These key interactions could not occur with Neu5Ac, thus explaining the marked preference for Neu5Gc. The mode of binding of the Neu5Gc moiety in the monosaccharide and the trisaccharide complex was identical. The C2 oxygen of Neu5Gc, the site of attachment to underlying sugar moieties, points outwards making no contacts with SubB. Nevertheless, the remaining two sugar moieties present in the trisaccharide are able to make some contact SubB, with the side chains of Thr_{107} and Glu_{108} hydrogen bonding with the N-acetyl carbonyl group on the GlcNAc at the reducing end, as well as water-mediated hydrogen bonds between the second and third sugar moieties with residues on the 106–108 loop region of SubB. The small number of additional interactions between SubB and the tertiary sugar in the trisaccharide is consistent with the Neu5Gc moiety driving the specificity and affinity for the interaction with SubB (Byres *et al.*, 2008).

The biological relevance of the structural analysis was confirmed by mutagenesis of key glycan-interacting residues in SubB. Mutagenesis of Ser_{12}, Glu_{36} and Tyr_{78} to Ala, Ala, and Phe, respectively, significantly reduced cell binding and specific cytotoxicity of the respective holotoxin. Of these, the Ser_{12} mutation had the greatest impact, reducing cytotoxicity by 99.98%. Importantly, mutagenesis of Tyr_{78}, which interacts only with the OH group unique to Neu5Gc, reduced cytotoxicity by 96.9% (Byres *et al.*, 2008).

The biological significance of the high specificity of SubAB for Neu5Gc-terminating glycans is also apparent from the observation that fluorescent-labelled SubAB bound to kidney tissue from wild type mice, but not CMP-N-acetylneuraminic acid hydroxylase (*Cmah*)-knock-out mice that can not make Neu5Gc due to the inability to convert CMP-Neu5Ac to CMP-Neu5Gc. Humans also lack this enzyme owing to a mutation in the *Cmah* gene that occurred after evolutionary separation of the *Hominin* lineage from the great apes (Varki, 2001), raising the possibility of human genetic resistance to the toxin. However, humans have been shown to be capable of assimilating Neu5Gc from dietary sources and incorporating it into glycoconjugates expressed on epithelial and endothelial surfaces (Tangvoranuntakul *et al.*, 2003). This would enable expression of high-affinity receptors on the cell surface, thereby conferring susceptibility to SubAB. Indeed, *in vitro* binding of SubAB to human gut epithelium and microvascular endothelium has recently been demonstrated (Byres *et al.*, 2008). Ironically, the foods that are the richest sources of Neu5Gc are red meat and dairy products, and these are also the commonest source of STEC contamination. This is a unique paradigm of bacterial pathogenesis, whereby humans may directly contribute to disease through dietary choices, simultaneously exposing themselves to the risk of STEC infection and sensitizing their tissues to SubAB.

Applications of SubAB and future directions

Interestingly, defects in chaperone function, particularly those affecting ER stress responses, are now strongly implicated in cellular senescence and a range of degenerative conditions including cataracts and Parkinson's and Alzheimer's diseases (Macario *et al.*, 2005). Through its capacity to rapidly and specifically abolish BiP function, SubAB provides a new tool in cell biology enabling *in vitro* modelling of key events in the pathogenesis of chaperonopathies. It has also already been proven to be a useful tool for examining the roles of BiP and ER stress in a range of other cellular processes, including ERAD (Lass *et al.*, 2008), T-cell activation and inflammatory responses of a variety of cell types (Takano *et al.*, 2007; Hayakawa *et al.*, 2008), as well as in certain viral infections, which involve perturbation of ER function (Buchkovich *et al.*, 2008). The mode of action of SubAB also raises the exciting possibility of its utility for treatment of certain cancers. Tumours up-regulate BiP in response to ER stress induced by rapid growth, and this is a crucial anti-apoptotic mechanism. BiP up-regulation is linked to metastatic po-

tential and resistance to chemotherapy (Li and Lee, 2006). Such tumours should be particularly susceptible to SubAB. If holotoxin were used, an appropriate targeting strategy would be required to prevent damage to normal tissues. Alternatively, the A subunit alone could be delivered via an appropriate targeting molecule, for example, epidermal growth factor (EGF) to specifically target EGF receptor- (EGFR-) positive tumours (including melanomas, gliomas, breast, prostate, and ovarian cancers). Recently, an EGF-SubA fusion protein has been synthesized, and this killed EGFR-positive rat glioma and human breast and prostate cancer cells at picomolar concentrations (Paton *et al.*, 2008). Cell lines expressing moderate to high levels of EGFR, which is associated with invasiveness and metastatic potential, were the most susceptible. Moreover, EGF-SubA acted synergistically with drugs such as thapsigargin that induce ER stress, enabling effective deployment at concentrations well below the cytotoxicity threshold for each component. Thus, EGF-SubA may provide a potential therapeutic avenue for potentially dangerous tumours, without the significant collateral damage and side effects that occur with conventional chemotherapy.

As might be expected given its comparatively recent discovery, much remains to be learned about the biological properties of SubAB, and its role in disease in both humans and animals. In particular, whilst we know that SubAB triggers ER stress signalling pathways, the precise molecular and cellular events whereby this leads to apoptosis remain to be elucidated. Understanding these events may be important for clinical management of affected patients. It is also interesting that to date, SubAB has been detected only in strains of *E. coli* that also produce Stx, and production of two distinct, highly potent AB_5 cytotoxins by a single pathogen raises fascinating questions regarding their relative contributions to pathogenesis. The mode of action of Stx involves direct inhibition of the elongation step of eukaryotic protein synthesis by StxA-mediated modification of eukaryotic 28S rRNA (Paton and Paton, 1998), while SubAB inhibits protein synthesis indirectly through activation of eIF2α, as described above. Thus, there is a possibility for either additive or synergistic inhibition of protein synthesis in cells targeted by both toxins. Conversely, SubAB-mediated BiP cleavage might interfere with retro-translocation of StxA into the cytosol, thereby antagonizing Stx function. An additional avenue for interaction between SubAB and Stx relates to effects on inflammatory signalling pathways. Stx is known to induce a ribotoxic stress response and expression of TNF-α, IL-1β, IL-6, and a number of CC and CXC chemokines in target cells, particularly epithelial cells and macrophages (Cherla et al., 2006). The resultant inflammation may play a significant role in pathogenesis of STEC disease. Effects of SubAB, with or without Stx, on these processes remain to be investigated.

References

Boyce, M., and Yuan, J. (2006). Cellular response to endoplasmic reticulum stress: a matter of life or death. Cell Death Differ. 13, 363–373.

Buchkovich, N.J., Maguire, T.G., Yu, Y., Paton, A.W., Paton, J.C., and Alwine, J.C. (2008). Human cytomegalovirus specifically controls the levels of the endoplasmic reticulum chaperone BiP/GRP78, which is required for virion assembly. J. Virol. 82, 31–39.

Byres, E., Paton, A.W., Paton, J.C., Löfling, J.C., Smith, D.F., Wilce, M.C.J., Talbot, U.M., Chong, D.C., Yu, H., Huang, S., Chen, X., Varki, N.M., Varki, A., Rossjohn, J., and Beddoe, T. (2008). Incorporation of a non-human glycan mediates human susceptibility to a bacterial toxin. Nature 456, 648–652.

Cergole-Novella, M.C., Nishimura, L.S., dos Santos, L.F., Irino, K., Vaz, T.M.I., Bergamini, A.M.M., and Guth, B.E.C. (2008). Distribution of virulence profiles related to new toxins and putative adhesins in Shiga toxin-producing Escherichia coli isolated from diverse sources in Brazil. FEMS Microbiol. Lett. 274, 329–334.

Cherla, R.P., Lee, S.Y., Mees, P.L., and Tesh, V.L. (2006). Shiga toxin 1-induced cytokine production is mediated by MAP kinase pathways and translation initiation factor eIF4E in the macrophage-like THP-1 cell line. J. Leukocyte Biol. 79, 397–407.

Chong, D.C., Paton, J.C., Thorpe, C.M., and Paton, A.W. (2008). Clathrin-dependent trafficking of subtilase cytotoxin, a novel AB5 toxin that targets the endoplasmic reticulum chaperone BiP. Cell. Microbiol. 10, 795–806.

Fan, E., Merritt, E.A., Verlinde, C.L.M.J., and Hol, W.G.J. (2000). AB(5) toxins: structures and inhibitor design. Curr. Opin. Struct. Biol. 10, 680–686.

Fujii, J., Matsui, T., Heatherly, D.P., Schlegel, K.H., Lobo, P.I., Yutsudo, T., Ciraolo, G.M., Morris, R.E., and Obrig, T. (2003). Rapid apoptosis induced by Shiga toxin in HeLa cells. Infect. Immun. 71, 2724–2735.

Gething, M. J. (1999). Role and regulation of the ER charperone BiP. Cell Devel. Biol. 10, 465–472.

Hayakawa, K., Hiramatsu, N., Okamura, M., Yao, J., Paton, A.W., Paton, J.C., and Kitamura, M. (2008). Blunted activation of NF-κB and NF-κB-dependent gene expression by geranylgeranylacetone: Involvement of unfolded protein response. Biochem. Biophys. Res. Comm. 365, 47–53.

Hendershot, L. M. (2004). The ER chaperone BiP is a master regulator of ER function. Mt. Sinai J. Med. 71, 289–297.

Karmali, M.A., Petric, M., Lim, C., Fleming, P.C., Arbus, G.S., and Lior, H. (1985). The asociation between idiopathic hemolytic uremic syndrome and infection by Verotoxin-producing Escherichia coli. J. Infect. Dis. 151, 775–782.

Khaitan, A., Jandhyala, D.M., Thorpe, C.M., Ritchie, J.M., and Paton, A.W. (2007). The operon encoding SubAB a novel cytotoxin is present in Shiga toxin-producing Escherichia coli isolates from the United States. J. Clin. Microbiol. 45, 1374–1375.

Izumiya, H., Iyoda, S., Terajima, J., Ohnishi, M., Yamasaki, S, and Watanabe, H. (2006). Distribution of the subA gene among LEE-negative STEC isolates in Japan. Abstract P07.2.11, pp. 84–85, Abstracts of the Sixth International Symposium and Workshop on Shiga toxin (Verocytotoxin) -producing E. coli Infections (VTEC 2006 Melbourne).

Lass, A., Kujawa, M., McConnell, E., Paton, A.W., Paton, J.C., and Wójcik, C. (2008). Decreased ER-associated degradation of α-TCR induced by Grp78 depletion with the SubAB cytotoxin. Int. J. Biochem. Cell Biol. In press.

Lencer, W.I., and Tsai, B. (2003). The intracellular voyage of cholera toxin: going retro. Trends Biochem. Sci. 28, 639–645.

Li, J., and Lee, A.S. (2006). Stress induction of GRP78/BiP and its role in cancer. Curr. Mol. Med. 6, 45–54.

Lindgren, S.W., Samuel, J.E., Schmitt, C.K., and O'Brien, A.D. (1994). The specific activities of Shiga-like toxin type II (SLT-II) and SLT-II-related toxins of enterohemorrhagic Escherichia coli differ when measured by Vero cell cytotoxicity but not by mouse lethality. Infect. Immun. 62, 623–631.

Ling, H., Bast, D. Brunton, J.L., and Read, R.J. (1998). Structure of the Shiga-like toxin I B-pentamer complexed with an analogue of its receptor Gb3. Biochemistry 37, 1777–1788.

Macario, A. J. L., and Conway de Macario, E. (2005). Sick chaperones, cellular stress and disease. N. Engl. J. Med. 353, 1489–1501.

Merritt, E.A., Sarfarty, S., Jobling, M.G., Chang, T., Holmes, R.K., Hirst, T.R., and Hol, W.G. (1997). Structural studies of receptor binding by cholera toxin mutants. Protein Sci. 6, 1516–1528.

Merritt, E.A., Sixma, T.K., Kalk, K.H., van Zanten, B.A., and Hol, W.G. (1994). Galactose-binding site in Escherichia coli heat-labile enterotoxin (LT) and cholera toxin (CT). Mol. Microbiol. 13, 745–753.

Morinaga, N., Yahiro, K., Matsuura, G., Watanabe, M., Nomura, F., Moss, J., and Noda, M. (2007). Two distinct cytotoxic activities of subtilase cytotoxin produced by Shiga-toxigenic Escherichia coli. Infect. Immun. 75, 488–496.

Morinaga, N., Yahiro, K., Matsuura, G., Moss, J., and Noda, M. (2008). Subtilase cytotoxin, produced by Shiga-toxigenic *Escherichia coli*, transiently inhibits protein synthesis of Vero cells via degradation of BiP and induces cell cycle arrest at G1 by down-regulation of cyclin D1. Cell. Microbiol. *10*, 921–929.

Murzin, A.G. (1993). OB (oligonucleotide/oligosaccharide binding)-fold: common structural and functional solution for non-homologous sequences. EMBO J. *12*, 861–867.

Noda, M., Yutsudo, T., Nakabayashi, N., Hirayama, T., and Takeda, Y. (1987). Purification and some properties of Shiga-like toxin from *Escherichia coli* O157:H7 that is immunologically identical to Shiga toxin. Microb. Pathog. 2, 339–349.

Paton, A.W., Paton, J.C., Backer, M., and Backer, J. (2008). Cleavage of BiP by Subtilase cytotoxin. Patent Cooperation Treaty Application WO2008/014574.

Paton, A.W., Beddoe, T., Thorpe, C.M., Whisstock, J.C., Wilce, M.C.J., Rossjohn, J., Talbot, U.M. and Paton J.C. (2006). AB₅ subtilase cytotoxin inactivates the endoplasmic reticulum chaperone BiP. Nature *443*, 548–552.

Paton, A.W., and Paton, J.C. (2005). Multiplex PCR for direct detection of Shiga toxigenic *Escherichia coli* producing the novel subtilase cytotoxin. J. Clin. Microbiol. *43*, 2944–2947.

Paton, A.W., Srimanote, P., Talbot, U.M., Wang, H., and Paton, J.C. (2004). A new family of potent AB₅ cytotoxins produced by Shiga toxigenic *Escherichia coli*. J. Exp. Med. *200*, 35–46.

Paton, A.W., Woodrow, M.C., Doyle, R.M., Lanser, J.A. and Paton J.C. (1999). Molecular characterization of a Shiga-toxigenic *Escherichia coli* O113:H21 strain lacking *eae* responsible for a cluster of cases of hemolytic-uremic syndrome. J. Clin. Microbiol. *37*, 3357–3361.

Paton, J.C., and Paton, A.W. (1998). Pathogenesis and diagnosis of Shiga toxin-producing *Escherichia coli* infections. Clin. Microbiol. Rev. *11*, 450–479.

Richardson, S.E., Rotman, T.A., Jay, V., Smith, C.R., Becker, L.E., Petric, M., Olivieri, N.F., and Karmali, M.A. (1992). Experimental Verocytotoxemia in rabbits. Infect. Immun. *60*, 4154–4167.

Siezen, R.J., and Leunissen, J.A.M. (1997). Subtilases: The superfamily of subtilisin-like serine proteases. Protein Science *6*, 501–523.

Srimanote, P., Paton A.W., and Paton, J.C. (2002). Characterization of a novel type IV pilus locus carried on the large plasmid of human-virulent strains of locus of enterocyte effacement-negative Shiga-toxigenic *Escherichia coli*. Infect. Immun. *70*, 3094–3100.

Stein, P.E., Boodhoo, A., Armstrong, G.D., Heerze, L.D., Cockle, S.A., Klein, M.H. & Read, R.J. (1994). Structure of a pertussis toxin-sugar complex as a model for receptor binding. Nat. Struct. Biol. *1*, 591–596.

Szegezdi, E., Logue, S.E., Gorman, A.M., and Samali, A. (2006). Mediators of endoplasmic reticulum stress-induced apoptosis. EMBO Reports 7, 880–885.

Takano, Y., Hiramatsu, N., Okamura, M., Hayakawa, K., Shimada, T., Kasai, A., Yokouchi, M., Shitamura, A., Yao, J., Paton, A.W., Paton, J.C., and Kitamura, M. (2007). Suppression of cytokine responses by GATA inhibitor K-7174: Implication for unfolded protein response. Biochem. Biophys. Res. Comm. *360*, 470–475.

Talbot, U.M., Paton, J.C., Paton, A.W. (2005). Protective immunization of mice with an active site mutant of subtilase cytotoxin of Shiga toxigenic *Escherichia coli*. Infect. Immun. 73, 4432–4436.

Tangvoranuntakul, P., Gagneux, P., Diaz, S., Bardor, M., Varki, N., Varki, A., and Muchmore, E. (2003). Human uptake and incorporation of an immunogenic nonhuman dietary sialic acid. Proc. Natl. Acad. Sci. USA *100*, 12045–12050.

Varki, A. (2001). Loss of N-glycolylneuraminic acid in humans: mechanisms, consequences, and implications for hominid evolution. Am. J. Phys. Anthropol. Suppl. *33*, 54–69.

Wang, H., Paton, J.C., and Paton, A.W. (2007). Pathologic changes in mice induced by subtilase cytotoxin, a potent new *Escherichia coli* AB₅ toxin that targets the endoplasmic reticulum. J. Infect. Dis. *196*, 1093–1101.

Wolfson, J., Thorpe, C.M., May, K.L., Paton, J.C., Jandhyala, D.M., and Paton, A.W. (2008). Subtilase cytotoxin activates PERK, ATF6 and IRE1 endoplasmic reticulum stress-signaling pathways. Cell. Microbiol. *10*, 1775–1786.

Yahiro, K., Morinaga, N., Satoh, M., Matsuura, G., Tomonaga, T., Nomura, F., Moss, J., and Noda, M. (2006). Identification and characterization of receptors for vacuolating activity of subtilase cytotoxin. Mol. Microbiol. *62*, 480–490.

Yu, M., and Haslam, D.B. (2005). Shiga toxin is transported from the endoplasmic reticulum following interaction with the luminal chaperone HEDJ/ERdj3. Infect. Immun. *73*, 2524–2532.

Yutsudo, T., Nakabayashi, N., Hirayama, T., and Takeda, Y. (1987). Purification and some properties of a Vero toxin from *Escherichia coli* O157:H7 that is immunologically unrelated to Shiga toxin. Microb. Pathog. 3, 21–30.

Pasteurella multocida Toxin

5

Joachim H.C. Orth

Abstract

Pasteurella multocida toxin (PMT) is the major pathogenic determinant of *Pasteurella multocida*. The species *Pasteurella multocida* leads to various diseases of animals and humans. The toxin is the causative agent of the economically important atrophic rhinitis in swine. Stimulation of several signalling pathways is induced by PMT. Most remarkable is a potent mitogenic effect. Phospholipase Cβ and the small GTPase Rho are activated due to stimulation of heterotrimeric G proteins of the $G\alpha_q$ and $G\alpha_{12/13}$ family. Here most recent results on studies of PMT are presented.

Introduction

PMT is the major virulence factor of *Pasteurella multocida*, a Gram-negative coccobacillus. *P. multocida* was used by Pasteur for his milestone vaccination studies in 1880. The pathogenicity is disclosed in the species name multo-cida (*multus* many, *-cidus* from *caedere* to kill) (Garrity *et al.*, 2005).

PMT given subcutaneously leads to dermonecrosis accompanied by congested vessels hepatic, splenic and testicular lesions (Elling *et al.*, 1988; Ackermann *et al.*, 1992). Thus, PMT belongs to the group of dermonecrotic toxins. These toxins are defined by their ability to induce dermonecrosis after subcutaneous administration. The group of dermonecrotic toxins consists of cytotoxic necrotizing factors CNF1 and CNF2 from pathogenic *Escherichia coli*, CNFy from *Yersinia pseudotuberculosis* and dermonecrotic toxin (DNT) from *Bordetella* subspecies *bronchiseptica*, *parapertussis* and *per-*tussis (Horiguchi, 2001; Lemonnier *et al.*, 2007). CNFs and DNT target Rho GTPases leading to their activation. Opposing PMT activates Rho only indirect due to activation of heterotrimeric G proteins.

Pathogenicity of *Pasteurella multocida*

Pasteurella multocida is a widespread pathogen in the animal kingdom but also in humans. It causes live-threatening illnesses of rabbits, cattle, sheep, and birds. Although *P. multocida* is an important pathogen it could occur as a commensal of the normal nasopharyngeal microflora of adult animals, e.g. cattle, rabbits, cats and dogs. The species *Pasteurella multocida* consists of four subspecies: *multocida*, *gallicida*, *septica* and *tigris*. *P. multocida* ssp. *multocida* in turn is divided into five serogroups (A, B, D, E and F) (Harper *et al.*, 2006; Capitini *et al.*, 2002), but PMT is produced only by serogroups A and D (Frandsen *et al.*, 1991).

Fowl cholera is most commonly caused by *P. multocida* subsp. *multocida* serogroup A. The site of infection is the respiratory tract ranging from acute to chronic infections. Acute infections are characterized by general septicaemic lesions. Further symptoms could be fever, anorexia and diarrhoea. In chronic infections suppurative lesions are widely spread, e.g. over the respiratory tract or the conjunctiva (Christensen and Bisgaard, 2000).

Rabbits are highly susceptible to infection with *P. multocida*. Pasteurellosis is the primary respiratory disease in domesticated animals, leading to characteristic 'snuffles'. The primary manifestation is rhinitis, sinusitis, conjunctivitis and dacryocystitis. A life-threatening haemorrhagic pneumonia may accompany in additionally stressed animals (Deeb and DiGiacomo, 2000).

In cattle *P. multocida* is associated with pneumonia and haemorrhagic septicaemia. The former is mainly induced by serogroup A and the later by serogroup B and E. Haemorrhagic septicaemia is characterized by growing oedema of the head and neck and swelled haemorrhagic lymph nodes. Pneumonia is triggered by concomitant stress factors such as shipping (therefore also called shipping fever), overcrowding or predisposing infections (Harper *et al.*, 2006; Dabo *et al.*, 2007).

P. multocida-induced infections in humans can be divided in various groups. Most human infections occur as local infections arising from cat and dog bites. Other animals implicated in bite wound infections are rats, horses or rabbits (Woolfrey *et al.*, 1985; Capitini *et al.*, 2002). A bite is not a prerequisite for infection: contact with saliva from colonized animals could also lead to infection (Kristinsson, 2007). Various biotypes of *P. multocida* could be isolated from human origin (Oberhofer, 1981). Complications of local infections are severe abscesses and tenosynovitis (Weber *et al.*, 1984). *P. multocida* colonized the respiratory tract can be interpreted as a normal commensal organism but it can also induce pneumonia, tracheobronchitis, lung abscess and empyema (Klein and Cunha, 1997). Infection of the respiratory tract has been described, e.g. for farmers, veterinarians or milkmen (Kristinsson, 2007). Further systemic and severe diseases are bacteraemia, meningitis, brain abscess, peritonitis or intra-abdominal abscess (Spencker *et al.*, 1979; Weber *et al.*, 1984). Like an opportunistic pathogen, *P. multocida* induces severe diseases in predisposed patients.

Atrophic rhinitis and bone resorption

Atrophic rhinitis is a nonfatal respiratory disease of pigs mainly characterized by loss of nasal turbinate bones, leading in the most severe form to a twisted and/or shortened snout. In the final stage notable destructions of tissue occur, which are associated with decalcification and destruction of turbinate bones and of bones forming the upper jaw of the snout. Early symptoms are those of an acute rhinitis with swelling of mucous membranes proceeding to difficulties in breathing and, in severe cases, to haemorrhage from the nostrils (Duthie, 1947; Pearce and Roe, 1966).

In pigs showing symptoms of atrophic rhinitis a retarded growth with decreased food intake is observed resulting in economical impact (van Diemen *et al.*, 1994a, 1995; Ackermann *et al.*, 1996). Reduced growth rate and food efficiency are not caused by reduced porcine growth hormone production (Ghoshal *et al.*, 1991).

The infectious nature of atrophic rhinitis was discussed for several years. Transmission of material from pigs affected with atrophic rhinitis to healthy animals e.g. by intranasal instillation were inconsistent (Duthie, 1947; Moynihan, 1947; Gwatkin *et al.*, 1949). But finally *P. multocida* was recognized in tissue material of pigs suffering from atrophic rhinitis (Heddleston *et al.*, 1954; Pearce and Roe, 1966). In addition strains of *Bordetella bronchiseptica* were isolated from nasal mucus of pigs affected with atrophic rhinitis. Early studies in pathogen-free piglets proposed a *B. bronchiseptica*-dependent model leading to turbinate atrophy, which could be increased in severity by coinoculation with *P. multocida* (Miniats and Johnson, 1980; Pedersen and Barford, 1981). Contribution of *B. bronchiseptica*- and *P. multocida*-derived toxins were revealed later on. Only toxigenic strains of *P. multocida* lead to severe stages of atrophic rhinitis. Inoculation with *B. bronchiseptica* strains expressing dermonecrotic toxin (DNT) or challenging with DNT alone supported colonization with *P. multocida* (Chanter *et al.*, 1989; Ackermann *et al.*, 1991; Elias *et al.*, 1992). However, the predisposing effect of DNT to infection with toxigenic *P. multocida* was not confirmed by comparison of a DNT-producing strain and an isogenic mutant of *B. bronchiseptica* that does not produce DNT. Pathogen-free pigs only infrequently were colonized by toxigenic *P. multocida* but both strains of *B. bronchiseptica* assisted to colonization with toxigenic *P. multocida* and turbinate atrophy occurred to the same extent (Brockmeier and Register, 2007). Consequently infection with *B. bronchiseptica* independently of DNT-expression predispose to secondary infection with *P. multocida*.

Moreover challenging of gnotobiotic pigs with purified PMT revealed that the toxin itself (given via different administration routs) is the causative agent of atrophic rhinitis. Toxin given intravenously, subcutaneously (Williams *et al.*, 1990; Ackermann *et al.*, 1993), intraperitoneally (Lax and Chanter, 1990), intramuscularly (Martineau-Doize *et al.*, 1990, 1991a,b) or intranasally via aerosols (Dominick and Rimler, 1986, 1988; van Diemen *et al.*, 1994a) leads to atrophy of nasal turbinates. Therefore PMT is the primary etiologic factor in the pathogenesis of atrophic rhinitis.

The mechanism of PMT-induced turbinate atrophy is not well understood. Data based on histological and ultrastructural studies were incoherent. On the one hand inhibition of bone formation by degeneration of osteoblasts and additional increase of osteoclastic osteolysis was reported (Dominick and Rimler, 1986, 1988). Other studies could not find changes in osteoblasts but detected an increase in the number of osteoclasts (Martineau-Doize *et al.*, 1990, 1991b, 1993). Toxin-induced ^{45}Ca release was reported from prelabelled mouse fetal long bones, which was accompanied by an increase in preosteoclasts and osteoclasts (Kimman *et al.*, 1987) and also from murine calvaria (Felix *et al.*, 1992). When rat osteoblastic osteosarcoma cells were directly targeted by the toxin, a reduced expression of osteoblastic markers was detected associated with reduced matrix mineralization (Sterner-Kock *et al.*, 1995). In contradiction, PMT stimulated DNA synthesis and cell proliferation in quiescent chicken osteoblasts. However, at the same time expression markers of osteoblast differentiation were down-regulated by PMT (Mullan and Lax, 1996). Analysis of the underlying signal pathways has still to be completed. Recent results indicate that down-regulation of osteoblast differentiation involves toxin-induced activation of the small GTPase Rho (Harmey *et al.*, 2004). Contribution of toxin-stimulated G_q signalling is not revealed yet, but recently constitutively active mutants of $G\alpha_q$ were found to impair differentiation of mouse osteoblastic MC3T3-E1 cells (Ogata *et al.*, 2007).

Two reports focused upon the origin of the increased number of osteoclasts. In mononuclear adherent mouse bone marrow cells (progenitor cells) PMT stimulated proliferation and differentiation into postmitotic osteoclasts (Jutras and Martineau-Doize, 1996). A porcine bone marrow cell model was used to investigate the effect of PMT on formation of osteoclasts and osteoblasts. PMT treatment of bone marrow cells resulted in earlier appearance and increased number of osteoclasts, while cultures developing into osteoblastic cells were retarded in cell growth due to PMT treatment. Interestingly, the effect of PMT on osteoclast development was also observed after addition of ultrafiltrate of PMT-intoxicated cells to fresh cultures. This could indicate that a soluble mediator and not the toxin directly is responsible for PMT action (Gwaltney *et al.*, 1997). This hypothesis of a mediator could be strengthening by a co-culture approach of osteoblasts and osteoclasts. Resorption of bone was not observed after toxin treatment of osteoclasts alone or co-cultures with cell–cell contact prevented. Nevertheless PMT induced bone breakdown in co-cultures of osteoblasts and osteoclasts with cell–cell contact permitted, indicating an osteoblast-dependent mechanism in PMT-induced bone loss (Mullan and Lax, 1998).

Vaccination

Because PMT was defined as the primary aetiological factor in atrophic rhinitis, it was conclusive to test the effect of vaccination of sows with toxin-containing vaccines. First Pederson and Barford vaccinated newborn piglets with a *P. multocida* vaccine containing PMT and challenged them with *B. bronchiseptica* and either a toxigenic or a non-toxigenic strain of *P. multocida* (Pedersen and Barfod, 1982).

The importance of PMT in developing atrophic rhinitis was also strengthened by comparing effects of vaccines containing detoxified preparations of either crude extract of toxigenic *P. multocida* or affinity-purified PMT. Both vaccines efficiently improved survival rates after challenging with pure toxin (Bording and Foged, 1991; Foged, 1991). In a rat model additional effects of PMT were tested. Vaccination with detoxified toxin not only prevented death but also hepatic necrosis or elevated leukocyte counts (Thurston et al., 1991). Interestingly vaccination of pregnant gilts with formaldehyde-detoxified PMT prevented the offspring in developing atrophic rhinitis (Foged et al., 1989). To optimize vaccination different strategies were studied. Non-toxic fragments of PMT were developed and used for vaccination of pigs and mice. It could be clearly demonstrated that these non-toxic PMT derivates efficiently protect against atrophic rhinitis (Nielsen et al., 1991; Petersen et al., 1991; Liao et al., 2006). Also genetically modified non-toxigenic PMT was used. Mutagenesis of Ser1164 to Ala or Cys1165 to Ser created a non-toxic PMT without impairment of the ability to induce protective immunity in pigs (To et al., 2005). The same mutation (Cys1165 to Ser) was used in a DNA vaccine approach. A DNA vaccine vector was used containing a signal sequence and encoding the entire toxin with mentioned mutation. Antibody response was tested in mice for optimization and additional in pigs but the ability of the vaccine to protect against disease remains unestablished (Register et al., 2007).

Another approach in protection pigs against atrophic rhinitis was either the combination of truncated PMT, B. bronchiseptica and adjuvant or expression of truncated PMT in B. bronchiseptica. Whereas the combinational approach was successful in respect to antibody production against PMT (Riising et al., 2002) the expression of truncated PMT in B. bronchiseptica lead to antibody response to B. bronchiseptica but no PMT specific antibodies were detected (Rajeev et al., 2003).

Intranasal administration of different preparations of detoxified PMT stimulated a protective response in rabbits, which could be enhanced by coadministration of the adjuvant cholera toxin (Suckow et al., 1995; Jarvinen et al., 1998, 2000).

Notable there is no antibody or cellular immune response detected in piglets challenged with PMT (van Diemen et al., 1994b). Only one-third of pigs suffering from atrophic rhinitis developed neutralizing antibodies against the toxin (Bechmann and Schoss, 1991). The lack of an immune response towards PMT might

be reasoned in low concentrations of PMT or in modification of the immune response by the toxin itself. To check the later possibility piglets were vaccinated intranasal with a combination of PMT without or with keyhole limpet haemocyanin, ovalbumin and tetanus toxoid. A lower total antibody response was detected in animals treated with PMT (van Diemen et al., 1996).

In vitro PMT induces activation of human monocyte-derived dendritic cells and mouse bone marrow-derived dendritic cells to mature but in vivo it is a poor adjuvant and it blocks the adjuvant effect of cholera toxin in mice (Bagley et al., 2005). Additionally, PMT inhibits the proper function of the motility machinery in dendritic cells to migrate to regional lymph nodes by activating Rho GTPases. This was proposed to limit the development of an adaptive immune response (Blöcker et al., 2006).

Detection of toxigenic P. multocida

Only toxigenic strains of P. multocida lead to production of atrophic rhinitis in pigs. Thus, detection of the toxin is of interest. First lethal tests in mice were described (Rutter, 1983). Culture filtrates of P. multocida were intraperitoneal injected in BALB/C mice. The same assay could be used as a serum neutralizing test to detect antibodies against PMT in animals. To avoid animal testing an in vitro cell culture assay was developed by two groups. Rutter and Luther showed a cell culture assay depending on a cytopathic effect of PMT on embryo bovine lung cells and was found to completely agree with results from mouse lethality test (Rutter and Luther, 1984). Pennings and Storm introduced a cell culture assay in Vero cells and compared it with the guinea pig skin test (Pennings and Storm, 1984). In later studies cell culture assays and enzyme-linked immunosorbent assays were used to detect the toxin (Oppling et al., 1994; Schimmel et al., 1994; Finco-Kent et al., 2001). Additional PCR methods were used for detection of toxigenic P. multocida (Vasfi et al., 1997; Townsend et al., 2000). Very recently an indirect enzyme-linked immunosorbent assays was introduced as an alternative for time-consuming neutralization tests (Takada-Iwao et al., 2007).

PMT and signal transduction

PMT stimulates various signal transduction pathways. G_q-dependent phospholipase C (PLC) β is activated to induce inositoltrisphosphate production and Ca^{2+} mobilization (Staddon *et al.*, 1991a). PMT also facilitates the effect of receptor agonists on phophatidylinositol-4,5-bisphosphate (PIP_2) hydrolysis (Murphy and Rozengurt, 1992). In addition focal contact assembly, paxillin phosphorylation and activation of the small GTPase Rho, leading to actin stress fibre formation, are observed after PMT treatment (Dudet *et al.*, 1996; Lacerda *et al.*, 1996; Thomas *et al.*, 2001). Moreover, PMT causes a strong mitogenic effect in different cell types (Rozengurt *et al.*, 1990). These effects are mainly connected to activation of heterotrimeric G protein of two different families, the $G\alpha_q$- and $G\alpha_{12/13}$-family.

Mitogenic effect

PMT was found to be an extremely strong mitogen for various cell types as Swiss 3T3, BALB/c, 3T6 or tertiary mouse embryo or human fibroblasts. Concentrations of PMT as low as 1 to 2 pM induced half-maximal effect on stimulation of DNA synthesis (Rozengurt *et al.*, 1990). The toxin stimulates anchorage-independent cell growth in Rat1 fibroblasts. Colony formation was stimulated at picomolar concentration and was observed after transient toxin exposure (Higgins *et al.*, 1992). Also proliferation of quiescent Swiss 3T3 cells, fetal bones (Dudet *et al.*, 1996) or osteoblasts (Mullan and Lax, 1996) was induced by PMT. In vivo nasal infection with *P. multocida* caused proliferation of bladder epithelium in gnotobiotic pigs (Hoskins *et al.*, 1997).

The pathways, which are perturbed by PMT to induce mitogenicity, were elucidated by Seo *et al.* The ability of the toxin to induce activation/phosphorylation of extracellular signal-regulated kinase (ERK) was compared with endogenous G protein-coupled receptors (Seo *et al.*, 2000). Inhibitors of G_q signalling blocked PMT- and GPCR-induced ERK phosphorylation. Depending on cell type, it has been shown that GPCR-induced ERK activation occurs via different pathways: Ras-dependent activation via receptor-tyrosine kinases called transactivation or Ras-independent via stimulation of PKC isoforms (Rozengurt, 2007). The effect of PMT was blocked by inhibitors of epidermal growth factor receptor but was not sensitive to inhibitors of PKC (Seo *et al.*, 2000). In the model of Seo *et al.* PMT employs G_q signalling to transactivate EGF receptor and to induce Ras-dependent ERK activation. The dependence on the cell type was shown by using cardiomyocytes and cardiac fibroblasts. In cardiac fibroblasts, activation of ERK depends on EGF receptor transactivation, whereas ERK activation was independent of EGF receptor in cardiomyocytes (Sabri *et al.*, 2002). However, when mitogenic signalling and cell cycle progression was studied in Swiss 3T3 and Vero cells, mitogenic signalling was down-regulated and cell cycle was arrested after an initial strong response (Wilson *et al.*, 2000). Interestingly, PMT-induced ERK activation was also found in $G\alpha_{q/11}$-deficient mouse embryonic fibroblasts (MEF) (Zywietz *et al.*, 2001), indicating the involvement of $G\alpha_q$-independent pathways. Using $G\alpha_{q/11}$- and $G\alpha_{12/13}$-deficient MEF, the participation of $G\alpha_{12/13}$ and $G\alpha_{q/11}$ in ERK signalling was demonstrated (Orth *et al.*, 2005).

Activation of Gq-dependent PLCβ

PMT increases inositol phosphates leading to elevation of intracellular Ca^{2+} (Rozengurt *et al.*, 1990; Staddon *et al.*, 1991a). Inositol phosphates are released by hydrolysis of PIP_2 due to activity of phosphatidylinositol-specific phospholipase C (PLC) isoforms. PLCγ is coupled to tyrosine-phosphorylated growth factor receptors and PLCβ to heterotrimeric G protein pathways. It has been shown that neuropeptide-induced (e.g. bombesin) increase in inositol phosphates is facilitated by PMT. By contrast, growth factor induced increase in inositol phosphates via tyrosine kinase receptor was not enhanced by the toxin (Murphy and Rozengurt, 1992). Therefore, it was suggested that PMT acts on the same pathway as agonists of G protein-coupled receptors. This idea was supported by studying Ca^{2+}-dependent Cl^- current in *Xenopus* oocytes (Wilson *et al.*, 1997). Wilson *et al.* injected PMT into oocytes, which led to a Cl^- current within

20 s after administration. Coinjection of specific antibodies against various PLCβ and PLCγ isoforms revealed that the PMT-induced response is mediated via PLCβ1. In line with this result antibody against PLCβ1-activating G proteins $G\alpha_{q/11}$ diminished PMT-induced Cl$^-$ current. Stimulation of PMT-induced Cl$^-$ current was only possible once, additional injection with PMT gave no further response.

It was shown by gene deletion of the α-subunits of G_q and G_{11} that PMT acts on PLCβ via $G\alpha_q$ but not via $G\alpha_{11}$ (Zywietz et al., 2001). This finding was surprising, because both G proteins are closely related and $G\alpha_q$-coupled receptors were constantly found to activate also $G\alpha_{11}$. Both G proteins share sequence identity of 89%. To determine the region, which is responsible for this unique specificity of PMT, $G\alpha_q/G\alpha_{11}$ deficient cells were used, which are unresponsive towards PMT in regard to PLCβ stimulation. Introduction of $G\alpha_q$ but not of $G\alpha_{11}$ into these cells reconstituted the sensitivity of these cells for PLCβ activation by PMT. Studies with chimeras of $G\alpha_q$ and $G\alpha_{11}$ showed that a small region, covering residues 105 through 113, defined the structural requirement of $G\alpha_q$ as compared to $G\alpha_{11}$ to mediate the PLCβ stimulatory effects of PMT (Orth et al., 2004). Exchange of Gln^{105} or Asn^{109} in $G\alpha_{11}$ to histidine rendered $G\alpha_{11}$ capable of mediating PMT-induced activation of PLCβ. But the mutation of histidines inside $G\alpha_q$ to the complementary amino acids of $G\alpha_{11}$ did not lead to unresponsive $G\alpha_q$ mutants, indicating more complex structural requirements for PMT-induced activation of G proteins. The importance of the helical domain, harbouring residues 105 and 109 of $G\alpha_q$ and of $G\alpha_{11}$, is surprising. So far almost all functions of $G\alpha_q$ are related to the Ras-like GTPase domain of the G proteins. The functional role of the helical domain is still largely enigmatic. However, these results indicate that this region especially helix αB is involved in the PMT effects. Moreover, the observation that exchange of one amino acid residue in this region largely effects the functional properties of the $G\alpha_{q/11}$ proteins suggests that minor structural changes may have major consequences in the function of G proteins.

During recent years various findings suggested that the all-helical domain of other G proteins might have a role in signal transduction. For example, it was shown by crystal structure analysis that the G$_o$LOCO motif (also called G protein regulatory motif (GRP)), which is found in some GTPase-activating proteins of the RGS (regulator of G protein signalling) family (e.g., RGS12 and RGS14), binds to the all-helical domain of G proteins (Kimple et al., 2002; Skiba et al., 1999). Also AGS proteins (activators of G protein signalling), which have GDI-like functions, possess a G$_o$LOCO domain (Takesono et al., 1999). AGS3 protein binds to the GDP-bound forms of all G$_i$ isoforms, to Gα_t and weakly to Gα_q. Recently, Natochin and coworkers (Natochin et al., 2002) identified residues 144–151 of Gα_i, which are located in the helical domain, to be essential for interaction with AGS3. However, it appears that AGS3 has activity only towards Gα_i and Gα_t. Comparing the activation of heterotrimeric G proteins with the mode of activation of small GTPases and GTP-binding elongation factors, Cherfils and Chabre, suggested a model of G-protein activation (Cherfils and Chabre, 2003), which depends on the interaction and contact area of Gβγ and Gα to stabilize a nucleotide-free complex. In contrast to the dissociation model of G protein activation, they proposed that upon activation, the N-terminus of the Gγ-subunit interacts in a hook-like manner with the helical domain, especially with the αA or αB helices of Gα. Moreover, by fluorescence resonance energy transfer (FRET)-based studies, it has been recently proposed that subunit rearrangement rather than dissociation is involved in G$_i$ protein activation (Bunemann et al., 2003). It is fascinating to speculate that the same region, e.g. αA or αB helix of Gα, which is important for G protein activation, is involved in PMT-induced activation of G$_q$.

Receptor-dependent activation of Gα_q is facilitated by PMT

PMT activates Gα_q and facilitates stimulation of inositol phosphate accumulation induced by agonists via G$_q$-coupled membrane receptors (Murphy and Rozengurt, 1992). Noteworthy, phosphorylation of a C-terminal Tyr356 was described to enhance the ability of Gα_q to activate PLCβ1 (Umemori et al., 1997). Baldwin et al. suggested that the enhancement of agonist-induced stimulation of G_q signalling by PMT is caused by tyrosine phosphorylation of $G\alpha_q$, an effect unrelated to the typical biological effect of the toxin, which results in activation of PLCβ (Baldwin *et al.*, 2003). This hypothesis was deduced from the observation that the inactive PMT mutant (PMT^{C1165S}) as PMT^{wt} stimulated tyrosine phosphorylation of $G\alpha_q$ and exhibited an enhancement of agonist-induced activation of G_q signalling. In other studies, however, the facilitating effect of PMT on GPCR signalling was strictly dependent on the full activity of PMT (Orth *et al.*, 2007b).

To clarify the role of tyrosine phosphorylation of $G\alpha_q$ in PMT-induced activation and enhancement of agonist stimulation, a $G\alpha_q$ mutant Tyr356Phe was generated. When this mutant was introduced into $G\alpha_{q/11}$-deficient MEF, PMT was able to stimulate inositol phosphate accumulation and facilitated agonist-induced inositol phosphate accumulation. Thus, the effect of PMT on agonist-induced G_q signalling should be independent of the C-terminal tyrosine residue (Orth *et al.*, 2007b).

Interaction of GPCR and Gq.

It has been shown that the C-terminal amino acid residues of Gα proteins are essential for interaction with GPCRs (Liu *et al.*, 1995, 2002;

Conklin *et al.*, 1993; 1996; Martin *et al.*, 1996; Hamm, 1998; Kostenis *et al.*, 2005). To study the role of the C-terminus of $G\alpha_q$ the C-terminal last five residues of $G\alpha_q$ were deleted and the effect of the deletion in $G\alpha_{q/11}$-deficient cells was studied (Orth *et al.*, 2007b). Although the toxin was able to stimulate inositol phosphate production in the presence of the C-terminal $G\alpha_q$ deletion, receptor agonists were not able to activate PLCβ via $G\alpha_q$. These findings indicate that the action of PMT is independent of the coupling of the G protein with GPCRs. In yeast it has been shown that unoccupied GPCR causes inhibition of heterologously expressed G_q. This inhibition is then released by occupation of the receptor by its specific ligand (Ladds *et al.*, 2007). To exclude the possibility that PMT acts by modifying a possible negative interaction of G_q protein with ligand-free receptor, the effects of the toxin we studied by using α_{1b}-adrenoceptor–$G\alpha_q$ fusion protein in $G\alpha_{q/11}$-deficient cells. Also under these conditions PMT activated the G_q, supporting the notion that the toxin acts independently of any G-protein–receptor interaction (Orth *et al.*, 2007b).

Activation of the small GTPase RhoA by Gα$_q$ and Gα$_{12/13}$

PMT activates RhoA signalling pathways (Fig. 5.1). The toxin induces reorganization of the

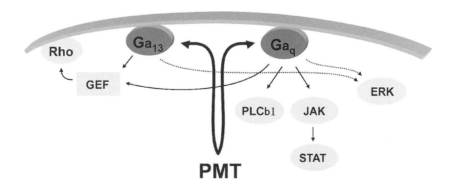

Figure 5.1 Sketch of signalling pathways activated by *Pasteurella multocida* toxin. After receptor-mediated endocytosis of PMT the toxin affects heterotrimeric G proteins to stimulate signalling of $G\alpha_q$ and $G\alpha_{13}$. PMT-induced activation of $G\alpha_q$ leads to activation of PLCβ1, which in turn hydrolyses PIP_2 to DAG and IP_3. The generated second messengers increase intracellular Ca^{2+} levels and activity of PKC. Further effects of $G\alpha_q$ activation is the stimulation of the JAK-STAT pathway manipulating gene expression. A third outcome of $G\alpha_q$ activation is the stimulation of the small GTPase Rho due to activation of a Rho guanine nucleotide exchange factor (GEF). Additionally, Rho is activated by PMT-induced activation of $G\alpha_{13}$. ERK phosphorylation occurs via $G\alpha_{13}$ and $G\alpha_q$ dependent pathways.

actin cytoskeleton and formation of stress fibres (Lacerda et al., 1996; Dudet et al., 1996; Essler et al., 1998). Moreover, the direct activation of RhoA induced by PMT was demonstrated using pull-down assays with Rho-binding effectors, which exclusively interact with the GTP-bound form of the GTPase (Zywietz et al., 2001).

PMT-induced increase in endothelial permeability was found to depend on Rho-activated Rho kinase, leading to inactivation of myosin light chain phosphatase. This results in an increase in phosphorylation of myosin light chain, causing endothelial cell retraction and increase in endothelial permeability (Essler et al., 1998). It was supposed that the damaging effects of PMT on endothelium could explain vascular effects of PMT (Aepfelbacher and Essler, 2001).

Several laboratories reported that RhoA is activated via signal pathways, involving various types of G proteins, including $G_{12/13}$ and $G_{q/11}$. First, Buhl et al. reported that $G\alpha_{12/13}$ is involved in RhoA-mediated stress fibre formation (Buhl et al., 1995). Later it was shown that the guanine nucleotide exchange factor p115RhoGEF directly interacts with $G\alpha_{13}$ to activate RhoA (Kozasa et al., 1998; Hart et al., 1998). In addition, the RhoA GEFs LARG and PDZ-RhoGEF were shown to activate RhoA mediated by $G_{12/13}$ (Suzuki et al., 2003; Fukuhara et al., 2001). Expression of dominant active (GTPase deficient) $G\alpha_{12}$ and $G\alpha_{13}$ induces RhoA activation in various cell types (Kranenburg et al., 1999; Chikumi et al., 2002). Moreover, activation of membrane receptors, which couple to $G_{12/13}$ and $G_{q/11}$ proteins by agonists like thrombin and lysophosphatidic acid, were shown to induce RhoA-dependent stress fibre formation even in $G_{q/11}$-deficient cells (Gohla et al., 1999). Other findings indicate that RhoA is also activated via $G_{q/11}$. This was most convincingly demonstrated in $G\alpha_{12/13}$-deficient cells. GPCRs, which couple to both $G_{q/11}$ and $G_{12/13}$ family G proteins were able to activate RhoA even in the absence of $G\alpha_{12}$ and $G\alpha_{13}$ (Vogt et al., 2003). Moreover, the GEF LARG was shown to interact not only with $G_{12/13}$ but also with $G_{q/11}$ (Booden et al., 2002). Recently, a novel guanine nucleotide exchange factor p63RhoGEF has been described that activates RhoA via $G_{q/11}$ (Lutz et al., 2005). Interestingly, this GEF protein appears

to be specific for $G_{q/11}$ and does not interact with $G_{12/13}$.

It was studied whether the PMT-induced activation of RhoA is mediated via $G_{q/11}$ and/or via $G_{12/13}$. This is an important question, because PMT is frequently used as a tool to specifically activate G_q signal pathways (Sleight et al., 2002; Sabri et al., 2002; Wilson and Ho, 2004). To differentiate between G_q- and $G_{12/13}$-mediated signalling the cyclic peptide YM-254890, which is a highly specific inhibitor of receptor-mediated activation of $G_{q/11}$ was used (Takasaki et al., 2004). YM-254890 completely blocked the activation of PLCβ by PMT (Orth et al., 2005), which is in line with the view that inositol phosphate accumulation induced by PMT is exclusively dependent on G_q (Orth et al., 2004).

YM-254890 and overexpression of RGS2, RGS16 and lscRGS partially inhibited PMT-stimulated serum response element (SRE)-dependent luciferase activity (Orth et al., 2005). RGS2 acts on G_i and $G_{q/11}$ but preferentially on $G_{q/11}$ (Heximer et al., 1999; Ingi et al., 1998). RGS16 acts on G_i and $G_{q/11}$ (Wieland and Chen, 1999; Hollinger and Hepler, 2002). lscRGS, the RGS domain of the Rho guanine nucleotide exchange protein lsc, appears to be specific for $G\alpha_{12/13}$. Thus, studies with specific inhibitors indicated that $G\alpha_{12/13}$ are involved in PMT-induced increase in luciferase activity. To further study this topic in more detail, cells deficient in the genes of $G\alpha_{12/13}$ and $G\alpha_{q/11}$ were used.

In $G\alpha_{q/11}$-deficient cells, Rho protein was strongly activated by PMT, which was shown by a rhotekin pull-down assay (Zywietz et al., 2001) and serum response element-dependent luciferase assay (Orth et al., 2005). In line with the proposed specificity of YM-254890, this activation of RhoA was not affected by compound YM-254890. Most importantly, when the G_q pathway was blocked by YM-254890, it was possible to reconstitute the RhoA-dependent luciferase activation by PMT in $G\alpha_{12/13}$-deficient cells after introducing $G\alpha_{13}$ into the cells (Orth et al., 2005). This finding is another strong indication for the hypothesis that $G_{12/13}$ is target of PMT.

Taken together, using the novel compound YM-254890, which specifically blocks $G_{q/11}$

activation, and by applying cells deficient in the genes for $G\alpha_{q/11}$ or $G\alpha_{12/13}$, it was able to dissect signalling pathways induced by PMT involving both subgroups of G proteins. PMT activates RhoA and RhoA-dependent pathways by stimulation of $G\alpha_q$ and by stimulation of $G\alpha_{12/13}$. These findings indicate that the effect of PMT is not specific for $G\alpha_q$ but also involves $G\alpha_{12/13}$.

Cancer development by PMT?

Recently, *P. multocida* has been linked to cancer, as natural infection with *P. multocida* or injection of its major virulence factor PMT cause proliferation of the bladder epithelium in the absence of an inflammatory reaction (Ward *et al.*, 1998). Although some of the pathways that are activated by PMT are linked to known oncogenes, such as RhoA, Src or Erk kinases, to date there are no data available whether chronic *P. multocida* infection may facilitate oncogenesis. Recently two studies focused on this topic and found different connections of PMT-induced signalling with cancer predisposition (Orth *et al.*, 2007a; Aminova and Wilson, 2007).

One report found that PMT induces activation of signal transducer and activator of transcription (STAT) proteins via Janus kinases (JAK). STAT1, STAT3 and STAT5 were identified as new targets of PMT-induced signalling. Using $G\alpha_{q/11}$-deficient MEF and the inhibitor of $G\alpha_q$ YM-254890 STAT activation could be linked to PMT-induced $G\alpha_q$ activation. Interestingly, PMT-dependent STAT phosphorylation remained constitutive for at least 18 h. A novel mechanism to maintain activation of STATs was supposed by the finding that PMT caused down-regulation of the expression of the suppressor of cytokine signalling (SOCS)-3. Moreover, stimulation of Swiss 3T3 cells with PMT increased transcription of the cancer-associated STAT-dependent gene cyclooxygenase (COX)-2 (Orth *et al.*, 2007a).

The second report studied the effect of PMT on cell signalling pathways during differentiation of 3T3-L1 adipocytes (Aminova and Wilson, 2007). PMT affected the differentiation to adipocytes preventing the expression of important adipocyte markers as PPARγ (peroxisome-proliferator-activated receptor γ) or C/EBPα (CAATT enhancer-binding protein α)

but stimulated cell cycle progression. Inhibition of differentiation was not only due to activation of the canonical PMT-induced $G\alpha_q$-PLCβ-Ca^{2+} pathway. Moreover two additional pathways were activated by PMT, Wnt/β-catenin and Notch1 pathway. PMT maintained β-catenin to block differentiation of preadipocytes to adipocytes. At the same time the expression of Notch1 was down-regulated as measured by mRNA levels. In adipocytes, PMT led to re-entry of cell cycle and down-regulation of PPARγ, C/EBPα and Notch1 (Aminova and Wilson, 2007). Mutation leading to high stabilization of β-catenin has been found, e.g. in endometrioid ovarian cancer or hepatoblastoma (Polakis, 2007). Also, Notch1 signalling plays a pivotal role in metazoan development. It is a main regulator how cells respond to intrinsic or extrinsic developmental signals deciding on differentiation, proliferation or apoptotic programs (Artavanis-Tsakonas *et al.*, 1999).

One possible connection between infections caused by bacteria and cancer is chronic inflammation that supports anti-apoptotic or chronic proliferative conditions eventually allowing tumour onset or promotion (Lax, 2005). Host cells infected with *P. multocida* that have a continuous production of toxin might therefore show constitutive STAT activation or long term effects by affecting the Notch1 and β-catenin pathway.

Oswald *et al.* named bacterial toxins that modulate the eukaryotic cell cycle cyclomodulins (Oswald *et al.*, 2005). It was suggested to group PMT to this family of bacterial toxins as it directly disturbs host cell signalling and cell cycle control (Orth *et al.*, 2007a). To date there is no established link between bacteria, such as *P. multocida*, with known mitogenic properties, and tumour formation (Lax and Thomas, 2002). However, the above mentioned studies (Orth *et al.*, 2007a; Aminova and Wilson, 2007) might indicate that the STAT pathway and/or the β-catenin/Notch1 pathway are missing links that could connect bacterial toxins to tumour development.

PMT as a tool

As PMT acts on various signal pathways it can be used as a tool to investigate signal transduction pathways.

PMT used as a mitogen

To confirm the selectivity of growth factor receptor inhibitors, PMT was used to stimulate DNA synthesis in Swiss 3T3 cells, tertiary mouse embryonic fibroblasts or rat-1 cells. Nordihydroguaiaretic acid was found to inhibit platelet-derived growth factor (PDGF)-stimulated DNA synthesis. Using PMT amongst others stimuli of DNA synthesis the authors found no inhibitory effect of nordihydroguaiaretic acid on PMT-induced DNA synthesis, excluding unspecific effects of the substance on proliferation (Domin et al., 1994).

The replication dependence of apoptosis induced by DNA methylating drugs (e.g. temezolomide) was also an utilization for PMT. The toxin increases proliferation rate, which in turn enhanced susceptibility of temezolomide-treated cells for apoptosis (Roos et al., 2007).

PMT used as an activator of Rho GTPase in plants

Also in plant cells PMT is used to decipher signal pathways. In Vigna unguiculata (L.) the involvement of Rho GTPases in root hair deformation was investigated. PMT activates the small GTPase Rho leading to deformation of root hair (Kelly-Skupek and Irving, 2006). Another bacterial toxin C3 toxin, a Rho ADP-ribosylating and thereby inhibiting toxin (Aktories et al., 1987), proofed the Rho dependence of the PMT-induced effect.

PMT used to target $G\alpha_q$

Most often PMT is adopted to target $G\alpha_q$ signalling pathways. In some studies PMT was used to activate $G\alpha_q$ but in others it was used to inhibit $G\alpha_q$ signalling. The inhibition of $G\alpha_q$ signalling goes back to observations of Wilson et al. that prolonged treatment of cells with PMT uncouples $G\alpha_q$ from its receptor (Wilson et al., 1997, 2000).

PMT was utilized for a set of studies in cardiomyocytes. The coupling function of α_1- and β-adrenergic receptors in cardiomyocytes cultured from mouse and rat ventricles was revealed. In mouse cardiomyocytes α_1-adrenergic receptor failed to activate PLCβ, ERK, p38-MAPK or to stimulate hypertrophy. Nevertheless in rat cardiomyocyte α_1-adrenergic receptor activated

$G\alpha_q$-dependent pathways leading to hypertrophy. PMT confirmed the proper function of $G\alpha_q$ and $G\alpha_q$-dependent pathway for PLCβ activation in mouse cardiomyocytes indicating a localization of the defect in PLC signalling on receptor-G protein coupling or on RGS proteins (Sabri et al., 2000).

Activation of $G\alpha_q$ in cardiomyocytes is characterized by hypertrophy in the case of modest overexpression or apoptosis in the case of intense activation. Treatment of cardiomyocytes with PMT leads to cardiomyocyte enlargement. Additionally novel PKC isoforms are activated and ERK is phosphorylated independently of EGFR transactivation. Interestingly in cardiac fibroblasts, PMT-induced ERK phosphorylation depends on EGFR transactivation. PMT induced a repression of basal and EGF-induced Akt phosphorylation. The enhanced susceptibility to apoptosis due to $G\alpha_q$ activation was therefore connected to the decreased Akt activation due to survival factors (Sabri et al., 2002).

PMT was utilized as a tool to study the involvement of Shc proteins in cardiac hypertrophy. Shc proteins are expressed as three isoforms and function as molecular adapters to link surface receptors to mitogenic response. It was shown that p66Shc, which was activated through $G\alpha_q$-dependent mechanism (protease-activated receptor-1 and PMT) presents a new candidate of hypertrophy-induced mediator of cardiomyocyte apoptosis and heart failure (Obreztchikova et al., 2006).

Platelet-derived growth factor receptor (PDGF)-induced activation of phosphatidylinositol (PI) 3-kinase is antagonized by $G\alpha_q$-coupled receptors. In Rat-1 fibroblasts stimulation of α_{1A}-adrenoceptor inhibits PDGF-induced binding of PI-3-kinase and phosphorylation of PDGF receptor at Tyr751. The activity of the protein tyrosine phosphatase SHP-2, which dephosphorylates Tyr751 was enhanced in cells treated with α_{1A}-adrenoceptor agonist. The same results were obtained by treatment with PMT indicating that activation of $G\alpha_q$ pathways lead to enhanced tyrosine dephosphorylation of PDGF receptor to block PI-3-kinase activation (Baldwin et al., 2003).

To study the cross-talk of insulin receptor with heterotrimeric G proteins, PMT was employed. Insulin and PMT were compared in their ability to generate inositol phosphates. The activation of $G\alpha_q$ by insulin was not further enhanced by additional treatment with PMT, indicating activation of identical pathways, i.e. $G\alpha_q$, by insulin and PMT (Sleight *et al.*, 2002).

Furthermore the crosstalk of different families of heterotrimeric G proteins was studied in more detail by PMT. In human mast cells the pathway, which couples adenosine A_{2B}-receptor to interleukin (IL)-4 secretion, was investigated. The secretion of IL-4 strictly depended on the activation of the $G\alpha_q$-PLCβ-Ca^{2+} pathway. In line with this finding treatment with PMT induced an increase in IL-4 secretion. Inhibitors of protein kinase A, an effector of $G\alpha_s$-cAMP pathway, only partial blocked IL-4 secretion. Surprisingly concomitant treatment of human mast cells with forskolin, an activator of adenylyl cylase, and PMT enhanced IL-4 secretion. These results indicate that the $G\alpha_s$ pathway is not mandatory for IL-secretion but it modulates $G\alpha_q$-dependent IL-4 secretion (Ryzhov *et al.*, 2006).

Not only the activation of $G\alpha_q$-coupled receptors but also expression of constitutive active mutants of $G\alpha_q$ are mimicked by PMT. The role of connexin (Cx)43-based gap junction channels in cell–cell coupling of Rat-1 cells was investigated. Disruption of gap junctional communication was observed after treatment with PMT or transfection with constitutive active $G\alpha_q$ but no other Gα families. It was revealed that the depletion of PIP_2 occurring after activation of $G\alpha_q$-PLCβ1 pathway is sufficient to close Cx43 channels (van Zeijl *et al.*, 2007).

The secretion of Cl^- in the intestine is triggered by stimulation of muscarinic receptors. Activation of muscarinic receptors leads to a biphasic effects: a short-lasting activation and a long-lasting inhibition. PMT treatment of mucosa–submucosa preparations of rats evoked a transient secretion and subsequently a down-regulation of muscarinic receptor-dependent secretion. The authors concluded that $G\alpha_q$ is involved in activation of Cl^- secretion by muscarinic receptors as well as in long term down-regulation of intestinal secretion. Additionally, activation of Cl^- secretion might contribute to mechanisms whereby *P. multocida* evoke diarrhoea (Hennig *et al.*, 2008).

In the above-mentioned studies PMT was adopted to confirm involvement of $G\alpha_q$-PLCβ1 pathways in various cellular effects. Inversely the toxin was used to exclude involvement of PMT-dependent pathways. Participation of PLCβ1 in histamine-induced catecholamine secretion from bovine adrenal chromaffin cells or in saccharin-induced current in frog rod taste cells were excluded by employment of PMT (Okada *et al.*, 2001; Donald *et al.*, 2002)

As mentioned before, PMT was used not only to stimulate $G\alpha_q$ pathway but also to block it after prolonged treatment (Wilson *et al.*, 1997; 2000). This special feature was mainly utilized to investigate the regulation of G protein-activated inwardly rectifying K^+ channels (GIRKs). GIRKs are activated in a pertussis toxin, i.e. G_i-dependent manner by binding βγ-subunits, which are released from $G\alpha_i$. Inhibition of GIRKs involves different classes of G protein-coupled receptors (e.g. G_i-coupled M_2 receptor or G_q-coupled α_1 adrenoceptor). To address the question whether both classes of receptor elicit their inhibitory effects on GIRKs by the same mechanism PMT was used. PMT action did not change M_2 receptor-induced inhibition of GIRKs indicating no effect of PMT on G_i in this system. However, α_1 adrenoceptor/$G\alpha_q$-mediated inhibition of GIRKs was blocked by prolonged PMT pretreatment (Tonello *et al.*, 1997; Bunemann *et al.*, 2000). The effect of prolonged PMT treatment to block G_q-dependent inhibition of GIRKs was also established in atrial myocytes (Meyer *et al.*, 2001).

Recently an equal approach was adopted to confirm the involvement of $G\alpha_q$ in M_1 receptor-mediated activation of TRP (transient receptor potential) channel 5. These channels are Ca^{2+}-permeable non-selective cation channels, which could be activated by G protein-coupled receptors, e.g. M_1 receptor. Prolonged PMT treatment inhibited M_1 receptor-induced currents indicating the participation of $G\alpha_q$ (Dattilo *et al.*, 2008).

Interestingly, the inhibition of $G\alpha_q$-dependent effects occurred even when $G\alpha_{11}$, which is not a target of PMT, was present. The question arises why $G\alpha_{11}$, which is similarly ubiquitously expressed as $G\alpha_q$, is not able to substitute the uncoupled $G\alpha_q$.

Cell binding and uptake

There is substantial evidence that PMT is an intracellularly acting toxin. After addition of toxin to cells a lag phase is observed (Rozengurt *et al.*, 1990). It is suggested that PMT binds to ganglioside-type receptors and is internalized by endocytic vesicles. Colloidal gold-labelled toxin was located within one minute after its addition to the plasma membrane of canine osteosarcoma and monkey kidney (Vero) cells. Competition with mixed gangliosides inhibited association of colloidal gold-labelled toxin with Vero cells (Pettit *et al.*, 1993) and counteracted its effect on DNA synthesis (Dudet *et al.*, 1996). Bafilomycin A1 an inhibitor of vacuolar ATPases, leading to acidification of endosomes, inhibited toxin activity suggesting a trafficking of PMT through an acidic compartment during uptake. Inhibition of toxin action was also observed after early (30 minutes after addition of toxin) but not late (two hours) treatment of cells with lysosomotrophic compounds (e.g. methylamine, ammonium chloride and chloroquine) (Staddon *et al.*, 1991a). Corresponding to the hypothesis that PMT is taken up via an acidic compartment, it was found that PMT undergoes conformational changes at low pH. These changes should occur during acidification of early endosome to elicit membrane translocation via the endosomal membrane to the cytosol (Smyth *et al.*, 1995, 1999). The uptake via an acidic compartment was also proofed by mimicking endosomal acidification by exposing toxin-bound cells to acidic medium. Under these conditions the toxin is directly translocated to the cytosol (Baldwin *et al.*, 2004).

From gene to protein

Before the toxin encoding gene was cloned, native PMT was purified by various methods, e.g. from sonic extracts of *P. multocida* using monoclonal antibodies or chromatographic techniques (Nakai *et al.*, 1984; Foged, 1988). A molecular size between 143 and 160 kDa was reported for native PMT. The gene encoding the toxic activity was found in a chromosomal DNA library of toxigenic type D strains of *P. multocida* (Petersen and Foged, 1989) and was named *toxA*. The *toxA* gene was cloned into plasmid vector and expressed under its own promoter in *Escherichia coli*. Recombinant PMT protein was purified by using monoclonal antibodies or standard chromatographic techniques (Petersen and Foged, 1989; Kamps *et al.*, 1990; Lax and Chanter, 1990). The recombinant and native expressed toxin are indistinguishable in respect to molecular mass, toxicity or antigenicity (Lax and Chanter, 1990; Higgins *et al.*, 1992) except an N terminal modification in native toxin, which does not occur in recombinant protein (Petersen, 1990). The *toxA* gene has been sequenced (Petersen, 1990; Buys *et al.*, 1990; Lax *et al.*, 1990). The primary structure of PMT consists of 1285 amino acids with a calculated mass of 146 kDa (Petersen, 1990; Buys *et al.*, 1990). Searching for homologue proteins was not successful at this time. Only some striking motifs were found: two extensive hydrophobic regions located near to each other were predicted at amino acid residues 385 to 469 suggesting a role in membrane targeting (Petersen, 1990), which could be verified some years later (Baldwin *et al.*, 2004). A proposed conserved motif of ADP-ribosylating toxins (Lax *et al.*, 1990) turned out inactive as PMT possesses no ADP-ribosyltransferase activity and mutation of the motif did not effect toxin activity (Staddon *et al.*, 1991b; Ward *et al.*, 1994). Petersen proposed that a cluster of cysteine and histidine residues between positions 1158 and 1229 could provide a site for metal ion binding (Petersen, 1990). Additionally, it has been reported that the production and the lethality of PMT is affected by different cations (Erler *et al.*, 1994; Reissbrodt *et al.*, 1994). Meanwhile the importance of His1205 was proofed by mutational studies (Orth *et al.*, 2003; Pullinger and Lax, 2007) but it contributes to a catalytic triad of a thiol protease (Kitadokoro *et al.*, 2007) (see below).

Later the complete genomic sequence of *P. multocida* was presented (May *et al.*, 2001) and the DNA sequence flanking the *toxA* gene was studied in more detail (Pullinger *et al.*, 2004). A homology to bacteriophage tail protein genes and a bacteriophage antirepressor was found. Therefore, it was proposed that PMT resides within a prophage. Mitomycin C induced lytic cycle of the temperate prophage indicating its functionality (Pullinger *et al.*, 2004). The presence of PMT gene within lytic prophage could be also a key to explain release of the toxin from

P. *multocida*. Ultrastructural studies localized PMT inside the bacterial cytoplasm. PMT is not secreted from *in vitro* cultures (Nakai *et al.*, 1985; iDali *et al.*, 1991; Erler, 1993). Therefore, a lytic cycle of the phage could enhance expression of PMT and release of PMT to intoxicate target cells. The expression of the toxin depends strongly on the medium used for the growth of *P. multocida* (Erler, 1993). A minimal medium avoiding the growth of *P. haemolytica* or *P. trehalosi* but supporting the growth of toxigenic and non-toxigenic *P. multocida* was described by Jablonski *et al.* and was used to produce PMT (Jablonski *et al.*, 1996).

Domain architecture

PMT can be expressed recombinantly in *E. coli* with high yields as his-tagged (Pullinger *et al.*, 2001) or GST-fusion protein (Busch *et al.*, 2001) (Fig. 5.2). PMT consists of 1285 amino acid residues and the primary structure of the N-terminus of the toxin shares over 20% of identical amino acids with the N-terminus of the cytotoxic necrotizing factors CNF1 and CNF2 of *Escherichia coli*, and CNFy of *Yersinia pseudotuberculosis*. CNF1, CNF2 and CNFy are deamidases acting on Rho proteins (Flatau *et al.*, 1997; Schmidt *et al.*, 1997). The catalytic domain of CNF1 is localized to the C-terminal part of

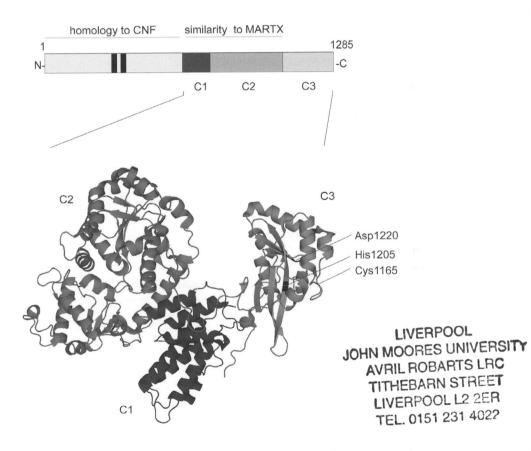

Figure 5.2 Domain architecture of *Pasteurella multocida* toxin. PMT consists of 1285 amino acid residues with a calculated mass of 146 kDa. PMT is a typical AB toxin, consisting of a receptor binding and translocation domain in the N-terminal part and an active domain in the C-terminal part. The primary structure shows a homology to cytotoxic necrotizing factors of *E. coli* and *Y. pseudotuberculosis* in the N terminus (residue 1–580) and a similarity to multifunctional auto-processing repeats-in-toxin toxins (MARTX) (residue 575–1104). Within the N terminus two hydrophobic helices (residues 402–423 and 437–457, indicated by black bars) are located essential for uptake procedure. The crystal structure of the C terminus revealed three domains named C1, C2 and C3. The C1 domain targets the toxin to the plasma membrane whereas the function of C2 domain is unknown. The C3 domain has similar folding to thiol proteases. The catalytic triad (Cys1165, His1205, Asp1220) is indicated in the ribbon diagram. Image of the crystal structure was generated with PyMol using PDB data file 2EBF (C-PMT).

the protein (Lemichez et al., 1997; Schmidt et al., 1998), whereas the N-terminal part harbours the receptor binding domain (Lemichez et al., 1997; Fabbri et al., 1999). In keeping with this domain structure of CNF1, a C-terminal amino acid of PMT was proposed to be important for biological activity of the toxin (Ward et al., 1998) and the receptor binding and translocation domain was localized in the N-terminus (Pullinger et al., 2001; Baldwin et al., 2004).

Active domain

It was observed that none of the toxin fragments studied induced a cytopathic effect when applied to the cell culture medium of intact cells (Wilson et al., 1999), no dermonecrotic activity was observed when guinea pigs were challenged with toxin fragments (Nakai and Kume, 1987) or anchorage-independent cell growth was not induced (Higgins et al., 1992). However, a C-terminal fragment encompassing amino acids 581 through 1285 provoked the same morphological changes and redistribution of the actin cytoskeleton as the holotoxin when introduced into embryo bovine lung cells by electroporation. Moreover, this fragment caused an increase in total inositol phosphates as observed with the holotoxin and was similarly efficient as the parent toxin (Busch et al., 2001). In other hands a slightly smaller fragment of the toxin (amino acids 681 to 1285) was expressed and tested for catalytic activity by microinjection into Swiss 3T3 cells (Pullinger et al., 2001). After microinjection this fragment induced changes in morphology and stimulated DNA synthesis equal to the full-length toxin.

Recently, a C-terminal part of PMT (amino acids 569 to 1285) was crystallized (Kitadokoro et al., 2007; Miyazawa et al., 2006). The C terminus of PMT consists of three distinct domains with a 'feed', 'body' and 'head' called like structure. The C1 domain (feed), encompassing amino acid 575 to 719, shows a structural homology to the N-terminal domain of *Clostridium difficile* toxin B. In *Clostridium difficile* toxin B this domain is suggested to localize the toxin to the lipid membrane (Reinert et al., 2005). In accordance with this homology fusion proteins of EGFP and the PMT-C1 domain were found to be localized at the plasma membrane (Kitadokoro et al., 2007). The C2 domain (body), encompassing residues 720–1104, revealed a homology to folylpolyglutamate synthetase. The third domain C3 (head), encompassing residues 1105–1285, consists of two subdomains providing a cleft between them. This cleft is supposed to be the catalytic centre of the toxin. Surprisingly a disulphide bond was found between cysteine-1159 and -1165. At least cysteine 1165 was known previously to be essential for the toxin activity (Ward et al., 1998; Busch et al., 2001) whereas the mutation of cysteine-1159 to serine has no major influence on toxin activity. Breaking the disulphide bond by mutation of cysteine 1159 to serine leads a reorientation of the free cysteine-1165. Comparison of this structure reveals a folding similar to thiol proteases or N-acetyltransferases with a catalytic triad consisting of the essential amino acids cysteine 1165 (Ward et al., 1998; Busch et al., 2001), histidine-1205 (Orth et al., 2003; Pullinger and Lax, 2007) and aspartic acid 1220. It was suggested by Kitadokoro et al. that the required breaking of the disulphide bond takes place during uptake of the toxin (Kitadokoro et al., 2007).

Recently, a similarity between C1 and C2 domain and multifunctional autoprocessing repeats-in-toxin toxins (MARTX), e.g. *Vibrio cholerae* toxin, was found (Satchell, 2007). MARTX harbour five putative activity domains and are autoprocessed upon cell uptake, releasing activity domains into the cytosol. In analogy to MARTX it was proposed that the putative catalytic active thiol protease is not the only activity carried by PMT. In fact the thiol protease-containing C3 domain was supposed to be required for autoproteolytic cleavage of PMT during uptake liberating C1 and C2 domain with actually unknown toxin activity (Satchell, 2007).

However, the assumption of Kitadokoro et al. that the C3 domain harbours the catalytic centre and toxic mechanism of the toxin was confirmed by exogenously expression of C-terminal fragments of PMT in mammalian cells. The last 180 amino acids encompassing the C3 domain induced serum response signalling and calcium-dependent signalling comparable to expressed full length toxin (Aminova et al., 2008). However, so far the substrates of the thiol protease PMT are not known and the mechanism of PMT-induced activation of heterotrimeric G proteins remains unclear.

Receptor binding and translocation domain

The sequence homology of PMT and CNF strongly suggests a similar molecular organization with an active domain in the C terminus and a receptor binding and translocation domain in the N terminus. In line with this hypothesis an N-terminal fragment of PMT (amino acids 1 to 506) competed with full-length toxin for binding to cell surface receptors (Pullinger *et al.*, 2001). Additional analysis of the N terminus predicted a transmembrane domain with hydrophobic and amphipathic helices similar to that of CNF1 and CNF2. In PMT two hydrophobic helices (helix one: amino acid 402 to 423, helix two: amino acid 437 to 457) are interrupted by a hydrophilic loop (amino acid 424 to 436). Mutation of acidic residues inside the hydrophilic loop leads to abrogated toxicity but did not to affected cell binding (Baldwin *et al.*, 2004). Thus, hydrophobic helices should play essential role in translocation of the toxin from the acidified endosome to the cytosol. Whether further processing, i.e. proteolytic cleavage, occurs as described for various bacterial toxins is not clear. However, toxin activity seems to be independent of processing at least if PMT is artificially introduced into cells, e.g. *Xenopus* oocytes (Wilson *et al.*, 1997).

Future perspectives

Over the last decade substantial progress was achieved in elucidating the action of PMT. The $G\alpha_q$ pathway was identified as a target of PMT and the toxins ability to distinguish between the highly related G proteins $G\alpha_q$ and $G\alpha_{11}$ was revealed. The latter one was the starting point to attribute a function to the helical domain of $G\alpha_q$, which is essential to serve as a target for PMT. Another milestone was the recognition of additional heterotrimeric G proteins as targets of PMT. Although $G\alpha_{13}$ was shown to be activated by PMT, it remains unclear whether the toxin dissects between $G\alpha_{12}$ and $G\alpha_{13}$ as described for $G\alpha_q/G\alpha_{11}$.

The major question, which is still not answered, is the molecular mechanism of PMT to activate G proteins. One breakthrough was the crystallization of the C terminus of PMT harbouring the toxin activity to stimulate heterotrimeric G proteins signalling. The crystal structure reveals a structural folding of the catalytic centre similar to thiol proteases or N-acetyltransferases. But the target, which is affected by PMT to evoke $G\alpha_q$ and $G\alpha_{13}$ signalling, has still to be established.

There is growing evidence for a connection of bacterial pathogens and carcinogenesis. Mainly the property to disturb cell cycle control and to activate signalling pathways, which are linked to known oncogenes, provide a basis for further research. Very recently initial studies revealed new PMT-activated pathways, i.e. JAK-STAT-, Notch1- or β-catenin pathway, which might present a link of PMT to carcinogenesis. But these investigations are at a very early stage.

References

Ackermann, M.R., Adams, D.A., Gerken, L.L., Beckman, M.J., and Rimler, R.B. (1993). Purified *Pasteurella multocida* protein toxin reduces acid phosphatase-positive osteoclasts in the ventral nasal concha of gnotobiotic pigs. Calcif. Tissue Int. 52, 455–459.

Ackermann, M.R., Register, K.B., Stabel, J.R., Gwaltney, S.M., Howe, T.S., and Rimler, R.B. (1996). Effect of *Pasteurella multocida* toxin on physeal growth in young pigs. Am. J. Vet. Res 57, 848–852.

Ackermann, M.R., Rimler, R.B., and Thurston, J.R. (1991). Experimental model of atrophic rhinitis in gnotobiotic pigs. Infect. Immun. 59, 3626–3629.

Ackermann, M.R., Tappe, J.P., Jr., Thurston, J.R., Rimler, R.B., Shuster, D.E., and Cheville, N.F. (1992). Light microscopic and ultrastructural pathology of seminiferous tubules of rats given multiple doses of *Pasteurella multocida* group D protein toxin. Toxicol. Pathol. 20, 103–111.

Aepfelbacher, M. and Essler, M. (2001). Disturbance of endothelial barrier function by bacterial toxins and atherogenic mediators: a role for Rho/Rho kinase. Cell Microbiol. 3, 649–658.

Aktories, K., Weller, U., and Chhatwal, G.S. (1987). *Clostridium botulinum* type C produces a novel ADP-ribosyltransferase distinct from botulinum C2 toxin. FEBS Lett. 212, 109–113.

Aminova, L.R., Luo, S., Bannai, Y., Ho, M., and Wilson, B.A. (2008). The C3 domain of *Pasteurella multocida* toxin is the minimal domain responsible for activation of Gq-dependent calcium and mitogenic signaling. Protein Sci. 17, 1–5.

Aminova, L.R. and Wilson, B.A. (2007). Calcineurin-independent inhibition of 3T3-L1 adipogenesis by *Pasteurella multocida* toxin: suppression of Notch1, stabilization of beta-catenin and pre-adipocyte factor 1. Cell Microbiol. 9, 2485–2496.

Artavanis-Tsakonas, S., Rand, M.D., and Lake, R.J. (1999). Notch signaling: cell fate control and signal integration in development. Science 284, 770–776.

Bagley, K.C., Abdelwahab, S.F., Tuskan, R.G., and Lewis, G.K. (2005). Pasteurella multocida toxin activates human monocyte-derived and murine bone marrow-derived dendritic cells in vitro but suppresses antibody production in vivo. Infect. Immun. 73, 413–421.

Baldwin, M.R., Lakey, J.H., and Lax, A.J. (2004). Identification and characterization of the Pasteurella multocida toxin translocation domain. Mol. Microbiol. 54, 239–250.

Baldwin, M.R., Pullinger, G.D., and Lax, A.J. (2003). Pasteurella multocida toxin facilitates inositol phosphate formation by bombesin through tyrosine phosphorylation. J. Biol. Chem. 278, 32719–32725.

Bechmann, G. and Schoss, P. (1991). [Detection of neutralizing antibodies against Pasteurella multocida toxin in swine with atrophic rhinitis]. Dtsch. Tierarztl. Wochenschr. 98, 310–312.

Blöcker, D., Berod, L., Fluhr, J.W., Orth, J., Idzko, M., Aktories, K., and Norgauer, J. (2006). Pasteurella multocida toxin (PMT) activates RhoGTPases, induces actin polymerization and inhibits migration of human dendritic cells, but does not influence macropinocytosis. Int. Immunol. 18, 459–464.

Booden, M.A., Siderovski, D.P., and Der, C.J. (2002). Leukemia-associated Rho guanine nucleotide exchange factor promotes G alpha q-coupled activation of RhoA. Mol. Cell Biol. 22, 4053–4061.

Bording, A. and Foged, N.T. (1991). Characterization of the immunogenicity of formaldehyde detoxified Pasteurella multocida toxin. Vet. Microbiol. 29, 267–280.

Brockmeier, S.L. and Register, K.B. (2007). Expression of the dermonecrotic toxin by Bordetella bronchiseptica is not necessary for predisposing to infection with toxigenic Pasteurella multocida. Vet. Microbiol. 125, 284–289.

Buhl, A.M., Johnson, N.L., Dhanasekaran, N., and Johnson, G.L. (1995). $G\alpha_{12}$ and $G\alpha_{13}$ stimulate Rho-dependent stress fiber formation and focal adhesion assembly. J. Biol. Chem. 270, 24631–24634.

Bunemann, M., Frank, M., and Lohse, M.J. (2003). Gi protein activation in intacT-cells involves subunit rearrangement rather than dissociation. Proc. Natl. Acad. Sci. USA 100, 16077–16082.

Bunemann, M., Meyer, T., Pott, L., and Hosey, M. (2000). Novel inhibition of gbetagamma-activated potassium currents induced by M(2) muscarinic receptors via a pertussis toxin-insensitive pathway. J. Biol. Chem. 275, 12537–12545.

Busch, C., Orth, J., Djouder, N., and Aktories, K. (2001). Biological activity of a C-terminal fragment of Pasteurella multocida toxin. Infect. Immun. 69, 3628–3634.

Buys, W.E.C., Smith, H.E., Kamps, A.M.I.E., and Smits, M.A. (1990). Sequence of the dermonecrotic toxin of Pasteurella multocida ssp. multocida. Nucleic Acids Res. 18, 2815–2816.

Capitini, C.M., Herrero, I.A., Patel, R., Ishitani, M.B., and Boyce, T.G. (2002). Wound infection with Neisseria weaveri and a novel subspecies of Pasteurella multocida in a child who sustained a tiger bite. Clin. Infect. Dis. 34, E74-E76.

Chanter, N., Magyar, T., and Rutter, J.M. (1989). Interactions between Bordetella bronchiseptica and toxigenic Pasteurella multocida in atrophic rhinitis of pigs. Res. Vet. Sci. 47, 48–53.

Cherfils, J. and Chabre, M. (2003). Activation of G-protein Galpha subunits by receptors through Galpha-Gbeta and Galpha-Ggamma interactions. Trends Biochem Sci. 28, 13–17.

Chikumi, H., Fukuhara, S., and Gutkind, J.S. (2002). Regulation of G protein-linked guanine nucleotide exchange factors for Rho, PDZ-RhoGEF, and LARG by tyrosine phosphorylation: evidence of a role for focal adhesion kinase. J. Biol. Chem. 277, 12463–12473.

Christensen, J.P. and Bisgaard, M. (2000). Fowl cholera. Rev. Sci Tech. 19, 626–637.

Conklin, B.R., Farfel, Z., Lustig, K.D., Julius, D., and Bourne, H.R. (1993). Substitution of three amino acids switches receptor specificity of Gq alpha to that of Gi alpha. Nature 363, 274–276.

Conklin, B.R., Herzmark, P., Ishida, S., Voyno-Yasenetskaya, T.A., Sun, Y., Farfel, Z., and Bourne, H.R. (1996). Carboxyl-terminal mutations of Gq alpha and Gs alpha that alter the fidelity of receptor activation. Mol. Pharmacol. 50, 885–890.

Dabo, S.M., Taylor, J.D., and Confer, A.W. (2007). Pasteurella multocida and bovine respiratory disease. Anim Health Res Rev. 8, 129–150.

Dattilo, M., Penington, N.J., and Williams, K. (2008). Inhibition of TRPC5 channels by intracellular ATP. Mol. Pharmacol. 73, 42–49.

Deeb, B.J. and DiGiacomo, R.F. (2000). Respiratory diseases of rabbits. Vet. Clin. North Am. Exot. Anim Pract. 3, 465-vii.

Domin, J., Higgins, T., and Rozengurt, E. (1994). Preferential inhibition of platelet-derived growth factor-stimulated DNA synthesis and protein tyrosine phosphorylation by nordihydroguaiaretic acid. J. Biol. Chem. 269, 8260–8267.

Dominick, M.A. and Rimler, R.B. (1986). Turbinate atrophy in gnotobiotic pigs intranasally inoculated with protein toxin isolated from type D Pasteurella multocida. Am. J. Vet. Res. 47, 1532–1536.

Dominick, M.A. and Rimler, R.B. (1988). Turbinate osteoporosis in pigs following intranasal inoculation of purified Pasteurella toxin: histomorphometric and ultrastructural studies. Vet. Pathol. 25, 17–27.

Donald, A.N., Wallace, D.J., McKenzie, S., and Marley, P.D. (2002). Phospholipase C-mediated signalling is not required for histamine-induced catecholamine secretion from bovine chromaffin cells. J. Neurochem. 81, 1116–1129.

Dudet, L.I., Chailler, P., Dubreuil, D., and Martineau-Doize, B. (1996). Pasteurella multocida toxin stimulates mitogenesis and cytoskeleton reorganization in swiss 3T3 fibroblasts. J. Cell. Physiol. 168, 173–182.

Duthie, R.C. (1947). Rhinitis of Swine: I. Chronic atrophic rhinitis and congenital deformity of the skull. Can. J. Comp Med. Vet. Sci 11, 250–259.

Elias, B., Albert, M., Tuboly, S., and Rafai, P. (1992). Interaction between Bordetella bronchiseptica and toxigenic *Pasteurella multocida* on the nasal mucosa of SPF piglets. J. Vet. Med. Sci *54*, 1105–1110.

Elling, F., Pedersen, K.B., Hogh, P., and Foged, N.T. (1988). Characterization of the dermal lesions induced by a purified protein from toxigenic *Pasteurella multocida*. APMIS *96*, 50–55.

Erler, W. (1993). [The localization of toxin in the cells of *Pasteurella multocida*]. Zentralbl. Mikrobiol. *148*, 83–87.

Erler, W., Jacob, B., and Schlegel, J. (1994). The influence of cations on the letality and on the formation of the toxin of *Pasteurella multocida*. Microbiol. Res. *149*, 89–93.

Essler, M., Hermann, K., Amano, M., Kaibuchi, K., Heesemann, J., Weber, P.C., and Aepfelbacher, M. (1998). *Pasteurella multocida* toxin increases endothelial permeability via rho kinase and myosin light chain phosphatase. J. Immunol. *161*, 5640–5646.

Fabbri, A., Gauthier, M., and Boquet, P. (1999). The 5′ region of *cnf1* harbours a translational regulatory mechanism for CNF1 synthesis and encodes the cell-binding domain of the toxin. Mol. Microbiol.*33*, 108–118.

Felix, R., Fleisch, H., and Frandsen, P.L. (1992). Effect of *Pasteurella multocida* toxin on bone resorption *in vitro*. Infect. Immun. *60*, 4984–4988.

Finco-Kent, D.L., Galvin, J.E., Suiter, B.T., and Huether, M.J. (2001). *Pasteurella multocida* toxin type D serological assay as an alternative to the toxin neutralisation lethality test in mice. Biologicals *29*, 7–10.

Flatau, G., Lemichez, E., Gauthier, M., Chardin, P., Paris, S., Fiorentini, C., and Boquet, P. (1997). Toxin-induced activation of the G protein p21 Rho by deamidation of glutamine. Nature *387*, 729–733.

Foged, N.T. (1988). Quantitation and purification of the *Pasteurella multocida* toxin by using monoclonal antibodies. Infect. Immun. *56*, 1901–1906.

Foged, N.T. (1991). Detection of stable epitopes on formaldehyde-detoxified *Pasteurella multocida* toxin by monoclonal antibodies. Vaccine *9*, 817–824.

Foged, N.T., Nielsen, J.P., and Jorsal, S.E. (1989). Protection against progressive atrophic rhinitis by vaccination with *Pasteurella multocida* toxin purified by monoclonal antibodies. Vet. Rec. *125*, 7–11.

Frandsen, P.L., Foged, N.T., Petersen, S.K., and Bording, A. (1991). Characterization of toxin from different strains of *Pasteurella multocida* serotype A and D. Zentralbl. Veterinarmed. B *38*, 345–352.

Fukuhara, S., Chikumi, H., and Gutkind, J.S. (2001). RGS-containing RhoGEFs: the missing link between transforming G proteins and Rho? Oncogene *20*, 1661–1668.

Garrity, G.M., Bell, J.A., and Lilburn, T. (2005). *Pasteurellales* ord. nov. In Bergey's Manual of Systematic Bacteriology, 2nd edn, Vol. 2B The Gammaproteobacteria, D.J.Brenner, N.R.Krieg, and J.T.Staley, eds. (New York: Springer), pp. 850–912.

Ghoshal, N.G., Niyo, Y., and Trenkle, A.H. (1991). Growth hormone concentrations in plasma of healthy pigs and pigs with atrophic rhinitis. Am. J. Vet. Res *52*, 1684–1687.

Gohla, A., Offermanns, S., Wilkie, T.M., and Schultz, G. (1999). Differential involvement of $G\alpha_{12}$ and $G\alpha_{13}$ in receptor-mediated stress fiber formation. J. Biol. Chem. *274*, 17901–17907.

Gwaltney, S.M., Galvin, R.J., Register, K.B., Rimler, R.B., and Ackermann, M.R. (1997). Effects of *Pasteurella multocida* toxin on porcine bone marrow cell differentiation into osteoclasts and osteoblasts. Vet. Pathol. *34*, 421–430.

Gwatkin, R., Plummer, P.J., Byrne, J.L., and Walker, R.V. (1949). Rhinitis of swine: III. Transmission to baby pigs. Can. J. Comp Med. Vet. Sci *13*, 15–28.

Hamm, H.E. (1998). The many faces of G protein signaling. J. Biol. Chem. *273*, 669–672.

Harmey, D., Stenbeck, G., Nobes, C.D., Lax, A.J., and Grigoriadis, A.E. (2004). Regulation of osteoblast differentiation by *Pasteurella multocida* toxin (PMT): A role for Rho GTPase in bone formation. J. Bone Miner. Res. *19*, 661–670.

Harper, M., Boyce, J.D., and Adler, B. (2006). *Pasteurella multocida* pathogenesis: 125 years after Pasteur. FEMS Microbiol. Lett. *265*, 1–10.

Hart, M.J., Jiang, X., Kozasa, T., Roscoe, W., Singer, W.D., Gilman, A.G., Sternweis, P.C., and Bollag, G. (1998). Direct stimulation of the guanine nucleotide exchange activity of p115 RhoGEF by $G\alpha_{13}$. Science *280*, 2112–2114.

Heddleston, K.L., SHUMAN, R.D., and EARL, F.L. (1954). Atrophic rhinitis. IV. Nasal examination for *Pasteurella multocida* in two swine herds affected with atrophic rhinitis. J. Am. Vet. Med. Assoc. *125*, 225–226.

Hennig, B., Orth, J., Aktories, K., and Diener, M. (2008). Anion secretion evoked by *Pasteurella multocida* toxin across rat colon. Eur. J. Pharmacol. *583*, 156–163.

Heximer, S.P., Srinivasa, S.P., Bernstein, L.S., Bernard, J.L., Linder, M.E., Hepler, J.R., and Blumer, K.J. (1999). G protein selectivity is a determinant of RGS2 function. J. Biol. Chem. *274*, 34253–34259.

Higgins, T.E., Murphy, A.C., Staddon, J.M., Lax, A.J., and Rozengurt, E. (1992). *Pasteurella multocida* toxin is a potent inducer of anchorage-independenT-cell growth. Proc. Natl. Acad. Sci. USA *89*, 4240–4244.

Hollinger, S. and Hepler, J.R. (2002). Cellular regulation of RGS proteins: modulators and integrators of G protein signaling. Pharmacol. Rev. *54*, 527–559.

Horiguchi, Y. (2001). *Escherichia coli* cytotoxic necrotizing factors and *Bordetella* dermonecrotic toxin: the dermonecrosis-inducing toxins activating Rho small GTPases. Toxicon *39*, 1619–1627.

Hoskins, I.C., Thomas, L.H., and Lax, A.J. (1997). Nasal infection with *Pasteurella multocida* causes proliferation of bladder epithelium in gnotobiotic pigs. Vet. Rec. *140*, 22.

iDali, C., Foged, N.T., Frandsen, P.L., Nielsen, M.H., and Elling, F. (1991). Ultrastructural localization of the *Pasteurella multocida* toxin in a toxin-producing strain. J. Gen. Microbiol. *137*, 1067–1071.

Ingi, T., Krumins, A.M., Chidiac, P., Brothers, G.M., Chung, S., Snow, B.E., Barnes, C.A., Lanahan, A.A., Siderovski, D.P., Ross, E.M., Gilman, A.G., and Worley, P.F. (1998). Dynamic regulation of RGS2 suggests a novel mechanism in G-protein signaling and neuronal plasticity. J. Neurosci. *18*, 7178–7188.

Jablonski, P.E., Jaworski, M., and Hovde, C.J. (1996). A minimal medium for growth of *Pasteurella multocida*. FEMS Microbiol. Lett. *140*, 165–169.

Jarvinen, L.Z., HogenEsch, H., Suckow, M.A., and Bowersock, T.L. (1998). Induction of protective immunity in rabbits by coadministration of inactivated *Pasteurella multocida* toxin and potassium thiocyanate extract. Infect. Immun. *66*, 3788–3795.

Jarvinen, L.Z., HogenEsch, H., Suckow, M.A., and Bowersock, T.L. (2000). Intranasal vaccination of New Zealand white rabbits against pasteurellosis, using alginate-encapsulated *Pasteurella multocida* toxin and potassium thiocyanate extract. Com Med. *50*, 263–269.

Jutras, I. and Martineau-Doize, B. (1996). Stimulation of osteoclast-like cell formation by *Pasteurella multocida* toxin from hemopoietic progenitor cells in mouse bone marrow cultures. Can. J. Vet. Res *60*, 34–39.

Kamps, A.M.I.E., Kamp, E.M., and Smits, M.A. (1990). Cloning and expression of the dermonecrotic toxin gene of *Pasteurella multocida* ssp. *multocida* in *Escherichia coli*. FEMS Microbiol. Lett. *67*, 187–190.

Kelly-Skupek, M.N. and Irving, H.R. (2006). Pharmacological evidence for activation of phospholipid and small GTP binding protein signalling cascades by Nod factors. Plant Physiol. Biochem. *44*, 132–142.

Kimman, T.G., Löwik, C.W.G.M., Van de Wee-Pals, L.J.A., Thesingh, C.W., Defize, P., Kamp, E.M., and Bijvoet, O.L.M. (1987). Stimulation of bone resorption by inflamed nasal mucosa, dermonecrotic toxin-containing conditioned medium from *Pasteurella multocida*, and purified dermonecrotic toxin from *P. multocida*. Infect. Immun. *55*, 2110–2116.

Kimple, R.J., Kimple, M.E., Betts, L., Sondek, J., and Siderovski, D.P. (2002). Structural determinants for GoLoco-induced inhibition of nucleotide release by Gα subunits. Nature *416*, 878–881.

Kitadokoro, K., Kamitani, S., Miyazawa, M., Hanajima-Ozawa, M., Fukui, A., Miyake, M., and Horiguchi, Y. (2007). Crystal structures reveal a thiol protease-like catalytic triad in the C-terminal region of *Pasteurella multocida* toxin. Proc. Natl. Acad. Sci. USA *104*, 5139–5144.

Klein, N.C. and Cunha, B.A. (1997). *Pasteurella multocida* pneumonia. Semin. Respir. Infect. *12*, 54–56.

Kostenis, E., Martini, L., Ellis, J., Waldhoer, M., Heydorn, A., Rosenkilde, M.M., Norregaard, P.K., Jorgensen, R., Whistler, J.L., and Milligan, G. (2005). A highly conserved glycine within linker I and the extreme C terminus of G protein alpha subunits interact cooperatively in switching G protein-coupled receptor-to-effector specificity. J. Pharmacol. Exp. Ther. *313*, 78–87.

Kozasa, T., Jiang, X., Hart, M.J., Sternweis, P.M., Singer, W.D., Gilman, A.G., Bollag, G., and Sternweis, P.C. (1998). p115 RhoGEF, a GTPase activating protein for Galpha12 and Galpha13. Science *280*, 2109–2111.

Kranenburg, O., Poland, M., van Horck, F.P., Drechsel, D., Hall, A., and Moolenaar, W.H. (1999). Activation of RhoA by lysophosphatidic acid and Galpha12/13

subunits in neuronal cells: induction of neurite retraction. Mol. Biol. Cell *10*, 1851–1857.

Kristinsson, G. (2007). *Pasteurella multocida* infections. Pediatr. Rev. *28*, 472–473.

Lacerda, H.M., Lax, A.J., and Rozengurt, E. (1996). *Pasteurella multocida* toxin, a potent intracellularly acting mitogen, induces p125[FAK] and paxillin tyrosine phosphorylation, actin stress fiber formation, and focal contact assembly in Swiss 3T3 cells. J. Biol. Chem. *271*, 439–445.

Ladds, G., Goddard, A., Hill, C., Thornton, S., and Davey, J. (2007). Differential effects of RGS proteins on G alpha(q) and G alpha(11) activity. Cell Signal. *19*, 103–113.

Lax, A.J. (2005). Bacterial toxins and cancer – a case to answer? Nat. Rev. Microbiol. *3*, 343–349.

Lax, A.J. and Chanter, N. (1990). Cloning of the toxin gene from *Pasteurella multocida* and its role in atrophic rhinitis. J. Gen. Microbiol. *136*, 81–87.

Lax, A.J., Chanter, N., Pullinger, G.D., Higgins, T., Staddon, J.M., and Rozengurt, E. (1990). Sequence analysis of the potent mitogenic toxin of *Pasteurella multocida*. FEBS Lett. *277*, 59–64.

Lax, A.J. and Thomas, W. (2002). How bacteria could cause cancer: one step at a time. Trends Microbiol. *10*, 293–299.

Lemichez, E., Flatau, G., Bruzzone, M., Boquet, P., and Gauthier, M. (1997). Molecular localization of the *Escherichia coli* cytotoxic necrotizing factor CNF1 cell-binding and catalytic domains. Mol. Microbiol. *24*, 1061–1070.

Lemonnier, M., Landraud, L., and Lemichez, E. (2007). Rho GTPase-activating bacterial toxins: from bacterial virulence regulation to eukaryotic cell biology. FEMS Microbiol. Rev. *31*, 515–534.

Liao, C.M., Huang, C., Hsuan, S.L., Chen, Z.W., Lee, W.C., Liu, C.I., Winton, J.R., and Chien, M.S. (2006). Immunogenicity and efficacy of three recombinant subunit *Pasteurella multocida* toxin vaccines against progressive atrophic rhinitis in pigs. Vaccine *24*, 27–35.

Liu, J., Conklin, B.R., Blin, N., Yun, J., and Wess, J. (1995). Identification of a receptor/G-protein contact site critical for signaling specificity and G-protein activation. Proc. Natl. Acad. Sci. USA *92*, 11642–11646.

Liu, S., Carrillo, J.J., Pediani, J.D., and Milligan, G. (2002). Effective information transfer from the alpha 1b-adrenoceptor to Galpha 11 requires both beta/gamma interactions and an aromatic group four amino acids from the C terminus of the G protein. J. Biol. Chem. *277*, 25707–25714.

Lutz, S., Freichel-Blomquist, A., Yang, Y., Rumenapp, U., Jakobs, K.H., Schmidt, M., and Wieland, T. (2005). The guanine nucleotide exchange factor p63RhoGEF, a specific link between Gq/11-coupled receptor signaling and RhoA. J. Biol. Chem. *280*, 11134–11139.

Martin, E.L., Rens-Domiano, S., Schatz, P.J., and Hamm, H.E. (1996). Potent peptide analogues of a G protein receptor-binding region obtained with a combinatorial library. J. Biol. Chem. *271*, 361–366.

Martineau-Doize, B., Caya, I., Gagne, S., Jutras, I., and Dumas, G. (1993). Effects of *Pasteurella multocida* toxin on the osteoclast population of the rat. J. Comp Pathol. *108*, 81–91.

Martineau-Doize, B., Dumas, G., Larochelle, R., Frantz, J.C., and Martineau, G.P. (1991a). Atrophic rhinitis caused by *Pasteurella multocida* type D: morphometric analysis. Can. J. Vet. Res *55*, 224–228.

Martineau-Doize, B., Frantz, J.C., and Martineau, G.P. (1990). Effects of purified *Pasteurella multocida* dermonecrotoxin on cartilage and bone of the nasal ventral conchae of the piglet. Anat. Rec. *228*, 237–246.

Martineau-Doize, B., Menard, J., Girard, C., Frantz, J.C., and Martineau, G.P. (1991b). Effects of purified *Pasteurella multocida* dermonecrotoxin on the nasal ventral turbinates of fattening pigs: histological observations. Can. J. Vet. Res *55*, 377–379.

May, B.J., Zhang, Q., Li, L.L., Paustian, M.L., Whittam, T.S., and Kapur, V. (2001). Complete genomic sequence of *Pasteurella multocida*, Pm70. Proc. Natl. Acad. Sci. USA *98*, 3460–3465.

Meyer, T., Wellner-Kienitz, M.C., Biewald, A., Bender, K., Eickel, A., and Pott, L. (2001). Depletion of phosphatidylinositol 4,5-bisphosphate by activation of phospholipase C-coupled receptors causes slow inhibition but not desensitization of G protein-gated inward rectifier K+ current in atrial myocytes. J. Biol. Chem. *276*, 5650–5658.

Miniats, O.P. and Johnson, J.A. (1980). Experimental atrophic rhinitis in gnotobiotic pigs. Can. J. Comp Med. *44*, 358–365.

Miyazawa, M., Kitadokoro, K., Kamitani, S., Shime, H., and Horiguchi, Y. (2006). Crystallization and preliminary crystallographic studies of the *Pasteurella multocida* toxin catalytic domain. Acta Crystallogr. Sect. F. Struct. Biol. Cryst. Commun *62*, 906–908.

Moynihan, I.W. (1947). Rhinitis of Swine-II: II. An Effort to Transmit Chronic Atrophic Rhinitis of Swine. Can. J. Comp Med. Vet. Sci *11*, 259–260.

Mullan, P.B. and Lax, A.J. (1996). *Pasteurella multocida* toxin is a mitogen for bone cells in primary culture. Infect. Immun. *64*, 959–965.

Mullan, P.B. and Lax, A.J. (1998). *Pasteurella multocida* toxin stimulates bone resorption by osteoclasts via interaction with osteoblasts. Calcif Tissue Int. *63*, 340–345.

Murphy, A.C. and Rozengurt, E. (1992). *Pasteurella multocida* toxin selectively facilitates phosphatidylinositol 4,5-bisphosphate hydrolysis by bombesin, vasopressin, and endothelin. Requirement for a functional G protein. J. Biol. Chem. *267*, 25296–25303.

Nakai, T. and Kume, K. (1987). Purification of three fragments of the dermonecrotic toxin from *Pasteurella multocida*. Res Vet. Sci *42*, 232–237.

Nakai, T., Sawata, A., and Kume, K. (1985). Intracellular locations of dermonecrotic toxins in *Pasteurella multocida* and in *Bordetella bronchispetica*. Am. J. Vet. Res. *46*, 870–874.

Nakai, T., Sawata, A., Tsuji, M., Samejima, Y., and Kume, K. (1984). Purification of dermonecrotic toxin from a sonic extract of *Pasteurella multocida* SP-72 serotype D. Infect. Immun. *46*, 429–434.

Natochin, M., Gasimov, K.G., and Artemyev, N.O. (2002). A GPR-protein interaction surface of Giα:

Implications for the mechanism of GDP-release inhibition. Biochemistry *41*, 258–265.

Nielsen, J.P., Foged, N.T., Sorensen, V., Barfod, K., Bording, A., and Petersen, S.K. (1991). Vaccination against progressive atrophic rhinitis with a recombinant *Pasteurella multocida* toxin derivative. Can. J. Vet. Res *55*, 128–138.

Oberhofer, T.R. (1981). Characteristics and biotypes of *Pasteurella multocida* isolated from humans. J. Clin. Microbiol. *13*, 566–571.

Obreztchikova, M., Elouardighi, H., Ho, M., Wilson, B.A., Gertsberg, Z., and Steinberg, S.F. (2006). Distinct signaling functions for Shc isoforms in the heart. J. Biol. Chem. *281*, 20197–20204.

Ogata, N., Kawaguchi, H., Chung, U.I., Roth, S.I., and Segre, G.V. (2007). Continuous activation of G alpha q in osteoblasts results in osteopenia through impaired osteoblast differentiation. J. Biol. Chem. *282*, 35757–35764.

Okada, Y., Fujiyama, R., Miyamoto, T., and Sato, T. (2001). Saccharin activates cation conductance via inositol 1,4,5-trisphosphate production in a subset of isolated rod taste cells in the frog. Eur. J. Neurosci. *13*, 308–314.

Oppling, V., V, Kusch, M., Rubmann, P., and Cussler, K. (1994). [Detection of antibodies against *Pasteurella multocida* toxin with a cell culture method and an enzyme immunoassay]. ALTEX. *11*, 62–67.

Orth, J.H., Aktories, K., and Kubatzky, K.F. (2007a). Modulation of host cell gene expression through activation of STAT transcription factors by *Pasteurella multocida* toxin. J. Biol. Chem. *282*, 3050–3057.

Orth, J.H., Blöcker, D., and Aktories, K. (2003). His1205 and His 1223 are essential for the activity of the mitogenic *Pasteurella multocida* toxin. Biochemistry *42*, 4971–4977.

Orth, J.H., Lang, S., and Aktories, K. (2004). Action of *Pasteurella multocida* toxin depends on the helical domain of Galphaq. J. Biol. Chem. *279*, 34150–34155.

Orth, J.H., Lang, S., Preuss, I., Milligan, G., and Aktories, K. (2007b). Action of *Pasteurella multocida* toxin on Galpha(q) is persistent and independent of interaction with G-protein-coupled receptors. Cell Signal. *19*, 2174–2182.

Orth, J.H., Lang, S., Taniguchi, M., and Aktories, K. (2005). *Pasteurella multocida* toxin-induced activation of RhoA is mediated via two families of G{alpha} proteins, G{alpha}q and G{alpha}12/13. J. Biol. Chem. *280*, 36701–36707.

Oswald, E., Nougayrede, J.P., Taieb, F., and Sugai, M. (2005). Bacterial toxins that modulate host cell-cycle progression. Curr. Opin. Microbiol. *8*, 83–91.

Pearce, H.G. and Roe, C.K. (1966). Infectious porcine atrophic rhinitis: a review. Can. Vet. J. *7*, 243–251.

Pedersen, K.B. and Barfod, K. (1982). Effect on the incidence of atrophic rhinitis of vaccination of sows with a vaccine containing *Pasteurella multocida* toxin. Nord. Vet. Med. *34*, 293–302.

Pedersen, K.B. and Barford, K. (1981). The aetiological significance of *Bordetella bronchiseptica* and *Pasteurella multocida* in atrophic rhinitis of swine. Nord. Vet. Med. *33*, 513–522.

Pennings, A.M. and Storm, P.K. (1984). A test in Vero cell monolayers for toxin production by strains of *Pasteurella multocida* isolated from pigs suspected of having atropic rhinitis. Vet. Microbiol. *9*, 503–508.

Petersen, S.K. (1990). The complete nucleotide sequence of the *Pasteurella multocida* toxin gene and evidence for a transcriptional repressor, TxaR. Mol. Microbiol. *4*, 821–830.

Petersen, S.K. and Foged, N.T. (1989). Cloning and expression of the *Pasteurella multocida* toxin gene, *toxA*, in *Escherichia coli*. Infect. Immun. *57*, 3907–3913.

Petersen, S.K., Foged, N.T., Bording, A., Nielsen, J.P., Riemann, H.K., and Frandsen, P.L. (1991). Recombinant derivatives of *Pasteurella multocida* toxin: candidates for a vaccine against progressive atrophic rhinitis. Infect. Immun. *59*, 1387–1393.

Pettit, R.K., Ackermann, M.R., and Rimler, R.B. (1993). Receptor-mediated binding of *Pasteurella multocida* dermonecrotic toxin to canine osteosarcoma and monkey kidney (Vero) cells. Labor. Invest. *69*, 94–100.

Polakis, P. (2007). The many ways of Wnt in cancer. Curr. Opin. Genet. Dev. *17*, 45–51.

Pullinger, G.D., Bevir, T., and Lax, A.J. (2004). The *Pasteurella multocida* toxin is encoded within a lysogenic bacteriophage. Mol. Microbiol. *51*, 255–269.

Pullinger, G.D. and Lax, A.J. (2007). Histidine residues at the active site of the *Pasteurella multocida* toxin. Open Biochem. J. *1*, 7–11.

Pullinger, G.D., Sowdhamini, R., and Lax, A.J. (2001). Localization of functional domains of the mitogenic toxin of *Pasteurella multocida*. Infect. Immun. *69*, 7839–7850.

Rajeev, S., Nair, R.V., Kania, S.A., and Bemis, D.A. (2003). Expression of a truncated *Pasteurella multocida* toxin antigen in *Bordetella bronchiseptica*. Vet. Microbiol. *94*, 313–323.

Register, K.B., Sacco, R.E., and Brockmeier, S.L. (2007). Immune response in mice and swine to DNA vaccines derived from the *Pasteurella multocida* toxin gene. Vaccine *25*, 6118–6128.

Reinert, D.J., Jank, T., Aktories, K., and Schulz, G.E. (2005). Structural basis for the function of *Clostridium difficile* toxin B. J. Mol. Biol. *351*, 973–981.

Reissbrodt, R., Erler, W., and Winkelmann, G. (1994). Iron supply of *Pasteurella multocida* and *Pasteurella haemolytica*. J. Basic Microbiol. *34*, 61–63.

Riising, H.J., van, E.P., and Witvliet, M. (2002). Protection of piglets against atrophic rhinitis by vaccinating the sow with a vaccine against *Pasteurella multocida* and *Bordetella bronchiseptica*. Vet. Rec. *150*, 569–571.

Roos, W.P., Batista, L.F., Naumann, S.C., Wick, W., Weller, M., Menck, C.F., and Kaina, B. (2007). Apoptosis in malignant glioma cells triggered by the temozolomide-induced DNA lesion O6-methylguanine. Oncogene *26*, 186–197.

Rozengurt, E. (2007). Mitogenic signaling pathways induced by G protein-coupled receptors. J. Cell Physiol. *213*, 589–602.

Rozengurt, E., Higgins, T., Chanter, N., Lax, A.J., and Staddon, J.M. (1990). *Pasteurella multocida* toxin:

potent mitogen for cultured fibroblasts. Proc. Natl. Acad. Sci. USA *87*, 123–127.

Rutter, J.M. (1983). Virulence of *Pasteurella multocida* in atrophic rhinitis of gnotobiotic pigs infected with *Bordetella bronchiseptica*. Res. Vet. Sci. *34*, 287–295.

Rutter, J.M. and Luther, P.D. (1984). Cell culture assay for toxigenic *Pasteurella multocida* from atrophic rhinitis of pigs. Vet. Rec. *114*, 393–396.

Ryzhov, S., Goldstein, A.E., Biaggioni, I., and Feoktistov, I. (2006). Cross-talk between G(s)- and G(q)-coupled pathways in regulation of interleukin-4 by A(2B) adenosine receptors in human masT-cells. Mol. Pharmacol. *70*, 727–735.

Sabri, A., Pak, E., Alcott, S.A., Wilson, B.A., and Steinberg, S.F. (2000). Coupling function of endogenous alpha(1)- and beta-adrenergic receptors in mouse cardiomyocytes. Circ. Res *86*, 1047–1053.

Sabri, A., Wilson, B.A., and Steinberg, S.F. (2002). Dual actions of the Galpha(q) agonist Pasteurella multocida toxin to promote cardiomyocyte hypertrophy and enhance apoptosis susceptibility. Circ. Res. *90*, 850–857.

Satchell, K.J. (2007). MARTX, multifunctional auto-processing repeats-in-toxin toxins. Infect. Immun. *75*, 5079–5084.

Schimmel, D., Erler, W., Hanel, I., I, and Muller, W. (1994). [Detection of *Pasteurella multocida* toxin – a comparison of *in vitro* and *in vivo* methods]. ALTEX *11*, 59–61.

Schmidt, G., Sehr, P., Wilm, M., Selzer, J., Mann, M., and Aktories, K. (1997). Gln63 of Rho is deamidated by *Escherichia coli* cytotoxic necrotizing factor 1. Nature *387*, 725–729.

Schmidt, G., Selzer, J., Lerm, M., and Aktories, K. (1998). The Rho-deamidating cytotoxic-necrotizing factor CNF1 from *Escherichia coli* possesses transglutaminase activity – cysteine-866 and histidine-881 are essential for enzyme activity. J. Biol. Chem. *273*, 13669–13674.

Seo, B., Choy, E.W., Maudsley, W.E., Miller, W.E., Wilson, B.A., and Luttrell, L.M. (2000). *Pasteurella multocida* toxin stimulates mitogen-activated protein kinase via G$_{q/11}$-dependent transactivation of the epidermal growth factor receptor. J. Biol. Chem. *275*, 2239–2245.

Skiba, N.P., Yang, C.-S., Huang, T., Bae, H., and Hamm, H.E. (1999). The α-helical domain of Gα$_t$ determines specific interaction with regulator of G protein signaling 9. J. Biol. Chem. *274*, 8770–8778.

Sleight, S., Wilson, B.A., Heimark, D.B., and Larner, J. (2002). G(q/11) is involved in insulin-stimulated inositol phosphoglycan putative mediator generation in rat liver membranes: co-localization of G(q/11) with the insulin receptor in membrane vesicles. Biochem. Biophys. Res. Commun. *295*, 561–569.

Smyth, M.G., Pickersgill, R.W., and Lax, A.J. (1995). The potent mitogen Pasteurella multocida toxin is highly resistant to proteolysis but becomes susceptible at lysosomal pH. FEBS Lett. *360*, 62–66.

Smyth, M.G., Sumner, I.G., and Lax, A.J. (1999). Reduced pH causes structural changes in the potent mitogenic toxin of *Pasteurella multocida*. FEMS Microbiol. Lett. *180*, 15–20.

Spencker, F.B., Handrick, W., Buhrdel, P., and Braun, W. (1979). [Meningitis in an infant caused by *Pasteurella multocida*. Case report and review of other manifestations (author's transl)]. Infection 7, 183–186.

Staddon, J.M., Barker, C.J., Murphy, A.C., Chanter, N., Lax, A.J., Michell, R.H., and Rozengurt, E. (1991a). *Pasteurella multocida* toxin, a potent mitogen, increases inositol 1,4,5-triphosphate and mobilizes Ca^{2+} in swiss 3T3 cells. J. Biol. Chem. 266, 4840–4847.

Staddon, J.M., Bouzyk, M.M., and Rozengurt, E. (1991b). A novel approach to detect toxin-catalyzed ADP-ribosylation in intact cells: its use to study the action of *Pasteurella multocida* toxin. J. Cell Biol. 115,No.4, 949–958.

Sterner-Kock, A., Lanske, B., Uberschar, S., and Atkinson, M.J. (1995). Effects of the *Pasteurella multocida* toxin on osteoblastic cells *in vitro*. Vet. Pathol. 32, 274–279.

Suckow, M.A., Bowersock, T.L., Nielsen, K., Chrisp, C.E., Frandsen, P.L., and Janovitz, E.B. (1995). Protective immunity to *Pasteurella multocida* heat-labile toxin by intranasal immunization in rabbits. Lab. Anim. Sci. 45, 526–532.

Suzuki, N., Nakamura, S., Mano, H., and Kozasa, T. (2003). Galpha 12 activates Rho GTPase through tyrosine-phosphorylated leukemia-associated RhoGEF. Proc. Natl. Acad. Sci. USA 100, 733–738.

Takada-Iwao, A., Uto, T., Mukai, T., Okada, M., Futo, S., and Shibata, I. (2007). Evaluation of an indirect enzyme-linked immunosorbent assay (ELISA) using recombinant toxin for detection of antibodies against *Pasteurella multocida* toxin. J. Vet. Med. Sci 69, 581–586.

Takasaki, J., Saito, T., Taniguchi, M., Kawasaki, T., Moritani, Y., Hayashi, K., and Kobori, M. (2004). A novel Galphaq/11-selective inhibitor. J. Biol. Chem. 279, 47438–47445.

Takesono, A., Cismowski, M.J., Ribas, C., Bernard, M., Chung, P., Hazard III, S., Duzic, E., and Lanier, S.M. (1999). Receptor-independent activators of heterotrimeric G-protein signaling pathways. J. Biol. Chem. 274, 33202–33205.

Thomas, W., Pullinger, G.D., Lax, A.J., and Rozengurt, E. (2001). *Escherichia coli* cytotoxic necrotizing factor and *Pasteurella multocida* toxin induce focal adhesion kinase autophosphorylation and Src association. Infect. Immun. 69, 5931–5935.

Thurston, J.R., Rimler, R.B., Ackermann, M.R., Cheville, N.F., and Sacks, J.M. (1991). Immunity induced in rats vaccinated with toxoid prepared from heat-labile toxin produced by *Pasteurella multocida* serogroup D. Vet. Microbiol. 27, 169–174.

To, H., Someno, S., and Nagai, S. (2005). Development of a genetically modified nontoxigenic *Pasteurella multocida* toxin as a candidate for use in vaccines against progressive atrophic rhinitis in pigs. Am. J. Vet. Res 66, 113–118.

Tonello, F., Schiavo, G., and Montecucco, C. (1997). Metal substitution of tetanus neurotoxin. Biochem. J. 322, 507–510.

Townsend, K.M., Hanh, T.X., O'Boyle, D., Wilkie, I., Phan, T.T., Wijewardana, T.G., Trung, N.T., and Frost, A.J. (2000). PCR detection and analysis of *Pasteurella multocida* from the tonsils of slaughtered pigs in Vietnam. Vet. Microbiol. 72, 69–78.

Umemori, H., Inoue, T., Kume, S., Sekiyama, N., Nagao, M., Itoh, H., Nakanishi, S., Mikoshiba, K., and Yamamoto-Osaki, T. (1997). Activation of the G protein Gq/11 through tyrosinie phosphorylation of the α subunit. Science 276, 1878–1881.

van Diemen, P.M., de Jong, M.F., de Vries, R.G., van der, H.P., and Schrama, J.W. (1994a). Intranasal administration of *Pasteurella multocida* toxin in a challenge-exposure model used to induce subclinical signs of atrophic rhinitis in pigs. Am. J. Vet. Res 55, 49–54.

van Diemen, P.M., de Vries, R.G., and Parmentier, H.K. (1994b). Immune responses of piglets to *Pasteurella multocida* toxin and toxoid. Vet. Immunol. Immunopathol. 41, 307–321.

van Diemen, P.M., de Vries, R.G., and Parmentier, H.K. (1996). Effect of *Pasteurella multocida* toxin on *in vivo* immune responses in piglets. Vet. Q. 18, 141–146.

van Diemen, P.M., Henken, A.M., Schrama, J.W., Brandsma, H.A., and Verstegen, M.W. (1995). Effects of atrophic rhinitis induced by *Pasteurella multocida* toxin on heat production and activity of pigs kept under different climatic conditions. J. Anim. Sci. 73, 1658–1665.

van Zeijl, L., Ponsioen, B., Giepmans, B.N., Ariaens, A., Postma, F.R., Varnai, P., Balla, T., Divecha, N., Jalink, K., and Moolenaar, W.H. (2007). Regulation of connexin43 gap junctional communication by phosphatidilinositol 4,5-bisphosphate. J. Cell Biol. 177, 881–891.

Vasfi, M.M., Harel, J., and Mittal, K.R. (1997). Identification by monoclonal antibodies of serotype D strains of *Pasteurella multocida* representing various geographic origins and host species. J. Med. Microbiol. 46, 603–610.

Vogt, S., Grosse, R., Schultz, G., and Offermanns, S. (2003). Receptor-dependent RhoA activation in G12/G13-deficient cells. J. Biol. Chem. 278, 28743–28749.

Ward, P.N., Higgins, T.E., Murphy, A.C., Mullan, P.B., Rozengurt, E., and Lax, A.J. (1994). Mutation of a putative ADP-ribosylation motif in the *Pasteurella multocida* toxin does not affect mitogenic activity. FEBS Lett. 342, 81–84.

Ward, P.N., Miles, A.J., Sumner, I.G., Thomas, L.H., and Lax, A.J. (1998). Activity of the mitogenic *Pasteurella multocida* toxin requires an essential C-terminal residue. Infect. Immun. 66, 5636–5642.

Weber, D.J., Wolfson, J.S., Swartz, M.N., and Hooper, D.C. (1984). *Pasteurella multocida* infections. Report of 34 cases and review of the literature. Medicine (Baltimore) 63, 133–154.

Wieland, T. and Chen, C.-K. (1999). Regulators of G-protein signalling: a novel protein family involved in timely deactivation and desensitization of signalling via heterotrimeric G proteins. Naunyn-Schmiedeberg's Arch. Pharmacol. 360, 14–26.

Williams, P.P., Hall, M.R., and Rimler, R.B. (1990). Host response to *Pasteurella multocida* turbinate atrophy toxin in swine. Can. J. Vet. Res 54, 157–163.

Wilson, B.A., Aminova, L.R., Ponferrada, V.G., and Ho, M. (2000). Differential modulation and subsequent blockade of mitogenic signaling and cell cycle progression by *Pasteurella multocida* toxin. Infect. Immun. *68*, 4531–4538.

Wilson, B.A. and Ho, M. (2004). *Pasteurella multocida* toxin as a tool for studying G(q) signal transduction. Rev. Physiol Biochem. Pharmacol. *152*, 93–109.

Wilson, B.A., Ponferrada, V.G., Vallance, J.E., and Ho, M.F. (1999). Localization of the intracellular activity domain of *Pasteurella multocida* toxin to the N terminus. Infection and Immunity *67*, 80–87.

Wilson, B.A., Zhu, X., Ho, M., and Lu, L. (1997). *Pasteurella multocida* toxin activates the inositol tri- phosphate signaling pathway in *Xenopus* oocytes via $G_q\alpha$-coupled phospholipase C-β1. J. Biol. Chem. *272*, 1268–1275.

Woolfrey, B.F., Quall, C.O., and Lally, R.T. (1985). *Pasteurella multocida* in an infected tiger bite. Arch. Pathol. Lab Med. *109*, 744–746.

Zywietz, A., Gohla, A., Schmelz, M., Schultz, G., and Offermanns, S. (2001). Pleiotropic effects of *Pasteurella multocida* toxin are mediated by Gq-dependent and -independent mechanisms. Involvement of Gq but not G11. J. Biol. Chem. *276*, 3840–3845.

The Multifunctional Autoprocessing RTX Toxins of Vibrios

Karla J. F. Satchell and Brett Geissler

Abstract

Multifunctional autoprocessing RTX toxins are a unique family of secreted protein toxins, predominantly produced by the *Vibrio* sp. The best-characterized of these toxins is produced by *V. cholerae*, for which aspects of the regulation, secretion, and mechanism of toxicity have been defined. Within the eukaryotic cell, this toxin has three distinct biochemical activities resulting in autoprocessing, covalent cross-linking of actin, and inactivation of Rho-family GTPases, ultimately resulting in destruction of the actin cytoskeleton. Related toxins produced by *V. vulnificus* and *V. anguillarum* have also been characterized and found to have some similar, but also distinct mechanisms of action. For each *Vibrio* sp., the toxins have been linked to virulence and it is possible that these toxins function to assist the bacterium to evade host immune defences in some cases.

Introduction

The Gram-negative bacterium *Vibrio cholerae* causes the diarrhoeal disease cholera. Cholera is typically found throughout developing countries, especially in Africa, and is often associated with conditions of poor sanitation where *V. cholerae* is present in contaminated water and shellfish. The primary virulence mechanism of *V. cholerae* is the production of Cholera Toxin (CT), an A-B subunit toxin that elicits secretion of excessive fluid from enterocytes by increasing intracellular cAMP levels via ADP-ribosylation of the heterotrimeric G protein α subunit. The overwhelming secretory action induced by CT results in the massive diarrhoea associated with severe cholera disease (Rodighiero and Lencer, 2003).

Based on the importance of CT as a virulence factor, it was proposed that *V. cholerae* mutants deleted of the *ctxA* and *ctxB* genes that encode CT would be effective live attenuated vaccines for protection against cholera disease. In an early test of this hypothesis, the strain JBK70, a derivative of O1 El Tor N16961 deleted of *ctx* genes, was given to human volunteers. The volunteers succumbed to mild diarrhoea, fever, cramps, emesis, and anorexia – symptoms collectively called 'reactogenicity' (Levine *et al.*, 1988). This clinical observation demonstrates that virulence factors other than CT are produced by *V. cholerae* and that these factors are linked to moderate diarrhoea in humans.

Similarly, naturally occurring *V. cholerae* O1 and non-O1 strains that have not acquired the *ctx* genes have been implicated in clinical cases of septicaemia, inflammatory enteritis, and extraintestinal infections (Blake *et al.*, 1980; Mandell *et al.*, 1995; Mathan *et al.*, 1995; Shuangshoti and Reinprayoon, 1995; Ninin *et al.*, 2000). Thus, *V. cholerae*-associated disease in the absence of CT can be distinct in pathology compared to epidemic cholera and can vary between strains. As a whole, these observations demonstrate that there must be additional virulence factors, other than CT, that contribute to *V. cholerae* pathogenesis. Secreted factors that have been identified as potentially contributing to pathogenesis of *V. cholerae* include the zinc-metalloprotease haemagglutinin/protease, a pore-forming haemolysin/cytotoxin, two phage coat proteins (Zot and Ace) that may have dual functions as elicitors of fluid secretion, and effector proteins of type III and type VI secretion systems (reviewed in Fullner, 2002; Boardman and Satchell, in press).

This review will focus on a single secreted virulence determinant first described in 1999 (Lin et al., 1999), the Multifunctional-Autoprocessing V. cholerae RTX toxin, also known as MARTX$_{Vc}$. This toxin contributes to virulence by helping the bacteria avoid clearance from the small intestine and is thus proposed to function during the early stages of host colonization (Fullner et al., 2002; Haines et al., 2005; Olivier et al., 2007a,b). Its broad distribution among environment isolates suggests MARTX$_{Vc}$ may also have a function in extraintestinal survival (Cordero et al., 2007; Rahman et al., 2008). In addition, virulence-associated toxins produced by other Vibrio species that are similar to MARTX$_{Vc}$ will be discussed.

The *rtxA* gene

The MARTX$_{Vc}$ toxin is encoded on the V. cholerae large chromosome by the *rtxA* gene, the largest gene of the V. cholerae genome. Based on whole genome sequence analysis of El Tor O1 strain N16961, the predicted polypeptide has two potential start sites resulting in full-length proteins either 4545 or 4558 amino acids (aa) (Lin et al., 1999; Heidelberg et al., 2000). The actual length of the peptide chain has not been determined experimentally but the length of 4545 aa is favoured because of the presence of a consensus Shine–Dalgarno sequence immediately upstream of the second start.

The presence of the gene was first recognized during genome annotation as a large open reading frame tightly linked to the integrated CTXΦ prophage that encodes CT (Lin et al., 1999). Insertional inactivation of this gene eliminates a tissue culture cell rounding activity associated with V. cholerae El Tor and O139 strains, an activity that had been previously ascribed to the secretory action of CT (Theriot and Gardel, 1995). Indeed, the study by Lin et al. (1999) reveals that deletion of the CT-encoding *ctx* genes in some El Tor and O139 vaccine strains also removes 1907 base pairs (bp) of the *rtxA* gene accounting for the cell rounding defect of these mutants.

Cell rounding is not associated with classical O1 strains since these isolates have a naturally occurring 7,869 bp deletion that removes a large portion of the *rtxA* gene locus (Lin et al., 1999).

Despite its absence in classical O1 strains, ongoing molecular epidemiology analyses show that almost all other V. cholerae isolates do carry the *rtxA* gene (Chow et al., 2001; Dalsgaard et al., 2001; Faruque et al., 2004; Cordero et al., 2007; Rahman et al., 2008), and many of these isolates have cell rounding activity (Chow et al., 2001; Cordero et al., 2007). These *rtx*+ strains include a broad variety of El Tor O1 strains as well as isolates from other O-antigen groups and incorporate both clinical and environmental isolates. The strong conservation in both environmental and clinical isolates suggests that maintenance of this gene is important for environmental fitness as well as pathogenesis (Cordero et al., 2007; Rahman et al., 2008).

A more detailed analysis examining 24 strains for cell rounding directly associated with actin cross-linking (see section below) shows that most strains confer the cross-linking activity. However, there are strains, although rare, that are positive for the *rtxA* gene by PCR that lack actin cross-linking activity (Cordero et al., 2007). This finding is consistent with genome sequence analysis showing that the *rtxA* gene from O135 strain RC385 does not carry the domain necessary for actin cross-linking, but does carry an adenylate cyclase domain revealing potential heterogeneity of *rtxA* genes among different V. cholerae isolates (Heidelberg, 2007; Satchell, 2007).

Transcriptional regulation of the *rtx* locus

The *rtxA* gene is located within a larger *rtx* locus that consists of six genes organized into two divergently transcribed operons (Boardman and Satchell, 2004). The gene *rtxA* (VC1451) is the third gene in an operon with two other genes, *rtxH* (VC1449) and *rtxC* (VC1450). *rtxH* is a conserved hypothetical gene of unknown function, while the *rtxC* gene encodes a putative acyl transferase. In other RTX toxins such as E. coli haemolysin and Bordetella pertussis adenylate cyclase, the acyl transferase is required to add long chain fatty acids to the toxin and this modification is required for activity of these toxins (Ludwig and Goebel, 2006). A mutation has not been constructed in the *V. cholerae rtxC* gene so it is unknown if post-translational acylation of MARTX$_{Vc}$ is essential for its activity. The divergently transcribed operon comprises three genes *rtxBDE* (VC1448, VC1447, and VC1446, respectively) that encode components of the toxin secretion system (Fig. 6.1).

The two divergent operons that constitute the *rtx* locus are regulated by growth phase with optimal expression in Luria broth at an optical density of 0.4–1.0 (Boardman *et al.*, 2007). Mapping of the start site and transcript studies show that the *rtxH, rtxC,* and *rtxA* genes are co-ordinately expressed on a single mRNA and are produced only during logarithmic phase growth (Boardman, 2007; Boardman *et al.*, 2007). These genes are categorized as highly expressed in the intestinal lumen with relative rankings of 564, 326, and 395, respectively (Bina *et al.*, 2003). Microarray studies demonstrate that expression is decreased in minimal medium by loss of the small RNA *ryhB* involved in iron regulation, but this difference is not detected when bacteria are grown in Luria broth (Davis *et al.*, 2005; Mey *et al.*, 2005). By contrast, expression of *rtxH, rtxC,* and *rtxA* are each increased in the absence of the small RNA chaperone Hfq (Ding *et al.*, 2004). A Q-PCR study of the same mutant reveals that Hfq modulates the expression of the *rtxA* operon during log phase growth but does not regulate expression during stationary phase growth (Boardman *et al.*, 2007). Hence, the over-all expression of *rtxHCA* may be regulated by a small RNA in conjunction with Hfq during log phase, but the mechanism for *rtxHCA* operon transcriptional suppression during stationary phase is still unknown.

Similar to the *rtxHCA* operon, the divergently transcribed *rtxBDE* operon is coordinately expressed as a single operon and is likewise coordinately regulated by repression during stationary phase, rather than activation during logarithmic phase. This negative regulation is not linked to the well-characterized quorum sensing system of *V. cholerae* but is possibly linked to stress responses via the stringent response alarmone ppGpp, RpoS or RpoN (Yildiz *et al.*, 2004; Boardman *et al.*, 2007).

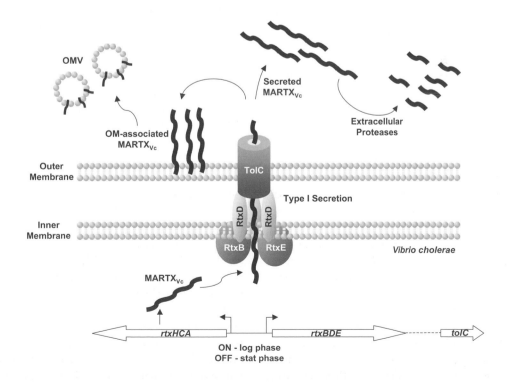

Figure 6.1 Regulation of the *rtx* locus and secretion of the MARTX$_{Vc}$ toxin. See text for details.

Type I secretion of MARTX$_{Vc}$

A characteristic shared by all members of the RTX family of exoproteins is secretion by a type I secretion system (T1SS). These bacterial membrane complexes generally consist of three protein components: a homodimer of an inner membrane transport ATPase, a trimer of a transmembrane fusion protein, and an outer membrane porin that is either a specialized porin or the common porin TolC (Ludwig and Goebel, 2006). The *rtxBDE* operon, described above, encodes the ATPase and transmembrane fusion components of the MARTX$_{Vc}$ T1SS, while the outer membrane porin is provided by the unlinked *tolC* gene (VC2436) (Bina and Mekalanos, 2001; Boardman and Satchell, 2004). Unlike other Type I secretion operons (Ludwig and Goebel, 2006), the *rtxBDE* operon includes genes for two transport ATPases (*rtxB* and *rtxE*) in addition to a gene for a transmembrane linker protein (*rtxD*) (Fig. 6.1). Mutagenesis studies reveal that extracellular export of MARTX$_{Vc}$ requires intact mononucleoside binding sites for both RtxB and RtxE, indicating ATP hydrolysis by both proteins is necessary for toxin export. This observation suggests the T1SS for MARTX$_{Vc}$ involves a heterodimeric ATPase, rather than a homodimer typical of T1SS. Both RtxD and TolC are also essential for toxin secretion (Boardman and Satchell, 2004). Indeed, similar to HlyD, the transmembrane fusion protein for export of *E. coli* haemolysis, the last four amino acids of RtxD are absolutely essential and probably play a similar role in contacting the outer membrane porin TolC (Schulein *et al.*, 1994; Boardman and Satchell, 2004).

Following secretion, the toxin is not a stable extracellular protein (Fig. 6.1). Within supernatant fluids, the toxin is rapidly inactivated by secreted proteases such that toxin activity can only be detected in supernatant fluids of protease-deficient mutants (Boardman *et al.*, 2007). Some secreted toxin seems to be associated with the bacterial outer membrane and/or outer membrane vesicles where it is protected from proteolysis (Boardman *et al.*, 2007). Whether newly expressed toxin or membrane-associated toxin is the primary active moiety against target cells remains to be tested in detail.

Repeat regions of MARTX$_{Vc}$: a structural element for cell binding and translocation?

After secretion from the bacterium, the next phase of cellular intoxication is binding of the toxin to the eukaryotic cell surface and translocation across the plasma membrane. For many toxin family groups, functions of receptor binding and translocation are linked to repetitive amino acid sequences.

Analysis of the primary sequence of MARTX$_{Vc}$ shows that both the N- and C-termini of the protein are composed predominantly of 18–20 aa glycine-rich repeats with a shared motif of G-7x-GxxN (Fig. 6.2). At the N-terminus between residues 73 and 1357 are found 50 repeats of two different consensus sequences. The A-repeats are 20 residue repeats featuring the central motif. These repeats occur in two groupings separated by 285 aa. Immediately downstream of the second group of A repeats are the 19 aa B repeats that share the central G-7x-GxxN motif but a distinct overall consensus sequence. There are 34 consecutive copies of this repeat (Lin *et al.*, 1999; Satchell, 2007).

At the C-terminus is found a third set of novel repeats. These are 18 aa repeats of which the first 9 aa are the same sequence as the nonapeptide calcium binding repeats that typify RTX toxins. MARTX$_{Vc}$ however carries a repeat distinct from other RTX toxins as it is 18 aa in length and shares the central G-7x-GxxN motif with the A and B repeats. Thus, these have been denoted as MARTX$_{Vc}$ C-repeats to distinguish this alternative primary structure (Lin *et al.*, 1999; Satchell, 2007). Overall, these three novel repeat structures are proposed to bind the eukaryotic cell surface and translocate the central ~1700 aa of the toxin to the cytosol, since this central region is composed of activity domains that function upon entry into the eukaryotic cell.

Cell rounding due to covalent cross-linking of cellular actin by MARTX$_{Vc}$

The action of the MARTX$_{Vc}$ toxin on eukaryotic cells results in rounding of tissue culture cells and loss of the integrity of the tight junctions between cells (Lin *et al.*, 1999; Fullner *et al.*, 2001). This rounding is due to actin depolymerization

MARTX$_{Vc}$

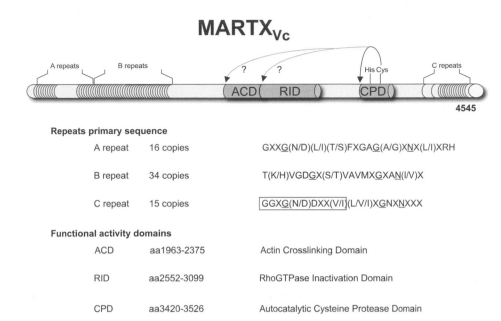

Figure 6.2 Structural features of MARTX$_{Vc}$. Repeat regions at the N and C termini are identified with the consensus sequences listed. The typical RTX GD-rich nonapeptide repeat is boxed. Location of domains described in text is shown. The known site of processing by the CPD is shown with a solid line and additional putative processing sites are indicated with dashed lines.

associated with a novel biochemical reaction that covalently cross-links the cellular actin into dimers, trimers, and larger oligomers (Fullner and Mekalanos, 2000). The cross-linking of monomers and oligomers occurs repeatedly, forming chains that can be resolved by SDS-PAGE up to at least 15-mers until ultimately all of the actin within the cell is cross-linked into an inactive form (Fig. 6.3) (Fullner and Mekalanos, 2000; Cordero *et al.*, 2006; Kudryashov *et al.*, 2008).

The cross-link is introduced in actin by an enzyme embedded within the MARTX$_{Vc}$ toxin at aa 1963–2375 (Sheahan *et al.*, 2004). This 'actin cross-linking domain (ACD)' can perform actin cross-linking in the absence of the remainder of the toxin when it is either transiently expressed within epithelial cells or fibroblasts as a fusion to green fluorescent protein (GFP) or when the ACD is delivered to cells by anthrax toxin protective antigen as a fusion to anthrax toxin lethal factor (Sheahan *et al.*, 2004; Cordero *et al.*, 2006). To demonstrate direct action of the enzyme, purified ACD has been shown to cross-link actin *in vitro* in a reaction composed of only actin, ATP, and Mg^{2+}. The reaction is cata-

lytic and progresses to completion in less than 10 minutes (Cordero *et al.*, 2006).

The ACD within MARTX$_{Vc}$ is 59% identical to the C-terminal portion of the *V. cholerae* protein VC1416, also known as VgrG-1. Transient expression of the C-terminal portion of VgrG-1 as a GFP fusion also causes cell rounding and actin cross-linking comparable to that generated with expression of ACD (Sheahan *et al.*, 2004). Purified VgrG-1 is also able to cross-link actin *in vitro* (Pukatzki *et al.*, 2007). Additional work by Pukatzki *et al.* (2006) shows that VC1416 is the first characterized effector of an IcmF-associated homologous protein cluster (IAHP), also known as type VI secretion, and is associated with rounding of J774 macrophages by some strains of *V. cholerae* (Pukatzki *et al.*, 2006; Pukatzki *et al.*, 2007). Thus, *V. cholerae* produces two proteins capable of inducing covalent cross-linking of actin, although they are delivered to cells by two distinct transport processes.

Polymerization of monomeric G-actin into F-actin filaments is dependent upon hydrolysis of ATP and occurs in the presence of Mg^{2+} (Korn *et al.*, 1987). Hence, the requirement of Mg^{2+}

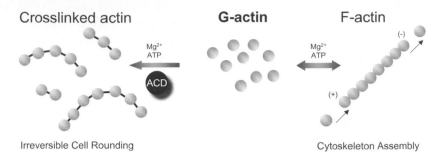

Figure 6.3 Actin occurs in two states in the cell, monomeric G-actin and polymerized F-actin. Assembly of actin is initiated by nucleation of actin followed by the rapid addition of monomers to the barbed (+) end. Elongation of the filaments depends on Mg^{2+} and ATP. The ACD utilizes G-actin as a substrate and the cross-linking occurs simultaneously with ATP hydrolysis and requires Mg^{2+}.

and ATP could indicate that actin needs to be polymerized prior to cross-linking. Contrary to this concept, it has been shown that monomeric G-actin, not polymerized F-actin, is the substrate for the cross-linking reaction both *in vitro* and *in vivo* (Fullner and Mekalanos, 2000; Cordero *et al.*, 2006; Kudryashov *et al.*, 2008). Moreover, it has been shown that the requirement for ATP and Mg^{2+} during ACD-mediated actin cross-linking is not to drive actin polymerization, but to energize the cross-linking reaction itself (Cordero *et al.*, 2006). Subsequent studies confirm that ACD itself is an ATPase that is stimulated to hydrolyze ATP in the presence of actin (Kudryashov *et al.*, 2008).

It is well-recognized that in the living cell, G-actin does not exist in a free form and is typically bound to one of the several actin binding proteins (Abs) (Pollard and Boris, 2003). Remarkably, the most abundant G-actin binding proteins, profilin, thymosin-β4, and gelsolin do not have an inhibitory effect on actin cross-linking and cofilin only partially inhibits cross-linking (Kudryashov *et al.*, 2008). Among the proteins tested, only DNaseI completely blocks the formation of actin oligomers, suggesting the cross-link reaction involves an interaction with the DNAseI binding loop (Kudryashov *et al.*, 2008). Thereby, cross-linking of actin by MARTX$_{Vc}$ likely progresses *in vivo* without interference by most ABPs, accounting for the ability of the enzyme to cross-link nearly every molecule of actin (Fullner and Mekalanos, 2000).

As G-actin is the substrate for cross-linking, higher order oligomers could arise by continued addition of G-actin monomers or by the joining of oligomers. In fact, the efficiency of cross-linking diminishes with the increasing size of oligomers, suggesting that formation of higher order oligomers occurs predominantly by the addition of monomers to linked oligomers, rather than the joining of oligomers (Kudryashov *et al.*, 2008).

In total, actin cross-linking by the MARTX$_{Vc}$ ACD or VgrG-1 ACD is proposed to drive rounding of cells by depleting cells of G-actin. This model is consistent with the sequestering model proposed for the action of fungal toxins latrunculin A and B to drive rounding of cells (Morton *et al.*, 2000). In the absence of G-actin, there would be a net depolymerization of actin filaments, ultimately resulting in the complete disassembly of the actin cytoskeleton. Overall, actin cross-linking seems to be an efficient mechanism for the permanent depolymerization of the actin cytoskeleton and destruction of the cell structure.

Cell rounding due to inactivation of RhoGTPases by MARTX$_{Vc}$

Deletion of the coding region for the ACD within the *rtxA* gene (RTXΔACD) on the *V. cholerae* chromosome eliminates actin cross-linking activity, further confirming that the ACD carries the cross-linking activity (Sheahan *et al.*, 2004). Unexpectedly, cells treated with *V. cholerae* expressing RTXΔACD still round, although cell rounding is somewhat slower, requiring 3–4 hours for complete rounding compared to 75–90 minutes when actin cross-linking occurs. The RTXΔACD rounded cells are phenotypically distinct from those rounded by wild-type toxin in that cells are less spherical and are not refractile to light. These results constitute the first evidence that MARTX$_{Vc}$ is a multifunctional toxin that causes cell rounding by at least two distinct mechanisms (Sheahan *et al.*, 2004).

This second cell rounding process is inhibited by constitutive activation of RhoGTPases by *E. coli* cytotoxic necrotizing factor (CNF1) indicating that cell rounding is due to RTXΔACD inactivation of the RhoGTPases (Sheahan and Satchell, 2007). Rho, Rac, and Cdc42 are three extensively characterized members of the RhoGTPase family that regulate assembly of actin stress fibres, lamellipodia and filopodia, respectively (Nobes and Hall, 1995). These GTPases cycle between an active membrane-localized GTP-bound state and an inactive GDP-bound cytosolic state dependent on guanine nucleotide exchange factors (GEFs) that mediate exchange of GDP for GTP and GTPase-activating proteins (GAPs) that stimulate the GTPase activity of the Rho family proteins. The inactive GDP-bound form can be sequestered in the cytosol in complex with guanine nucleotide dissociation inhibitors (GDIs)

that prevent dissociation of GDP adding an additional level of regulation for RhoGTPase activation (Fig. 6.4). Release of RhoGTPases from GDIs requires GDI displacement factors which allow RhoGTPases to recycle to the membrane for reactivation (DerMardirossian and Bokoch, 2005).

In cells treated with either the MARTX$_{Vc}$ holotoxin or RTXΔACD, no GTP-bound active Rho can be detected, although the overall content of Rho remains unchanged. Furthermore, Rho is found nearly exclusively within the cytosol (Sheahan and Satchell, 2007), consistent with sequestering of inactive Rho to the cytosol through its interaction with Rho-GDI proteins (DerMardirossian and Bokoch, 2005). Additional studies show that ~20% of Rac is relocalized to the cytosol while ~10% of CDC42 is inactivated (Sheahan and Satchell, 2007) demonstrating that Rho is a major target for this toxin with additional affects on related RhoGTPases.

The exact mechanism for inactivation of the RhoGTPases has not as yet been determined although several lines of evidence suggest the mechanism is indirect, affecting the vast regulatory network that controls the activation of RhoGTPases by GEFs and deactivation by GAPs. First, inactivation of Rho is reversible by constitutive activation of the RhoGTPases with CNF1. Second, Rho is reactivated after removal

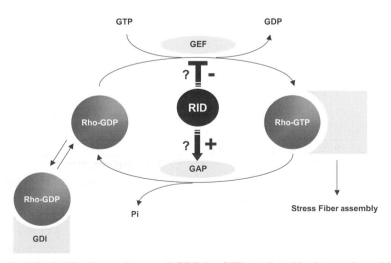

Figure 6.4 Rho is activated by the exchange of GDP for GTP catalysed by interaction with GEFs. GAPs deactivate Rho-GTP by increasing the rate of GTP hydrolysis by Rho. Rho-GDP can also be sequestered to the cytosol by its interaction with GDI proteins. Evidence suggests the RID of MARTX$_{Vc}$ does not inactivate Rho by a direct mechanism, but rather either deactivates host cell GEFs or activates a host cell GAP.

of toxin, even in the presence of the eukaryotic protein synthesis inhibitor cycloheximide. (Sheahan and Satchell, 2007). Together, these studies indicate that Rho itself is not affected by an irreversible event such as covalent modification or proteolysis.

The domain responsible for Rho inactivation on the MARTX$_{Vc}$ toxin is located within the central portion of the toxin at aa 2552–3099. Transient expression of this region causes cell rounding; and rounding and inactivation of Rho are observed when this region of MARTX$_{Vc}$ is delivered to cells as a fusion to anthrax toxin lethal factor (Sheahan and Satchell, 2007). Together, these results demonstrate this region of MARTX$_{Vc}$ is responsible for inactivation of RhoGTPases and it thereby has been denoted as the RhoGTPase-inactivation domain (RID). Sequence analyses demonstrate that this domain does not share characteristics with any other previously characterized bacterial toxins, suggesting that it employs a novel mechanism for RhoGTPase inactivation, likely activating a host cell GAP or inactivating host cell GEFs (Fig. 6.4) (Sheahan and Satchell, 2007).

Autocatalytic cleavage of MARTX$_{Vc}$ by an inositol hexakisphosphate-stimulated cysteine protease

Although much of the MARTX$_{Vc}$ toxin comprises sequences unique to the MARTX toxins, aa 3376–3625 is found in all MARTX toxins and is similar to a region found within *Clostridium* glucosyl transferase toxins (CGTs) including *C. difficile* toxin B (TcdB) (Fullner and Mekalanos 2000). All total, this region can be identified in 24 known and putative secreted bacterial proteins with 13 residues nearly 100% conserved (Prochazkova and Satchell, 2008). Transient expression of aa 3376–3637 fused to GFP is toxic to cells resulting in cells that are rounded, necrotic, and occasionally have condensed nuclei. A Western blot of cell lysates with anti-GFP antibody reveals the fusion protein is approximately 9 kDa smaller than expected. By contrast, transfection of a mutant plasmid carrying either a His3519A or C3568S mutation is not cytotoxic and these mutant fusion proteins maintain the expected size of 56.5 kDa. These re-

sults indicate that this domain is an autocatalytic endopeptidase with a His-Cys catalytic dyad and this domain has therefore been designated the cysteine protease domain (CPD). This cysteine protease is absolutely essential to the function of the toxin in cell rounding since a *V. cholerae* strain in which the catalytic cysteine has been altered shows a severe defect in actin cross-linking activity and Rho inactivation (Sheahan *et al.*, 2007).

In an *in vitro* reaction, recombinant CPD is only autoprocessed after addition of eukaryotic cell cytosol extract with processing occurring at the N-terminal side of the CPD, between residues L3428 and A3429 (Sheahan *et al.*, 2007). The requirement for cell cytosol in the reaction suggests that the CPD must be activated prior to autoprocessing. Indeed, the nucleotide GTP is a low efficiency stimulator that activates CPD upon binding (Sheahan *et al.*, 2007).

Research by Reineke *et al.* (2007) on TcdB illustrates that inositol hexakisphophate (InsP$_6$) is a component in eukaryotic cell lysates that stimulates processing of TcdB and CGTs and this InsP$_6$-induced processing is mediated by a domain similar to CPD (Egerer *et al.*, 2007). Indeed, mutation of the catalytic cysteine and histidine residues in TcdB also decreases cell rounding due to the inactivation of Rho, confirming that autoprocessing is essential for activity of this toxin as well (Barroso *et al.*, 1994; Egerer *et al.*, 2007). These findings suggest that these two families of toxins share a common mechanism for autoprocessing.

Consistent with findings on TcdB, InsP$_6$ is indeed a much more potent stimulator of *V. cholerae* CPD activity than GTP. InsP$_6$ activates CPD autoprocessing by binding with a $K_d = 0.6 \mu M$, a concentration about 100-fold below the physiological concentration of InsP$_6$. Binding of InsP$_6$ requires interaction between its negatively charged phosphate groups and positively charged lysines and arginines on CPD, particularly K3482 and K3611 (Prochazkova and Satchell, 2008). In total, it is proposed that upon translocation to the cytosol, CPD is activated by InsP$_6$ present within the cytosol to process the MARTX$_{Vc}$ toxin, thereby releasing the ACD and RID to intoxicate the cell. Similarly, despite different translocation strategies and catalytic mechanisms of action, the *Clostridial* toxins and possibly other secreted bacterial toxins share this same mechanism for InsP$_6$-induced autoprocessing after translocation.

MARTX toxins of other *Vibrios*

Using the repeat regions in a BLAST search, other MARTX toxins have been identified in seven different *Vibrio* species including animal, aquatic, and insect pathogens. Each of these genes is found in a similar two operon arrangement with *rtxHCA* divergently transcribed from *rtxBDE*. Of particular note, despite the strong conservation of the A, B, and C repeat regions and a CPD region, the central cores of the MARTX toxins are quite dissimilar. Indeed, as many as 10 distinct activity domains can be annotated (reviewed in Satchell, 2007). Among these other toxins, only those from *V. vulnificus* (MARTX$_{Vv}$) and *V. anguillarum* (MARTX$_{Va}$) have been studied experimentally.

MARTX$_{Vv}$

V. vulnificus causes gastrointestinal disease in healthy humans, but also causes a severe disease characterized by dermonecrotic skin lesions and septic shock in immunocompromised persons, particularly those with chronic liver disease. *V. vulnificus* infections can be particularly aggressive, with death occurring in as many as 75% of cases and as soon as 24 hours after contact with the bacterium (Gulig *et al.*, 2005). Comparison of the MARTX$_{Vv}$ toxin with MARTX$_{Vc}$ shows

that this toxin has extensive regions of similarity (Fig. 6.5). Despite this extensive homology, *V. vulnificus* does not have actin cross-linking activity, consistent with its lack of an ACD (Sheahan *et al.*, 2004; Kim *et al.*, 2008). Indeed, this observation forms part of the original rationale for the discovery of the ACD (Sheahan *et al.*, 2004). Despite the lack of actin cross-linking, *V. vulnficus* does induce cell rounding within 50 minutes dependent upon the MARTX$_{Vv}$, indicating an alternate mechanism for actin depolymerization (Dhakal *et al.*, 2006; Sheahan and Satchell, 2007; Kim *et al.*, 2008). Additional analysis of MARTX$_{Vv}$ reveals that this toxin has Rho-inactivating activity, likely tied to the RID that is shared with the MARTX$_{Vc}$ toxin (Fig. 6.5) (Sheahan and Satchell, 2007).

Unlike *V. cholerae*-intoxicated cells, cells incubated with *V. vulnificus* undergo cell lysis shortly after cell rounding is observed, usually within 70–180 minutes after the addition of bacteria (Lee *et al.*, 2007; Liu *et al.*, 2007; Kim *et al.*, 2008). Three independent studies establish that this potent cytotoxicity is associated with the MARTX$_{Vv}$ toxin. The first report links MARTX$_{Vv}$ with virulence in mice and cytotoxicity of tissue culture cells through characterization of the regulatory protein HlyU (Liu *et al.*, 2007). Previous studies show that *hlyU* controls both virulence and cytotoxicity of *V. vulnificus* (Kim *et al.*, 2003). However, when the two genes known to be regulated by HlyU, a

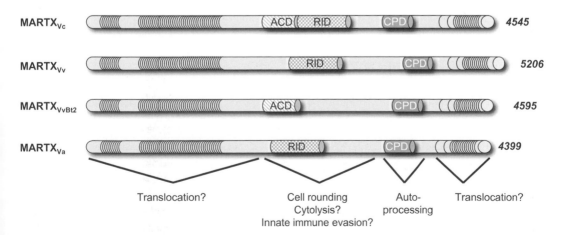

Figure 6.5 Relative domain structure of four MARTX toxins from vibrios: Vc, *V. cholerae*; Vv, *V. vulnificus*; VvBt2; *V. vulnificus* Biotype 2; Va, *V. anguillarum*. Numbers at right indicate size of the protein in aa. Domain designations are as described in Fig. 6.2. Known and putative functional regions of the protein are indicated as discussed in the text.

cytolysin and a protease, are disrupted, no defect in virulence and only a decrease in cytotoxicity are observed (Fan *et al.*, 2001). Comparative gene expression studies have shown that *rtxA* is down-regulated in the absence of *hlyU*, as are *rtxH* and *rtxC* within the same operon. That HlyU directly binds the *rtxHCA* promoter provides further evidence that HlyU regulates *rtxA* (Liu *et al.*, 2007). Disruption of *rtxA* decreases both virulence and cytotoxicity, thereby identifying MARTX$_{Vv}$ as the HlyU-regulated virulence factor responsible for the loss of virulence in *hlyU* mutant. Interestingly, deletion of *rtxH* along with *rtxC* decreases virulence in mice, but has no effect on cytotoxicity indicating that acylation of MARTX$_{Vv}$ and the unknown function of RtxH are not required for cell lysis, but may have a role in virulence (Liu *et al.*, 2007).

A concurrent study also shows a connection of MARTX$_{Vv}$ with virulence and cytotoxicity. In this study, transposon insertions in *rtxA* or the secretion gene *rtxE* reduce both the virulence and the cytotoxicity of *V. vulnificus* (Lee *et al.*, 2007; Lee *et al.*, 2008a). In addition, this study shows that the *rtx* genes are partly regulated by cell contact such that gene expression is increased when bacteria are added to tissue culture cells (Lee *et al.*, 2008a). These findings are confirmed by a third group (Kim *et al.*, 2008) that initiated studies of the *rtxA* gene based on its presence in the genome and its close homology to MARTX$_{Vc}$. They observe that expression of MARTX$_{Vv}$ is highest in *V. vulnificus* cells tightly associated with cells. Using single, double, and triple toxin gene knockouts, this group then shows that deletion of *rtxA* in combination with a cytolysin deletion abrogates the haemolytic activity of *V. vulnificus*, but the MARTX$_{Vv}$ toxin has a dominating role both in cytotoxicity and virulence. This group also demonstrates an important role for MARTX$_{Vv}$ in the progression of disease to septicaemia (Kim *et al.*, 2008). Altogether, by comparing these studies with those of the cytolysin and protease (Fan *et al.*, 2001), it is possible that MARTX$_{Vv}$ is the single most important secreted virulence-associated factor of *V. vulnificus*.

MARTX$_{VvBt2}$

V. vulnificus is not only a pathogen of humans, but some strains, known as the Biotype 2 strains, are pathogenic for eels and other aquatic species (Biosca *et al.*, 1991). Sequence analysis of plasmids carried by these strains reveal they carry *rtxA* genes that would produce a toxin that represents a hybrid between the actin depolymerizing MARTX$_{Vc}$ and the cytolytic MARTX toxin of *V. vulnificus* Biotype 1 strains (Lee *et al.*, 2008b), with Biotype 2 (MARTX$_{VvBt2}$) differing in part by the presence of an ACD (Fig. 6.5). A similarly arranged MARTX toxin has also been identified in *Aeromonas hydrophila* (Seshadri *et al.*, 2006; Satchell, 2007). A copy of the cytolytic Biotype 1 MARTX$_{Vv}$ is also expressed from the chromosome of the Biotype 2 strains. Yet, despite the presence of two active MARTX toxins, a plasmid-cured Biotype 2 mutant or an *rtxA*$_{VvBt2}$ mutant has reduced virulence in eels demonstrating that the plasmid-encoded ACD-containing copy of MARTX$_{VvBt2}$ contributes to virulence independent of the other MARTX toxin (Lee *et al.*, 2008b). In total, it seems *V. vulnificus* produces two distinct MARTX toxins depending on the biotype and that these genes are directly tied to pathogenesis in both mammalian and aquatic animals and may function independently during infection.

MARTX$_{Va}$

V. anguillarum is an aquatic pathogen that causes vibriosis, a fatal haemorrhagic septicaemic infection that affects fish, crustaceans, and bivalves. Similar to *V. vulnificus*, severe infection is typified by necrotic lesions (Austin and Austin, 1999). *V. anguillarum* produces a secreted haemolysin, but deletion of the haemolysin gene *vah1* has little effect on virulence in Atlantic salmon and the strain remains haemolytic (Rock and Nelson, 2006). Within a *vah1* deletion background, mutants that disrupt either the *rtxA* or the *rtxB* gene are non-haemolytic, identifying the MARTX$_{Va}$ toxin as the second *V. anguillarum* haemolysin. These mutants are also defective for cytotoxicity in Atlantic salmon kidney cells showing that, similar to MARTX$_{Vv}$, MARTX$_{Va}$ is a cytolytic toxin. Under reduced inoculation conditions, tissue culture cells treated with *V. anguillarum* are observed to round prior to lysis. This rounding is wholly dependent upon the *rtxA* gene, consistent with this toxin containing a RID. Finally, the *rtxA* mutant is avirulent in juvenile salmon (Li *et al.*, 2008), suggesting that, similar to *V. vulnificus*, the MARTX$_{Va}$ toxin is the most important factor associated with this lethal disease.

Innate immunity as a target for MARTX toxins?

As discussed above, it is clear that MARTX toxins have severe effects on eukaryotic cells and are important for virulence in animal models. However, the exact role of these toxins in disease progression remains unclear. MARTX$_{Vc}$ along with other accessory toxins, haemolysin and haemagglutinin/protease, have been demonstrated to be important to prevent clearance of the bacteria from the site of infection (Haines *et al.*, 2005; Olivier *et al.*, 2007b). Specifically, *V. cholerae* that do not produce accessory toxins (MARTX$_{Vc}$, haemolysin, and haemagglutinin/protease) fail to colonize the small intestine beyond 24 hr, consistent with a role in establishing a prolonged infection (Olivier *et al.*, 2007b). This effect suggests these toxins effectively block clearance by slowing or completely deactivating phagocytosis by immune cells that are the earliest responders to infection. Consistent with this model, MARTX$_{VvBt2}$ promotes survival of *V. vulnificus* Biotype 2 strains in eel blood, but not serum, suggesting it inactivates phagocytes (Lee *et al.*, 2008b). Furthermore, the actin crosslinking Type VI secretion effector VgrG-1 also affects phagocytic macrophage cells and amoebae (Pukatzki *et al.*, 2006). Thus, it seems possible that the function of the actin cross-linking toxins could be to directly block bacterial engulfment by phagocytic cells.

By contrast, MARTX$_{Vv}$ and MARTX$_{Va}$ have been shown to have cytolytic activity (Lee *et al.*, 2007; Liu *et al.*, 2007; Kim *et al.*, 2008; Li *et al.*, 2008). One study suggests that translocation of *V. vulnificus* from the gastrointestinal lumen to the bloodstream occurs more slowly in mice inoculated with an *rtxA* mutant indicating that cytolytic activity could be involved in destruction of the intestinal epithelial barrier to promote dissemination of *V. vulnificus* (Kim *et al.*, 2008). However, the necrotizing enteritis found in mice orally inoculated with *V. vulnificus* has been previously associated with the haemolysin (Fan *et al.*, 2001) leading to the conclusion that

disruption to the barrier to facilitate progression to septicaemia could be multifactorial.

In all, there are many intriguing questions remaining regarding the function of the *Vibrio* MARTX toxins during pathogenesis. It seems as if these toxins may have commonalities, but also have variances optimized for environmental survival and virulence of each specific species and strain.

References

Austin, B., and Austin, D. (1999). Bacterial fish pathogens: Disease of farmed and wild fish, 4th edn (London, UK, Springer Praxis Publishing).

Barroso, L.A., Moncrief, J.S., Lyerly, D.M., and Wilkins, T.D. (1994). Mutagenesis of the *Clostridium difficile* toxin B gene and effect on cytotoxic activity. Microb. Pathog. *16*, 297–303.

Bina, J., Zhu, J., Dziejman, M., Faruque, S., Calderwood, S., and Mekalanos, J. (2003). ToxR regulon of *Vibrio cholerae* and its expression in vibrios shed by cholera patients. Proc. Natl. Acad. Sci. USA *100*, 2801–2806.

Bina, J.E., and Mekalanos, J.J. (2001). *Vibrio cholerae tolC* is required for bile resistance and colonization. Infect. Immun. *69*, 4681–4685.

Biosca, E.G., Amaro, C., Alcaide, E., and Garay, E. (1991). First record of *Vibrio vulnificus* biotyype 2 from disease European eel, *Anguilla anguilla* L. J. Fish Dis. *14*, 103–109.

Blake, P.A., Weaver, R.E., and Hollis, D.G. (1980). Disease of humans (other than cholera) cause by vibrios. Annu. Rev. Microbiol. *34*, 341–367.

Boardman, B. (2007). Export of *Vibrio cholerae* RTX toxin. In Microbiology-Immunology (Chicago, IL, Northwestern University), Proquest document ID 1299821141, pp. 146.

Boardman, B.K., Meehan, B.M., and Satchell, K.J. (2007). Growth phase regulation of *Vibrio cholerae* RTX toxin export. J. Bacteriol. *189*, 1827–1835.

Boardman, B.K., and Satchell, K.J. (2004). *Vibrio cholerae* strains with mutations in an atypical type I secretion system accumulate RTX toxin intracellularly. J. Bacteriol. *186*, 8137–8143.

Boardman, B.K., and Satchell, K.J. Secreted proteins of *Vibrio cholerae*. In Bacterial Protein Secretion, K. Wooldridge, ed. (Norwich, UK, Horizon Scientific Press) in press.

Chow, K.H., Ng, T.K., Yuen, K.Y., and Yam, W.C. (2001). Detection of RTX Toxin Gene in *Vibrio cholerae* by PCR. J. Clin. Microbio.l *39*, 2594–2597.

Cordero, C.L., Kudryashov, D.S., Reisler, E., and Satchell, K.J. (2006). The actin cross-linking domain of the *Vibrio cholerae* RTX toxin directly catalyzes the covalent cross-linking of actin. J. Biol. Chem. *281*, 32366–32374.

Cordero, C.L., Sozhamannan, S., and Satchell, K.J. (2007). RTX toxin actin cross-linking activity in clinical and environmental isolates of *Vibrio cholerae*. J Clin Microbiol *45*, 2289–2292.

Dalsgaard, A., Serichantalergs, O., Forslund, A., Lin, W., Mekalanos, J., Mintz, E., Shimada, T., and Wells, J.G. (2001). Clinical and environmental isolates of *Vibrio cholerae* serogroup O141 carry the CTX phage and the genes encoding the toxin-coregulated pili. J. Clin. Microbiol. *39*, 4086–4092.

Davis, B.M., Quinones, M., Pratt, J., Ding, Y., and Waldor, M.K. (2005). Characterization of the small untranslated RNA RyhB and its regulon in *Vibrio cholerae*. J. Bacteriol. *187*, 4005–4014.

DerMardirossian, C., and Bokoch, G.M. (2005). GDIs: central regulatory molecules in Rho GTPase activation. Trends Cell Biol *15*, 356–363.

Dhakal, B.K., Lee, W., Kim, Y.R., Choy, H.E., Ahnn, J., and Rhee, J.H. (2006). *Caenorhabditis elegans* as a simple model host for *Vibrio vulnificus* infection. Biochem. Biophys. Res. Comm. *346*, 751–757.

Ding, Y., Davis, B.M., and Waldor, M.K. (2004). Hfq is essential for *Vibrio cholerae* virulence and down-regulates sigma expression. Mol. Microbiol.*53*, 345–354.

Egerer, M., Giesemann, T., Jank, T., Satchell, K.J., and Aktories, K. (2007). Auto-catalytic cleavage of *Clostridium difficile* toxins A and B depends on cysteine protease activity. J. Biol. Chem. *282*, 25314–25321.

Fan, J.J., Shao, C.P., Ho, Y.C., Yu, C.K., and Hor, L.I. (2001). Isolation and characterization of a *Vibrio vulnificus* mutant deficient in both extracellular metalloprotease and cytolysin. Infect. Immun. *69*, 5943–5948.

Faruque, S.M., Chowdhury, N., Kamruzzaman, M., Dziejman, M., Rahman, M.H., Sack, D.A., Nair, G.B., and Mekalanos, J.J. (2004). Genetic diversity and virulence potential of environmental *Vibrio cholerae* population in a cholera-endemic area. Proc. Natl. Acad. Sci. USA *101*, 2123–2128.

Fullner, K.J. (2002). Toxins of *Vibrio cholerae*: consensus and controversy. In Microbial pathogenesis and the intestinal epithelial cell, G. Hecht, ed. (Washington DC, ASM Press), pp. 481–502.

Fullner, K.J., Boucher, J.C., Hanes, M.A., Haines, G.K., 3rd, Meehan, B.M., Walchle, C., Sansonetti, P.J., and Mekalanos, J.J. (2002). The contribution of accessory toxins of *Vibrio cholerae* O1 El Tor to the proinflammatory response in a murine pulmonary cholera model. J. Exp. Med. *195*, 1455–1462.

Fullner, K.J., Lencer, W.I., and Mekalanos, J.J. (2001). *Vibrio cholerae*-induced cellular responses of polarized T84 intestinal epithelial cells dependent of production of cholera toxin and the RTX toxin. Infect. Immun. *69*, 6310–6317.

Fullner, K.J., and Mekalanos, J.J. (2000). In vivo covalent crosslinking of actin by the RTX toxin of *Vibrio cholerae*. EMBO J *19*, 5315–5323.

Gulig, P.A., Bourdage, K.L., and Starks, A.M. (2005). Molecular Pathogenesis of *Vibrio vulnificus*. J Microbiol (Korea) *43 Spec No*, 118–131.

Haines, G.K., 3rd, Sayed, B.A., Rohrer, M.S., Olivier, V., and Satchell, K.J. (2005). Role of toll-like receptor 4 in the proinflammatory response to *Vibrio cholerae* O1 El tor strains deficient in production of cholera toxin and accessory toxins. Infect. Immun. *73*, 6157–6164.

Heidelberg, J. (2007). *Vibrio cholerae* RC385 whole genome shotgun sequencing project, GenBank Accession Number AAKH02000000 (The Institute for Genomic Research).

Heidelberg, J.F., Eisen, J.A., Nelson, W.C., Clayton, R.A., Gwinn, M.L., Dodson, R.J., Haft, D.H., Hickey, E.K., Peterson, J.D., Umayam, L., *et al.* (2000). DNA sequence of both chromosomes of the cholera pathogen *Vibrio cholerae*. Nature *406*, 477–484.

Kim, Y.R., Lee, S.E., Kim, C.M., Kim, S.Y., Shin, E.K., Shin, D.H., Chung, S.S., Choy, H.E., Progulske-Fox, A., Hillman, J.D., *et al.* (2003). Characterization and pathogenic significance of *Vibrio vulnificus* antigens preferentially expressed in septicemic patients. Infect. Immun. *71*, 5461–5471.

Kim, Y.R., Lee, S.E., Kook, H., Yeom, J.A., Na, H.S., Kim, S.Y., Chung, S.S., Choy, H.E., and Rhee, J.H. (2008). *Vibrio vulnificus* RTX toxin kills host cells only after contact of the bacteria with host cells. Cell. Microbiol. *10*, 848–862.

Korn, E.D., Carlier, M.F., and Pantaloni, D. (1987). Actin polymerization and ATP hydrolysis. Science *238*, 638–644.

Kudryashov, D.S., Cordero, C.L., Reisler, E., and Satchell, K.J. (2008). Characterization of the enzymatic activity of the actin cross-linking domain from the *Vibrio cholerae* MARTXVc toxin. J. Biol. Chem. *283*, 445–452.

Lee, B.C., Lee, J.H., Kim, M.W., Kim, B.S., Oh, M.H., Kim, K.S., Kim, T.S., and Choi, S.H. (2008a). *Vibrio vulnificus rtxE* is important for virulence, and its expression is induced by exposure to host cells. Infect. Immun. *76*, 1509–1517.

Lee, C.T., Amaro, C., Wu, K.M., Valiente, E., Chang, Y.F., Tsai, S.F., Chang, C.H., and Hor, L.I. (2008b). A common virulence plasmid in biotype 2 *Vibrio vulnificus* and its dissemination aided by a conjugal plasmid. J. Bacteriol. *190*, 1638–1648.

Lee, J.H., Kim, M.W., Kim, B.S., Kim, S.M., Lee, B.C., Kim, T.S., and Choi, S.H. (2007). Identification and characterization of the *Vibrio vulnificus rtxA* essential for cytotoxicity *in vitro* and virulence in mice. J. Microbiol. (Korea) *45*, 146–152.

Levine, M.M., Kaper, J.B., Herrington, D., Losonsky, G., Morris, J.G., Clements, M.L., Black, R.E., Tall, B., and Hall, R. (1988). Volunteer studies of deletion mutants of *Vibrio cholerae* O1 prepared by recombinant techniques. Infect. Immun. *56*, 161–167.

Li, L., Rock, J.L., and Nelson, D.R. (2008). Identification and characterization of a repeat-in-toxin gene cluster in *Vibrio anguillarum*. Infect. Immun. *76*, 2620–2632.

Lin, W., Fullner, K.J., Clayton, R., Sexton, J.A., Rogers, M.B., Calia, K.E., Calderwood, S.B., Fraser, C., and Mekalanos, J.J. (1999). Identification of a *Vibrio cholerae* RTX toxin gene cluster that is tightly linked to the cholera toxin prophage. Proc. Natl. Acad. Sci. USA *96*, 1071–1076.

Liu, M., Alice, A.F., Naka, H., and Crosa, J.H. (2007). The HlyU protein is a positive regulator of *rtxA1*, a gene responsible for cytotoxicity and virulence in the human pathogen *Vibrio vulnificus*. Infect. Immun. *75*, 3282–3289.

Ludwig, A., and Goebel, W. (2006). Structure and mode of action of RTX toxins. In The comprehensive sourcebook of bacterial protein toxins, J.E. Alouf, and M.R. Popoff, eds. (Burlington, MA, Academic press).

Mandell, G.L., Bennett, J.E., and Dolin, R., eds. (1995). Principles and practices of infectious diseases (New York, NY, Churchill Livingstone).

Mathan, M.M., Chandy, G., and Mathan, V.I. (1995). Ultrastructural changes in the upper small intestinal mucosa in patients with cholera. Gastroenterology *109*, 422–430.

Mey, A.R., Craig, S.A., and Payne, S.M. (2005). Characterization of *Vibrio cholerae* RyhB: the RyhB regulon and role of *ryhB* in biofilm formation. Infect. Immun. 73, 5706–5719.

Morton, W.M., Ayscough, K.R., and McLaughlin, P.J. (2000). Latrunculin alters the actin-monomer subunit interface to prevent polymerization. Nature Cell Biol 2, 376–378.

Ninin, E., Caroff, N., Kouri, E., Espaze, E., Richet, H., Quilici, M.L., and Fournier, J.M. (2000). Nontoxigenic *Vibrio cholerae* O1 bacteremia: case report and review. Eur J Clin Microbiol Infect Dis *19*, 488–491.

Nobes, C.D., and Hall, A. (1995). Rho, Rac, and Cdc42 GTPases regulate the assembly of multimolecular focal complexes associated with actin stress fibers, lamellipodia, and filopodia. Cell *81*, 53–62.

Olivier, V., Haines, G.K., 3rd, Tan, Y., and Satchell, K.J. (2007a). Hemolysin and the multifunctional autoprocessing RTX toxin are virulence factors during intestinal infection of mice with *Vibrio cholerae* El Tor O1 strains. Infection and immunity 75, 5035–5042.

Olivier, V., Salzman, N.H., and Satchell, K.J. (2007b). Prolonged colonization of mice by *Vibrio cholerae* El Tor O1 depends on accessory toxins. Infect. Immun. 75, 5043–5051.

Pollard, T.D., and Borisy, G.G. (2003). Cellular motility driven by assembly and disassembly of actin filaments. Cell *112*, 453–465.

Prochazkova, K., and Satchell, K.J. (2008). Structure–function analysis of inositol hexakisphosphate-induced autoprocessing of the *Vibrio cholerae* multifunctional-autoprocessing RTX toxin. J. Biol. Chem. *283*, 23656–23664.

Pukatzki, S., Ma, A.T., Revel, A.T., Sturtevant, D., and Mekalanos, J.J. (2007). Type VI secretion system translocates a phage tail spike-like protein into target cells where it cross-links actin. Proc. Natl. Acad. Sci. USA *104*, 15508–15513.

Pukatzki, S., Ma, A.T., Sturtevant, D., Krastins, B., Sarracino, D., Nelson, W.C., Heidelberg, J.F., and Mekalanos, J.J. (2006). Identification of a conserved bacterial protein secretion system in *Vibrio cholerae* using the *Dictyostelium* host model system. Proc. Natl. Acad. Sci. USA *103*, 1528–1533.

Rahman, M.H., Biswas, K., Hossain, M.A., Sack, R.B., Mekalanos, J.J., and Faruque, S.M. (2008). Distribution of genes for virulence and ecological fitness among diverse *Vibrio cholerae* population in a cholera endemic area: tracking the evolution of pathogenic strains. DNA Cell Biol. 27, 347–355.

Reineke, J., Tenzer, S., Rupnik, M., Koschinski, A., Hasselmayer, O., Schrattenholz, A., Schild, H., and von Eichel-Streiber, C. (2007). Autocatalytic cleavage of *Clostridium difficile* toxin B. Nature *446*, 415–419.

Rock, J.L., and Nelson, D.R. (2006). Identification and characterization of a hemolysin gene cluster in *Vibrio anguillarum*. Infect. Immun. 74, 2777–2786.

Rodighiero, C., and Lencer, W.I. (2003). Trafficking of cholera toxin and related bacterial enterotoxins: pathways and endpoints. In Microbial pathogenesis and the intestinal epithelial cell, G. Hecht, ed. (Washington DC, ASM Press), pp. 385–401.

Satchell, K.J. (2007). MARTX: Multifunctional-autoprocessing RTX toxins. Infect. Immun. 75, 5079–5084.

Schulein, R., Gentschev, I., Schlor, S., Gross, R., and Goebel, W. (1994). Identification and characterization of two functional domains of the hemolysin translocator protein HlyD. Mol. Gen. Genet. *245*, 203–211.

Seshadri, R., Joseph, S.W., Chopra, A.K., Sha, J., Shaw, J., Graf, J., Haft, D., Wu, M., Ren, Q., Rosovitz, M.J., et al. (2006). Genome sequence of *Aeromonas hydrophila* ATCC 7966T: jack of all trades. J. Bacteriol. *188*, 8272–8282.

Sheahan, K.L., Cordero, C.L., and Satchell, K.J. (2004). Identification of a domain within the multifunctional *Vibrio cholerae* RTX toxin that covalently cross-links actin. Proc. Natl. Acad. Sci. USA *101*, 9798–9803.

Sheahan, K.L., Cordero, C.L., and Satchell, K.J. (2007). Autoprocessing of the *Vibrio cholerae* RTX toxin by the cysteine protease domain. EMBO J. *26*, 2552–2561.

Sheahan, K.L., and Satchell, K.J. (2007). Inactivation of small Rho GTPases by the multifunctional RTX toxin from *Vibrio cholerae*. Cell. Microbiol. 9, 1324–1335.

Shuangshoti, S., and Reinprayoon, S. (1995). Pathologic changes of gut in non-O1 *Vibrio cholerae* infection. J. Med. Assoc. Thai. 78, 204–209.

Theriot, J., and Gardel, C. (1995). *Vibrio cholerae* colonizing human cells; http://cmgm.stanford.edu/theriot/movies.htm#Hits.

Yildiz, F.H., Liu, X.S., Heydorn, A., and Schoolnik, G.K. (2004). Molecular analysis of rugosity in a *Vibrio cholerae* O1 El Tor phase variant. Mol. Microbiol. 53, 497–515.

Helicobacter pylori VacA Toxin

7

Timothy L. Cover and John C. Atherton

Abstract

Helicobacter pylori, a Gram-negative bacterium that colonizes the human stomach, secretes a toxin known as VacA. This toxin was initially identified based on its ability to cause vacuolation in cultured gastric epithelial cells. More recent studies have shown that VacA also causes several other alterations in gastric epithelial cells and that VacA targets multiple types of immune cells. Most VacA-induced cellular alterations are attributable to insertion of the toxin into cellular membranes and the formation of membrane channels. In this chapter, we highlight recent progress in understanding three features of VacA: (i) structural properties of VacA, (ii) targeting of T lymphocytes by VacA, and (iii) diversity among VacA proteins expressed by different *H. pylori* strains.

Introduction

Helicobacter pylori is a Gram-negative bacterium that persistently colonizes the human stomach (Suerbaum and Michetti, 2002; Blaser and Atherton, 2004). Most persons colonized with *H. pylori* never develop symptoms attributable to this infection. However, *H. pylori* infection is a risk factor for the development of peptic ulcer disease, gastric MALT lymphoma, and gastric adenocarcinoma (Suerbaum and Michetti, 2002; Atherton, 2006).

In 1988, Leunk *et al.* reported that *H. pylori* secretes a factor that can cause the formation of large intracellular vacuoles in cultured cells (Leunk *et al.*, 1988) (Fig. 7.1). The factor responsible for this activity was subsequently identified as an 88-kDa protein, designated VacA (Cover and Blaser, 1992). The primary amino acid sequence of VacA does not have any strong homology to other known proteins. It is now recognized that, in addition to causing cell vacuolation, VacA can cause numerous additional alterations in gastric epithelial cells, including activation of mitogen-activated protein (MAP) kinases, increased plasma membrane permeability, increased mitochondrial membrane permeability, reduced barrier function of polarized epithelial monolayers, and apoptosis (reviewed in de Bernard *et al.*, 2004; Gebert *et al.*, 2004; Cover and Blanke, 2005) (Fig. 7.2). Many of these effects are dependent on the insertion of VacA into cell membranes to form channels, and therefore, VacA has been classified as a pore-forming toxin. In contrast to most pore-forming bacterial toxins, which act mainly by forming pores in the plasma membrane of cells, VacA can cause cellular alterations by acting in an intracellular site (de Bernard *et al.*, 1997; Ye *et al.*, 1999).

Recent studies have shown that VacA not only targets gastric epithelial cells, but also can cause alterations in several types of immune cells. T-cells, B cells, eosinophils, macrophages, mast cells, and neutrophils are all susceptible to VacA (Molinari *et al.*, 1998b; Supajatura *et al.*, 2002; Boncristiano *et al.*, 2003; Gebert *et al.*, 2003; Zheng and Jones, 2003; de Bernard *et al.*, 2005; Kim *et al.*, 2007; Torres *et al.*, 2007). Thus far, the effects of VacA on T lymphocytes have been investigated in the most detail.

Early studies demonstrated that production of vacuolating toxin activity was a property of

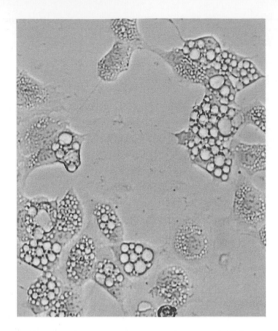

Figure 7.1 Epithelial cell vacuolation induced by VacA. Incubation of AGS gastric epithelial cells with broth culture supernatant from an *H. pylori* strain expressing type s1/i1/m1 VacA results in cell vacuolation. A similar effect is produced by purified VacA. The effect can be blocked by anti-VacA antibodies, and is not seen when cells are incubated with supernatant from a *vacA* null isogenic mutant strain.

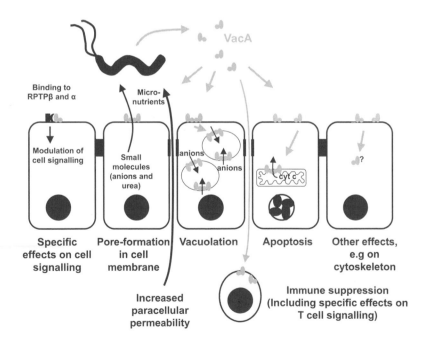

Figure 7.2 Multiple cellular alterations induced by VacA. Binding of VacA to gastric epithelial cells can result in activation of signal transduction pathways (including activation of MAP kinases), formation of channels in the plasma membrane, increased permeability of gastric epithelial monolayers, vacuolation, cytoskeletal alterations, and apoptosis. In addition, VacA can cause alterations in several types of immune cells, including T lymphocytes.

some *H. pylori* strains, but not others (Leunk et al., 1988; Figura et al., 1989; Cover et al., 1990). The basis for variation among strains in vacuolating toxin activity is determined in large part by variation in the amino acid sequences of VacA proteins produced by different strains. VacA proteins expressed by different *H. pylori* strains have been classified into several distinct families (Atherton et al., 1995; Rhead et al., 2007). Importantly, *H. pylori* strains expressing particular types of VacA proteins are associated with varying levels of risk for the development *H. pylori*-related diseases (Atherton et al., 2001; Cover and Blanke, 2005).

In this chapter, we highlight recent progress in understanding three features of the *H. pylori* VacA toxin: (i) structural properties of VacA, (ii) targeting of T lymphocytes by VacA, and (iii) diversity among VacA proteins expressed by different *H. pylori* strains. Thus far, most studies of VacA structure and function have been performed using s1/m1 forms of VacA. Therefore, in the initial sections of this chapter, we focus on structural and functional properties of type s1/m1 VacA. Other forms of VacA are discussed in detail in a later section of this chapter.

Overview of VacA secretion and VacA effects on gastric epithelial cells

Secretion of VacA via an autotransporter pathway

The *vacA* gene encodes a 140-kDa protoxin, which undergoes several proteolytic processing steps to yield the active secreted 88-kDa form of the toxin (Cover and Blaser, 1992; Cover et al., 1994; Schmitt and Haas, 1994; Telford et al., 1994; Nguyen et al., 2001) (Fig. 7.3). The 88-kDa VacA toxin is secreted as a soluble protein into the extracellular space (Cover and Blaser, 1992), and in addition, VacA can remain localized on the surface of *H. pylori* in distinct toxin-rich domains (Ilver et al., 2004). The mechanisms that govern release of VacA as a soluble protein and retention of VacA on the bacterial surface have not been elucidated.

The proteolytic processing and secretion of VacA are characteristic of members of the autotransporter (type Va) family of Gram-negative bacterial secreted proteins (Dautin and Bernstein, 2006). Bacterial proteins secreted by this pathway typically are processed to yield an amino-terminal signal sequence, a carboxy-terminal

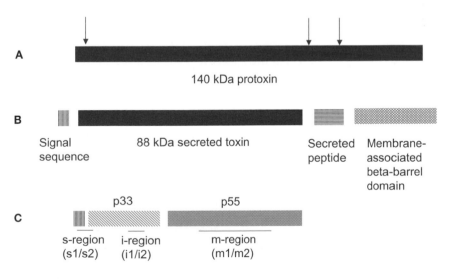

Figure 7.3 Proteolytic processing of VacA and regions of VacA diversity. **A** VacA is initially translated as a 140-kDa protoxin. **B** The protoxin undergoes proteolytic processing in at least 3 sites (arrows) to yield a signal sequence, a mature 88-kDa toxin, a small secreted peptide, and a carboxy-terminal domain with a beta-barrel structure that associates with the bacterial outer membrane. **C** VacA proteins can be classified into families based on diversity in three regions (s-region, i-region, and m-region). Within each region, two main types of proteins are recognized (s1 or s2, i1 or i2, m1 or m2).

domain with a beta-barrel structure, and a secreted passenger domain. The signal sequence contributes to translocation of the protein across the inner membrane and the carboxy-terminal domain facilitates translocation of the passenger domain across the outer membrane.

In the case of VacA, the first processing step is proteolytic cleavage of a 33-amino-acid amino-terminal signal sequence. VacA undergoes further proteolytic processing by one or more unidentified proteases, resulting in the 88-kDa amino-terminal 'passenger domain' (which is the active form of the toxin), a 12-kDa secreted peptide, and a ~33-kDa carboxyl-terminal domain that remains associated with the bacterial outer membrane (Schmitt and Haas, 1994; Telford et al., 1994; Nguyen et al., 2001; Bumann et al., 2002). Analogous to other members of the autotransporter protein family, the VacA carboxyl-terminal domain contains a phenylalanine-containing motif at the C-terminus and is predicted to have a beta-barrel structure (Cover et al., 1994; Schmitt and Haas, 1994). The carboxyl-terminal domain of VacA can transport the cholera toxin B subunit to the external face of the H. pylori outer membrane (Fischer et al., 2001), thereby providing experimental evidence that the C-terminal VacA domain has an important role in protein secretion.

Interactions of VacA with gastric epithelial cells

Several putative receptors for VacA on the surface of gastric epithelial cells have been identified, including receptor-like protein tyrosine phosphatases (RPTPα and β) and sphingomyelin (Yahiro et al., 1999; Padilla et al., 2000; Yahiro et al., 2003; Gupta et al., 2008). There is evidence that each of these receptors contributes to VacA-induced alterations of various types of epithelial cells. When administered intragastrically, VacA causes gastric injury in wild-type mice but not in RPTPβ-null mutant mice (Fujikawa et al., 2003). Following binding of VacA to the surface of cells, VacA can be internalized by cells via a clathrin-independent pinocytotic pathway, and then localizes to endosomal compartments or mitochondria (Galmiche et al., 2000; Willhite and Blanke, 2004; Gauthier et al., 2005, 2006,

2007). Lipid rafts have been implicated in the process leading to VacA internalization (Ricci et al., 2000; Patel et al., 2002; Schraw et al., 2002; Gauthier et al., 2004).

Upon addition of VacA to cells, alterations in plasma membrane anion permeability are detectable by patch clamping, and VacA-treated cells undergo depolarization (Szabo et al., 1999). These findings suggest that the toxin inserts directly into the plasma membrane to form anion-selective membrane channels. The formation of membrane channels by VacA has been studied in detail via the use of planar lipid bilayers, and such channels exhibit a moderate selectivity for anions over cations (Czajkowsky et al., 1999, 2005; Iwamoto et al., 1999; Tombola et al., 1999a,b, 2001). Glycosylphosphatidylinositol-anchored proteins have an important role in the process by which VacA forms membrane channels (Ricci et al., 2000; Patel et al., 2002; Schraw et al., 2002; Gauthier et al., 2004).

In addition to forming channels in the plasma membrane, it is possible that VacA may form channels in endosomal membranes. Conductance of chloride ions through VacA channels in the membranes of endosomes is thought to be an essential step in the process by which VacA causes the formation of cell vacuoles (Czajkowsky et al., 1999, 2005; Iwamoto et al., 1999; Tombola et al., 1999a,b; Montecucco and Rappuoli, 2001; Tombola et al., 2001; Genisset et al., 2007). It is unclear whether VacA can insert directly into endosomal membranes to form channels, or whether the channels form exclusively at the plasma membrane level and then are endocytosed.

Structural properties of the 88-kDa secreted VacA protein

Assembly of VacA into oligomeric structures

The secreted VacA protein is an 88-kDa monomer under denaturing conditions (Cover and Blaser, 1992), but under non-denaturing conditions, VacA monomers can assemble into large water-soluble oligomeric structures. By using a variety of microscopic techniques, it has been possible to visualize multiple types of oligomeric

VacA structures, including single layered forms (containing 6–9 subunits), bilayered forms (containing 12 or 14 subunits), and two-dimensional crystals (Lupetti *et al.*, 1996; Cover *et al.*, 1997; Lanzavecchia *et al.*, 1998; Czajkowsky *et al.*, 1999; Adrian *et al.*, 2002; El-Bez *et al.*, 2005) (Fig. 7.4). The bilayered forms have a flower-like appearance, and consist of a central ring with multiple peripheral arms or petals. Upon exposure to either acidic or alkaline pH, VacA oligomeric complexes dissociate into monomeric components (Cover *et al.*, 1997; Molinari *et al.*, 1998a; Yahiro *et al.*, 1999). VacA monomers can reassemble into oligomeric structures following neutralization of the pH (Cover *et al.*, 1997; Molinari *et al.*, 1998a; Yahiro *et al.*, 1999).

Images of membrane-associated VacA indicate that the toxin forms hexagonal ring-shaped structures (Czajkowsky *et al.*, 1999; Adrian *et al.*, 2002; Geisse *et al.*, 2004), similar in appearance to water-soluble VacA oligomers. Presumably these membrane-associated oligomeric structures correspond to VacA membrane channels. Binding of VacA to lipid-containing membranes is enhanced by acid-activation of the water-soluble oligomeric toxin (Czajkowsky *et al.*, 1999; Schraw *et al.*, 2002; Geisse *et al.*, 2004), and similarly, membrane channel formation by VacA requires acid activation of the toxin (Czajkowsky *et al.*, 1999; Iwamoto *et al.*, 1999; Geisse *et al.*, 2004). Channel formation probably results from interaction of VacA monomers with the membrane, followed by subsequent oligomerization and membrane insertion.

Crystal structure of the p55 VacA domain

The secreted 88-kDa toxin can undergo limited proteolytic nicking to yield two fragments, designated p33 and p55 (corresponding to amino acids 1–312 and 313 to 821, respectively) (Telford *et al.*, 1994; Cover *et al.*, 1997; Ye *et al.*, 1999; Nguyen *et al.*, 2001; Willhite *et al.*, 2002; Torres *et al.*, 2004; Torres *et al.*, 2005). There is no evidence that proteolytic cleavage into p33 and p55 fragments is required for VacA activity, but these two fragments are considered to represent two domains or subunits of VacA. The addition of the p33 domain alone or the p55 domain alone to cultured cells does not result in cell vacuolation. However, addition of a mixture of p33 and p55 to cells causes cell vacuolation (Torres *et al.*, 2004, 2005), and intracellular co-expression of p33 and p55 results in cell vacuolation (Ye *et al.*, 1999; Willhite *et al.*, 2002; Ye and Blanke, 2002).

A 2.4-Å crystal structure of the VacA p55 domain was recently determined (Gangwer *et al.*, 2007). The structure is predominantly a right-handed parallel beta-helix, and has a small globular domain at the C- terminus with mixed alpha and beta secondary structure elements (Fig. 7.5). The β-helix fold is composed of multiple ≈25-amino-acid quasirepeats, each of which forms a single coil of the helix.

Crystal structures have previously been determined for two other passenger domains that are secreted by type Va autotransporter pathways. These include pertactin (an adhesin from

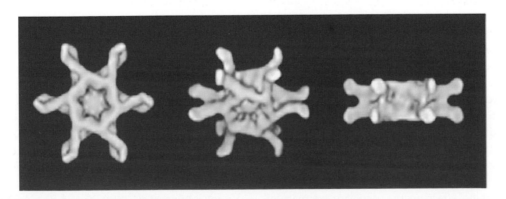

Figure 7.4 Assembly of VacA into oligomeric structures. This figure shows a 3-dimensional model of the structure of a VacA dodecamer, based on images obtained by cryo-negative staining electron microscopy (El-Bez *et al.*, 2005).

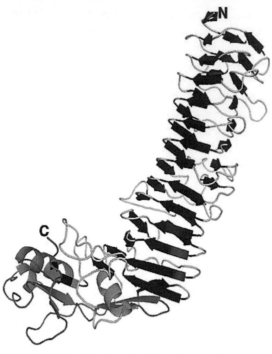

Figure 7.5 Crystal structure of the VacA p55 domain. The p55 structure consists primarily of beta-helical elements and contains a C-terminal globular domain (Gangwer *et al.*, 2007).

Bordetella pertussis) and Hbp (a haemoglobin protease from *E. coli*) (Emsley *et al.*, 1996; Otto *et al.*, 2005). VacA, pertactin, and Hbp do not share sequence similarities, but all contain a beta-helix fold (Gangwer *et al.*, 2007). Conservation of the β-helix fold among autotransporter passenger domains suggests that this structural feature is required for efficient secretion across the outer membrane and folding. A C-terminal β-helix cap may be important as a nucleus and/or chaperone for secretion and folding. The end of the VacA p55 β-helix is capped by a β-hairpin, similar to that observed in the structure of pertactin. A feature of the p55 structure that differs from the two other autotransporter passenger domain structures is the presence of multiple kinks that disrupt what would otherwise be continuous β-sheets (Gangwer *et al.*, 2007).

The secreted VacA passenger domain contains a single closely spaced pair of cysteine residues near the C-terminus (Cover *et al.*, 1994). Mutation of either of these cysteines in VacA to serine results in a decrease in toxin secretion but has no effect on the vacuolating activity of the toxin (Letley *et al.*, 2006). Analysis of the VacA p55 crystal structure indicates that these cysteines form a disulphide bond (Gangwer *et al.*, 2007). In contrast, disulphide bonds are not present in the structures of two previously determined autotransporter passenger domains (Emsley *et al.*, 1996; Otto *et al.*, 2005). In general, autotransporter sequences are characterized by a low cysteine content, and current models suggest that autotransporter passenger domains translocate across the outer membrane in an unfolded conformation. Since the periplasm is an oxidizing environment, it is predicted that the VacA passenger domain is translocated across the outer membrane with an intact disulphide bond.

No crystal structure is yet available for the VacA p33 domain (residues 1–311). However, a program known as BetaWrapPro, which identifies the β-helix motif with high sensitivity and selectivity, predicts that a large portion of p33 will adopt the β-helix fold (Gangwer *et al.*, 2007). The N-terminal portion of p33 is predicted to contain a strongly hydrophobic region near the amino-terminus (de Bernard *et al.*, 1998; Vinion-Dubiel *et al.*, 1999; Ye and Blanke, 2000; McClain *et al.*, 2003), and secondary structure prediction analyses suggest that p33 has α-helical structural elements at its N-terminus (residues 1 to 71) (Kim *et al.*, 2004; Gangwer *et al.*, 2007).

To understand the p55 structure in the context of VacA oligomers, the p55 structure was docked into the 19-Å cryo-EM map of the wild-type dodecamer. The elongated shape of the β-helix and the curve of the C-terminal foot allowed for unambiguous placement of 12 p55 subunits into the petal-like features of the map (Gangwer *et al.*, 2007). Thus, p55 is localized to the arms of the dodecameric complexes, and it is predicted that p33 elements are localized near the centre of dodecameric complexes. In support of this prediction, a comparison of wild-type and Δ6–27 dodecamer structures by cryo-EM suggests that the N-terminal domain of p33 is located in the central density of the structure (El-Bez *et al.*, 2005). A model proposes that VacA oligomerization is mediated by contacts between p33 and the N-terminus of p55 from a neighbouring subunit (Gangwer *et al.*, 2007; Ivie *et al.*, 2008).

Structure–function analysis of VacA

Several studies have shown that the p55 domain has an important role in binding of VacA to host cells (Garner and Cover, 1996; Pagliaccia *et al.*, 1998; Reyrat *et al.*, 1999; Wang and Wang, 2000; Wang *et al.*, 2001; Roche *et al.*, 2007). Specific VacA residues that mediate toxin binding to receptors on the surface of host cells have not been identified. However, deletion of a ~100 amino acid region at the C-terminus of VacA results in decreased binding of the toxin to cells (Wang and Wang, 2000). Moreover, antiserum against p55 inhibits binding of VacA to cells (Garner and Cover, 1996). Binding of VacA to cells is probably not mediated exclusively by the p55 domain, because individual recombinant p33 and p55 domains can each bind to cells (Reyrat *et al.*, 1999; Torres *et al.*, 2005), and individual p33 and p55 fragments each demonstrate affinity for lipids (Moll *et al.*, 1995). A mixture of p33 and p55 proteins exhibits markedly enhanced binding to the plasma membrane of mammalian cells in comparison to the individual p33 or p55 domains (Torres *et al.*, 2005).

A predicted hydrophobic region near the amino-terminus of the p33 domain is required for many of the toxin's cell-modulating effects (de Bernard *et al.*, 1998; Vinion-Dubiel *et al.*, 1999; Ye and Blanke, 2000; McClain *et al.*, 2003). The N-terminal hydrophobic region contains three tandem GXXXG motifs (defined by glycines at positions 14, 18, 22, and 26) (McClain *et al.*, 2001a; McClain *et al.*, 2003; Kim *et al.*, 2004), which are characteristic of transmembrane dimerization sequences. Mutagenesis of several residues within the amino-terminal hydrophobic region of VacA (including G14 and G18) abolishes the capacity of VacA to form membrane channels in planar lipid bilayers (McClain *et al.*, 2003) and also abolishes vacuolating toxin activity (de Bernard *et al.*, 1998; Ye and Blanke, 2000; McClain *et al.*, 2003). These data provide evidence that the hydrophobic region at the amino-terminus of VacA plays a role in membrane channel formation, and that membrane channel formation is required for VacA-induced cell vacuolation. Based on analysis of VacA associated with a lipid membrane, one study detected multiple additional protease-protected VacA fragments, which may represent additional regions of VacA that interact with the membrane (Wang *et al.*, 2000).

If VacA is added to the medium overlying cultured cells, the toxin binds to the cell surface, is internalized, and then causes cell vacuolation. Intracellular expression of VacA in transiently transfected HeLa cells also results in cell vacuolation (de Bernard *et al.*, 1997; Ye *et al.*, 1999). About 422 amino acids at the amino-terminal end of VacA (which includes the p33 domain and about 100 amino acids of the p55 domain) are sufficient to induce cell vacuolation if VacA is expressed intracellularly in transiently transfected cells (de Bernard *et al.*, 1997; de Bernard *et al.*, 1998; Ye *et al.*, 1999). Thus, the intracellularly active portion of VacA comprises a region required for insertion into membranes (residues 1–26) and also comprises regions (in both p33 and p55) that are required for VacA oligomerization.

Effects of VacA on T-lymphocytes

Initially, most studies of VacA activity focused on analysis of toxin-induced alterations in gastric epithelial cells or other epithelial cell types. More recently, there has been increasing recognition that VacA can have effects on various types of immune cells (Molinari et al., 1998b; Supajatura et al., 2002; Boncristiano et al., 2003; Gebert et al., 2003; Zheng and Jones, 2003; de Bernard et al., 2005; Kim et al., 2007; Torres et al., 2007). Effects of VacA on T lymphocytes have been studied in the most detail.

VacA causes multiple alterations in T lymphocytes (Boncristiano et al., 2003; Gebert et al., 2003; Sundrud et al., 2004; Oswald-Richter et al., 2006; Algood et al., 2007). When added to Jurkat T-cells, VacA inhibits production of interleukin 2 (IL-2) (a factor required for T-cell viability and proliferation), down-regulates surface expression of the IL-2 receptor, and causes numerous other transcriptional changes (Boncristiano et al., 2003; Gebert et al., 2003; Sundrud et al., 2004). These effects are attributable to the capacity of VacA to inhibit activation of nuclear factor of T-cells (NFAT) (Boncristiano et al., 2003; Gebert et al., 2003), a transcription factor that acts as a global regulator of immune response genes and that is required for optimal T-cell activation. The mechanism by which VacA inhibits NFAT activation may involve blocking calcium influx into cells from the extracellular milieu, thereby inhibiting the activity of the Ca^{2+}-calmodulin-dependent phosphatase calcineurin (an enzyme that dephosphorylates NFAT) (Boncristiano et al., 2003; Gebert et al., 2003). This VacA activity resembles the actions of cyclosporine A and FK506 (Tacrolimus), two immunosuppressive drugs that inhibit the NFAT phosphatase calcineurin. Other effects of VacA on T lymphocytes include activation of mitogen-activated protein kinases (p38 and MKK3/6) and activation of the Rac-specific nucleotide exchange factor, Vav (Boncristiano et al., 2003; Oswald-Richter et al., 2006). All of these effects of VacA on T-cells are reported to occur without a substantial increase in apoptosis or cell death.

VacA treatment of primary human CD4[+] T-cells results in inhibition of activation-induced proliferation, mitochondrial depolarization, ATP depletion, and cell cycle arrest (Sundrud et al., 2004; Oswald-Richter et al., 2006; Torres et al., 2007). In contrast to what is observed with Jurkat cells, VacA has relatively little if any inhibitory effect on IL-2 expression if primary human CD4[+] cells are activated with CD3/

CD28 antibodies (Sundrud et al., 2004; Oswald-Richter et al., 2006; Torres et al., 2007). VacA inhibits IL-2-induced proliferation of activated primary human CD4[+] T-cells due to a G1 cell cycle arrest, without affecting IL-2-dependent survival. The inhibitory effect of VacA on IL-2-driven cell cycle progression resembles the action of the immunosuppressive drug rapamycin, but in contrast to rapamycin, VacA does not have any detectable effect on p70 S6 kinase (Sundrud et al., 2004; Oswald-Richter et al., 2006; Torres et al., 2007).

The mechanisms by which VacA inhibits proliferation of primary human T-cells are not yet entirely clear, but VacA-induced alterations in mitochondrial function may be an important contributory factor (Torres et al., 2007). There is evidence that the effects of VacA on T-cells may occur via more than one mechanism. Some effects on T-cells are dependent on the formation of VacA channels in cell membranes (Boncristiano et al., 2003; Sundrud et al., 2004; Oswald-Richter et al., 2006). Other effects are the result of activation of intracellular signalling in T-cells, via a channel-independent mechanism (Boncristiano et al., 2003).

A recent study demonstrated that β2 integrin (CD18) mediates entry of VacA into human T-cells (Sewald et al., 2008). Integrins are heterodimeric transmembrane receptors, comprising α- and β-subunits, that mediate an assortment of cellular adhesive interactions and transmit signals across the plasma membrane. Numerous bacteria, viruses, and bacterial toxins are known to enter eukaryotic cells via processes that require integrins. T-cells can express about 12 different types of integrin heterodimers, including several leukocyte-specific β2 integrins (Hogg et al., 2003). Leukocyte function-associated antigen-1 (LFA-1; CD11a/CD18), the most abundant of the T-cell β2 integrins, has an important role in T-cell adherence to vascular endothelium and in formation of the immunologic synapse between T-cells and antigen-presenting cells. LFA-1 is typically in an inactive state in circulating T-cells, but it becomes rapidly activated when T-cells encounter various stimuli. Microscopic imaging of activated primary human CD4[+] T-cells showed that VacA colocalized with CD18 or CD11a at the uropod of cells, and after incubation with cells for 1 hour, VacA colocalized with LFA-1 in small intracellular vesicles (Sewald *et al.*, 2008). Immunoprecipitation experiments demonstrated that VacA interacted with CD18 in both cultured T-cell lines and primary human CD4[+] T-cells. LFA-1-deficient Jurkat T-cells were resistant to VacA-induced vacuolation and inhibition of IL-2 transcription, and genetic complementation of the defect restored VacA sensitivity. Moreover, expression of human β2 integrin heterodimers in Chinese hamster ovary (CHO) cells, which lack human integrins, enhanced the susceptibility of these cells to VacA-induced vacuolation (Sewald *et al.*, 2008).

In contrast to primary human CD4[+] T-cells, primary murine spleen CD4[+] T-cells and cells from a murine T lymphoblast cell line (EL4) are resistant to the actions of VacA (Algood *et al.*, 2007; Sewald *et al.*, 2008), and VacA was not internalized by PMA-activated murine CD4[+] cells (Sewald *et al.*, 2008). Expression of human LFA-1 in murine EL4 T-cells increased the susceptibility of these cells to VacA actions (Sewald *et al.*, 2008). Thus, the differences in susceptibility of mouse and human T-cells to VacA can be attributed to differences in the integrins expressed by these cell types. Since murine T-cells are resistant to VacA, it may be presumed that current mouse models of *H. pylori* infection provide an incomplete view of VacA actions *in vivo*.

It is hypothesized that VacA enters the lamina propria via disruptions in the gastric epithelial layer (Papini *et al.*, 1998; Amieva *et al.*, 2003), and thereby intoxicates T-cells. In addition, it is possible that VacA may target intraepithelial T lymphocytes (Bedoya *et al.*, 2003). It has been suggested that down-regulation of T-cell responses by VacA may contribute to *H. pylori* persistence (Boncristiano *et al.*, 2003; Gebert *et al.*, 2003; Sundrud *et al.*, 2004).

In addition to having effects on primary human CD4[+] T-cells, VacA also inhibits stimulation-induced proliferation of human CD8[+] T-cells and B cells (Torres *et al.*, 2007). Most of the effects of VacA on T lymphocytes are expected to result in localized immunosuppression. However, VacA also stimulates expression of cyclooxygenase-2, a proinflammatory enzyme, in T lymphocytes, which is expected to have a proinflammatory effect (Boncristiano *et al.*, 2003). Thus, the effects of VacA on lymphocytes are complex, and are characterized by both immunostimulatory and immunosuppressive actions.

Diversity among VacA proteins expressed by different *H. pylori* strains

Discovery of VacA diversity

In early studies of *H. pylori*, it was noted that there was considerable variation among strains in production of vacuolating toxin activity (Leunk *et al.*, 1988; Cover *et al.*, 1990). *H. pylori* infection is common and most infected persons do not develop *H. pylori*-associated diseases such as peptic ulceration or gastric adenocarcinoma. Differential toxicity could potentially explain why some *H. pylori*-infected persons develop disease whereas others do not. As discussed further below, many clinical studies have now shown an association between *H. pylori* strains expressing toxigenic forms of VacA and disease, although VacA is recognized as only one of several important disease-associated virulence factors (Atherton, 2006).

Important clues regarding the underlying basis of differential toxin activity became apparent during efforts to clone and sequence the toxin gene, *vacA*. Nucleotide sequence analyses revealed marked variation in *vacA* alleles from different strains, and importantly, these sequence variations were associated with differences in vacuolating toxin activity (Cover *et al.*, 1994; Atherton *et al.*, 1995). The most prominent differences have been localized to three regions of *vacA*, known as the signal (s), mid (m), and intermediate (i) regions (Fig. 7.3). Sequence variations in these regions correlate with toxin activity (Cover *et al.*, 1994; Atherton *et al.*, 1995; Atherton *et al.*, 1997; Rhead *et al.*, 2007), and exchange mutagenesis experiments have confirmed the biological importance of all three regions for toxicity (Pagliaccia *et al.*, 1998; Ji *et al.*, 2000; Letley and Atherton, 2000; McClain *et al.*, 2001b; Letley *et al.*, 2003; Rhead *et al.*, 2007). Moreover, numerous clinical studies have linked specific polymorphisms in these regions with different risks for development of gastroduodenal disease (Atherton *et al.*, 2001; Cover and Blanke, 2005; Rhead *et al.*, 2007).

Genetic variation among vacA alleles

The most obvious differences among *vacA* alleles are localized within an ≈800 bp 'mid' region, which encodes part of the p55 domain of VacA (Atherton *et al.*, 1995). There are two main mid region types, which have been termed m1 and m2. Nucleotide sequence differences between m1 and m2 alleles are present throughout the mid-region, but a particularly notable difference is a 75 bp insertion found in m2 alleles. The second major region of difference among *vacA* alleles is the 'signal' region, which encodes part of the VacA signal peptide, the signal peptide cleavage site and the N-terminus of the 88-kDa VacA toxin (Atherton *et al.*, 1995). The s2 type contains a 27 bp insertion, and the presence of this insertion alters the site of signal peptide cleavage, resulting in a 12-amino-acid extension at the N-terminus of type s2 VacA proteins (Atherton *et al.*, 1995; Letley and Atherton, 2000; McClain *et al.*, 2001b). Recently a third *vacA* region containing naturally occurring polymorphisms, termed the intermediate (i) region, has been described (Rhead *et al.*, 2007). This is a region of 111 bp encoding part of the p33 domain. The (i) region is characterized by minor differences clustered into two internal regions that are 30 and 39 bp in length.

Multiple minor variations within *vacA* families have been described, some large and consistent enough to encourage sub-classification. For example, m2a and m2b *vacA* families have been differentiated in *H. pylori* strains isolated in China (Pan *et al.*, 1998; Strobel *et al.*, 1998; van Doorn *et al.*, 1998a,b; Wang *et al.*, 1998; Owen and Xerry, 2007). Similarly, the s1 *vacA* type has been sub-divided into s1a, s1b and s1c types (Atherton *et al.*, 1995; van Doorn *et al.*, 1998b). These s1 subfamilies are distributed unequally throughout the world and are potentially useful for epidemiological purposes (Wang *et al.*, 2003; Yamazaki *et al.*, 2005). For example, the s1b type is common on the Iberian peninsula and in South America and the s1c type is common in East Asia (Van Doorn *et al.*, 1999). At present, there is no evidence that these *vacA* subfamilies

are associated with differences in levels of toxicity or differences in disease risk.

Recombination among vacA alleles

As described above, *vacA* alleles can be classified into s1 or s2, i1 or i2, and m1 or m2 forms, based on analysis of three discrete regions of the *vacA* gene. *H. pylori* has one of the highest recombination rates of all common bacterial species (Suerbaum *et al.*, 1998), and therefore, in principle, all combinations of these *vacA* forms can theoretically arise. Indeed, *vacA* alleles with three combinations of signal and mid region types are commonly found: s1/m1, s1/m2 and s2/m2 (Atherton *et al.*, 1995; Atherton *et al.*, 1997). The s2/m1 combination also occurs, but is rare (Letley *et al.*, 1999; Morales-Espinosa *et al.*, 1999). This suggests that there may be negative selective pressure against strains containing s2/m1 *vacA* alleles. Only *vacA* s1/m2 strains commonly vary in their i region type (Rhead *et al.*, 2007). Thus, the most commonly encountered *vacA* alleles are s1/i1/m1, s1/i1/m2, s1/i2/m2 and s2/i2/m2. The main s1/m1, s1/m2 and s2/m2 genotypes are found in varying proportions in most human populations (Van Doorn *et al.*, 1999; Owen and Xerry, 2007). An exception is Japan, where the toxigenic s1/m1 type is found almost universally (Ito *et al.*, 1998; Van Doorn *et al.*, 1999).

In bacteria with high rates of recombination, recombination occurs commonly in chromosomal regions of high sequence similarity and occurs rarely where sequences differ markedly. Thus, as expected, comparison of *vacA* alleles shows that recombination has occurred very commonly in regions between the signal and mid regions but less commonly within these regions. Although natural s1/s2 or m1/m2 hybrid sequences are uncommon, the latter have occasionally been found, particularly amongst Chinese strains (Pan *et al.*, 1998; Wang *et al.*, 1998; Atherton *et al.*, 1999). A potential reason for the rarity of strains containing s1/s2 or m1/m2 hybrid sequences is that the resultant proteins may have altered functions, and there may be negative selection against strains encoding such proteins. Analysis of non-synonymous (coding): synonymous (non-coding) substitution rates suggests that there are functional constraints on sequences in the m region, and supports the hypothesis that m1 and m2 forms of VacA have different functions (Atherton *et al.*, 1999). Similar analyses are problematic for the s and i regions because these are much smaller than the m region. However, recombination has been detected within the i region, resulting in i1/i2 hybrid sequences (Rhead *et al.*, 2007).

Several cases have been reported in which *H. pylori* isolates taken from a single human stomach are nearly identical but differ in *vacA* type, one having the s1/m1 and one the s1/m2 *vacA* type (Owen and Xerry, 2003; Aviles-Jimenez *et al.*, 2004; Prouzet-Mauleon *et al.*, 2005; Argent *et al.*, 2008a). Such 'microevolution' also occurs within families: adults and siblings who share the same 'strain' of *H. pylori* (and thus likely passed it among themselves) sometimes harbour variants with different *vacA* genotypes and phenotypes (Argent *et al.*, 2008a). Whether this type of microevolution occurs more frequently in developing countries (where infection with multiple strains is thought to be common) than in developed countries, remains to be determined.

Relationship between VacA types and VacA functional properties

Variation among H. pylori *strains in VacA activity and* vacA *transcription*

Numerous studies have shown that various types of VacA differ in the ability to cause vacuolation of cultured eukaryotic cells. As described in further detail below, important roles of the signal, intermediate and mid-regions in determining vacuolating cytotoxic activity have each been shown by exchange mutagenesis experiments. Importantly, toxin activity is not dependent on genetic elements outside *vacA*. In one experiment, replacement of type s2/m2 *vacA* in a non-toxigenic strain with type s1/m1 *vacA* from a toxigenic strain conferred full toxic activity (Letley *et al.*, 2003). Levels of *vacA* transcription vary by more than twentyfold among strains, and genetic differences 5′ to the *vacA* open reading frame (within a *cysS-vacA* intergenic region which contains the *vacA* promoter) do not obviously explain this (Forsyth *et al.*, 1998; Forsyth and Cover, 1999; Ayala *et al.*, 2004). Neverthe-

less, *vacA* transcriptional differences between some strains can be modulated by exchanging the *cysS–vacA* intergenic region (Forsyth *et al.*, 1998; Letley *et al.*, 2003). It remains unclear whether the level of *vacA* transcription in some strains is controlled by transcriptional regulators encoded by genetic loci located outside the *vacA* locus.

Although uncommon, *vacA* alleles can be truncated by partial gene deletion or frameshift mutations. As expected, such mutations lead to complete loss of VacA production and toxin activity (Ito *et al.*, 1998). In Japan, where virtually all *H. pylori* strains contain type s1/i1/m1 *vacA* alleles, deletion or frameshift mutations are the most common cause of non-toxicity (Ito *et al.*, 1998). Similar mutations are increasingly described in strains from Western populations (Falush *et al.*, 2001; Argent *et al.*, 2008a).

Functional differences between m1 and m2 forms of VacA

Most differences in toxicity among *H. pylori* strains are caused by natural polymorphisms in the signal, intermediate or mid-regions of *vacA*, and the large mid-region has been studied most extensively. Exchange of m1 and m2 mid regions between strains leads to changes in toxicity (Pagliaccia *et al.*, 1998; Ji *et al.*, 2000; Letley *et al.*, 2003) and shows that the first 148 amino acids of the mid-region contain the determinants of differential toxicity (Ji *et al.*, 2000). Both m1 and m2 forms of *vacA* are toxic to some cell lines (such as RK13 cells and primary gastric epithelial cells), but the m1 form is much more toxic than the m2 form to other cell lines (such as HeLa cells and some gastric epithelial cell lines, including AGS) (Pagliaccia *et al.*, 1998). This is attributable to differential binding of the m1 and m2 toxins to cells (Pagliaccia *et al.*, 1998). Notably, the differential binding is not due to differences in VacA binding to the VacA receptor RPTPβ: both m1 and m2 forms of VacA bind to this receptor (Skibinski *et al.*, 2006). The alternate receptor RPTPα is present in a different form on HeLa than on some other cell lines, and the HeLa form of RPTPα binds type m2 VacA less well than m1 VacA; this differential binding may explain the reduced susceptibility of these cells to m2 VacA (De Guzman *et al.*, 2005).

Recent VacA structural studies have provided clues about regions within the p55 domain that may be important for differential binding of m1 and m2 VacA proteins to host cells. Specifically, analysis of the crystal structure of the p55 VacA subunit indicates that nearly all of the amino acids that are divergent in m1 and m2 forms of VacA are surface-exposed, and thus are predicted to be accessible for contact with host cells (Gangwer *et al.*, 2007).

Functional differences between s1 and s2 forms of VacA

s2 forms of VacA have minimal or no vacuolating activity on cultured epithelial cell lines (Atherton *et al.*, 1995). The experimentally determined signal peptide cleavage sites differ between type s1 and s2 forms of VacA, such that s2 VacA has a 12 amino acid hydrophilic N-terminal extension (Atherton *et al.*, 1995; Letley and Atherton, 2000; McClain *et al.*, 2001b). Exchange mutagenesis experiments have shown that addition of the 12-amino-acid extension to type s1 VacA blocks vacuolating activity (Letley and Atherton, 2000; McClain *et al.*, 2001b), and removal of this 12-amino-acid extension from s2 VacA confers activity (Letley *et al.*, 2003). As described in a previous section of this chapter, the N-terminal hydrophobic region of VacA is required for membrane channel formation and vacuolating toxin activity (Ye and Blanke, 2000; Ye and Blanke, 2002; McClain *et al.*, 2003). Consistent with this, the hydrophilic extension found on s2 VacA alters membrane channel formation, although interestingly does not abolish it entirely (McClain *et al.*, 2001b). The s2 form of VacA is widespread in *H. pylori* strains cultured from many human populations, and therefore, it seems plausible that the s2 form of VacA has an important biological function. However, the function of s2 forms of VacA has not yet been elucidated.

Role of the intermediate (i) region

Recently, a third naturally occurring polymorphic region, the 'intermediate' (i) region has been described, and polymorphisms within this region affect VacA activity (Rhead *et al.*, 2007). Only

vacA s1/m2 strains commonly vary in i region type: i1 strains are fully toxic on cell lines sensitive to type m2 VacA, whereas i2 strains do not cause vacuolation of these cell lines. Exchange mutagenesis experiments showed that exchanging i2 elements into type i1 strains abolished toxic activity and exchanging i1 elements into i2 strains conferred activity, although not fully (Rhead *et al.*, 2007). The mechanisms underlying differences in the activity of i1 and i2 VacA proteins have not yet been investigated.

Associations of VacA types with disease

Approaches for studying the relationships between VacA types and disease

H. pylori is the main cause of both peptic ulceration and gastric adenocarcinoma (the second most common cause of cancer deaths worldwide) but most infections do not result in any disease (Suerbaum and Michetti, 2002; Atherton, 2006). Several early studies investigated a potential association between the vacuolating phenotype of *H. pylori* strains and disease, and most of these studies showed a positive association between toxicity and peptic ulceration (reviewed in Atherton *et al.*, 2001). However, such associations were not absolute. Many ulcer patients are infected with apparently non-vacuolating strains, and many non-ulcer patients are infected with vacuolating strains. One limitation of such studies is that phenotyping strains is very time-consuming, so the number of strains and subjects studied in most reports has generally been rather small. In addition, most studies have focused on analysing the vacuolating activity of *H. pylori* strains, but it seems possible that other activities of VacA might be more relevant to the development of disease.

A second approach has been to study antibody responses to VacA in *H. pylori*-infected persons. Some studies have shown a weak association between presence of an anti-VacA antibody response and peptic ulceration or gastric cancer (Atherton *et al.*, 2001; Sezikli *et al.*, 2006; Janulaityte-Gunther *et al.*, 2007), whereas other studies have not detected such an association (Atherton *et al.*, 2001). The problem with these serological studies is that an antibody response to VacA does not provide any information about the toxic activity of the colonizing strain (i.e. both active and inactive forms of VacA may elicit an antibody response). Attempts have been made to develop serum-based tests specific for VacA m1 or m2 infections but current serological assays are poorly discriminatory (Perez-Perez et al., 1999; Ghose et al., 2007).

A third approach, which has been used in the vast majority of clinical association studies, has been to examine potential associations of *vacA* s, m or, recently, i genotypes with disease. These studies have the advantage that simple, highly reproducible, PCR-based testing is possible, most easily on DNA from isolated strains but also on DNA extracted directly from gastric biopsy specimens (Gunn et al., 1998; Rudi et al., 1999).

Problems and pitfalls with disease association studies

There are several inherent problems with studies attempting to link VacA activity or *vacA* genotypes with disease. Peptic ulcers wax and wane, and consequently an ulcer patient in remission may be misclassified as a non-ulcer patient. Both ulcers and gastric cancers have causes other than *H. pylori*: in the case of ulcers, in developed countries non-steroidal anti-inflammatory drugs and aspirin ingestion are becoming nearly as common a cause of ulcers as *H. pylori*, and NSAID use is often surreptitious or unrecorded. Sometimes patients are colonized by more than one strain of *H. pylori* and this may include both cytotoxic and non-cytotoxic strains (Gonzalez-Valencia et al., 2000). Apart from the problem of knowing how to classify such dual infections, there are likely to be many unrecognized examples of multiple infection where a single strain is cultured but others are present in the stomach. Given these and other problems with association studies (some of which are further described below), it is not surprising that associations between toxigenic *vacA* genotypes and disease are often weak and sometimes not found. That they are found at all perhaps implies much stronger true associations between specific *vacA* genotypes and disease.

Associations between VacA and other virulence factors

A major problem in linking specific *vacA* types with disease has been the association of genes encoding active forms of VacA with genes encoding other virulence factors, particularly a 40-kb region of chromosomal DNA known as the *cag* pathogenicity island (PAI). The *cag* PAI encodes a type IV secretion system that translocates a bacterial effector protein, CagA, into gastric epithelial cells, and is widely regarded as the most clinically important *H. pylori* virulence factor (reviewed in (Atherton, 2006)). Strains possessing the *cag* PAI are usually *vacA* s1/m1 or s1/m2 and strains lacking the island are usually *vacA* s2/m2 (Atherton et al., 1995). There appears to be a relationship between *cag* PAI-encoded factors and VacA in modulating epithelial cellular phenotypes. Inactivation of either VacA or genes in the *cag* PAI slightly enhances the activity of the non-mutated system (Asahi et al., 2003; Yokoyama et al., 2005; Argent et al., 2008b). This is due in part to opposing actions of VacA and *cag* proteins on the signalling molecular NFAT in epithelial cells (Yokoyama et al., 2005) and in part due to reciprocal cytoskeletal effects (Argent et al., 2008b). Thus, the activities of individual virulence factors may be counterbalanced by the activities of other virulence factors. It has been suggested that *H. pylori* strains that are either very interactive with the human gastric mucosa (i.e. *cag* positive and toxigenic) or non-interactive (*cag* negative and non-toxigenic) may each have a niche in the gastric environment (Blaser and Atherton, 2004), but there may not be a niche for intermediate forms. Thus, the first two forms of *H. pylori* may have a selective advantage, whereas intermediate forms may be selected against. This is supported by the observation that genes encoding several non-*cag* virulence factors, including adhesins, often are found in strains that express active forms of the vacuolating toxin (Gerhard et al., 1999).

Whatever the reason for the association between *H. pylori* virulence factors, such an association creates a problem when analysing the relationship between specific virulence factors and the risk of gastroduodenal disease. In most clinical association studies, the number of strains discordant for specific virulence factors is too few to tease apart the relative importance of individual virulence factors. Multiple virulence factors (including products of the *cag* PAI, VacA, and various adhesins) are likely to be largely equivalent markers of disease-associated or disease-causing strains.

Association between vacA signal region types and disease

Numerous studies have reported an association between *H. pylori* strains containing type s1 *vacA* alleles and peptic ulcer disease or gastric cancer (reviewed in Atherton *et al.*, 2001; Cover and Blanke, 2005). However, other studies have failed to show an association and there may be publication bias in favour of positive studies. In most populations, approximately 80% of strains are the active *vacA* s1 type and although there is a significant association of *vacA* s1 strains with disease, many of these strains are not disease-associated. Thus, it may be more useful to think of the inactive *vacA* s2 type strains as non-pathogenic (not disease-associated) and the *vacA* s1 strains as able to cause disease but not always doing so.

Association of vacA mid region type with disease

Because *vacA* s1/m1 and s1/m2 strains are both usually *cag* positive, it is possible to analyse associations between *vacA* mid-region type and disease without confounding associations between *cag* status and disease (Bolek *et al.*, 2007). Many studies have shown positive associations of *vacA* m1 forms with disease, but interestingly, these associations appear more consistent and stronger with gastric adenocarcinoma than with duodenal ulceration (Atherton *et al.*, 2001). In support of this, the *vacA* m1 type is also associated with the cancer precursor condition, atrophic gastritis (Con *et al.*, 2007), and with the corpus-predominant gastric inflammation and *H. pylori* colonization pattern thought to predispose to gastric cancer (Soltermann *et al.*, 2007). *H. pylori* is well-recognized as causing gastric adenocarcinoma through first inducing gastric hypochlorhydria and gastric atrophy (Correa, 1992). That infection with strains containing *vacA* s1/m1 alleles is associated with atrophy and hypochlorhydria (Argent *et al.*, 2008a) suggests that such infections may increase cancer risk by increasing the risk of these precursor conditions. Studies of adult siblings with hypochlorhydria show that type s1/m1 strains were acquired in childhood (Argent *et al.*, 2008a). Thus, infection with *vacA* s1/m1 strains is likely to have occurred well before hypochlorhydria arose, implying a causal relationship.

Associations of vacA intermediate region type with disease

Strains with the recently described toxigenic intermediate region type (i1) were strongly associated with gastric cancer in an Iranian population, and multivariate analysis showed that the i1 genotype was a better marker of cancer-associated strains than s1 or m1 genotypes (Rhead *et al.*, 2007). This is likely to be due in part to the i1 region being a good marker of toxigenicity: all non-toxic s2 strains are i2, and all maximally toxic s1/m1 strains are i1. Thus, the i1 region appears to differentiate toxic from non-toxic s1/m2 strains. Reports on other populations are emerging and support the association between i1 region and disease. Both Japanese and Italian populations demonstrate associations between *vacA* i1 type and gastric cancer (Ito *et al.*, 2008) (Basso *et al.*, 2008), although in the latter study this became non-significant when CagA phosphorylation status was entered into the multivariate model. However, the Italian study also showed an independent association between i region type and the corpus-predominant gastric inflammatory pattern known to predispose to gastric cancer (Basso *et al.*, 2008). Gastric cancer and gastric ulcer are thought to arise on a similar background, so it is consistent that a positive association has been described between *vacA* i1 strains and gastric ulceration (Hussein *et al.*, 2008). More surprisingly, in the Italian study, multivariate analysis showed an association between *vacA* i1 strains and both gastric and duodenal ulcers, and this was independent of *cag*, including CagA phosphorylation status (Basso *et al.*, 2008). Type *vacA* i1 strains are much more common than i2 strains in Asian populations, but statistical associations between i1 forms and disease cannot readily be demonstrated due to the low number of i2 strains in these populations (Ogiwara *et al.*, 2008).

Future trends

Although VacA has been the topic of investigation in hundreds of publications, there are still many important gaps in our knowledge about this toxin. One important goal of future studies will be to elucidate the structural features of the VacA p33 domain that are required for membrane channel formation. Another goal will be to investigate in further detail the binding, internalization, and intracellular trafficking of VacA within cells, and to investigate the molecular mechanisms by which VacA causes multiple alterations in gastric epithelial cells and immune cells. It will be important to investigate differences in the structure and function of different forms of VacA in further detail. Finally, it will be important to understand more clearly the role of VacA *in vivo*, both in facilitating *H. pylori* colonization of the stomach and in contributing to gastroduodenal disease.

Acknowledgements

Supported in part by the National Institutes of Health (R01 AI39657), the Department of Veterans Affairs, Cancer Research UK, the UK Medical Research Council and the UK National Institute of Health Research Nottingham Biomedical Research Unit in GI and liver disease. We thank members of the Cover and Atherton laboratories for numerous scientific contributions and stimulating discussions.

References

Adrian, M., Cover, T.L., Dubochet, J., and Heuser, J.E. (2002). Multiple oligomeric states of the *Helicobacter pylori* vacuolating toxin demonstrated by cryoelectron microscopy. J. Mol. Biol. *318*, 121–133.

Algood, H.M., Torres, V.J., Unutmaz, D., and Cover, T.L. (2007). Resistance of primary murine CD4+ T-cells to *Helicobacter pylori* vacuolating cytotoxin. Infect. Immun. *75*, 334–341.

Amieva, M.R., Vogelmann, R., Covacci, A., Tompkins, L.S., Nelson, W.J., and Falkow, S. (2003). Disruption of the epithelial apical-junctional complex by *Helicobacter pylori* CagA. Science *300*, 1430–1434.

Argent, R.H., Thomas, R.J., Aviles-Jimenez, F., Letley, D.P., Limb, M.C., El-Omar, E.M., and Atherton, J.C. (2008a). Toxigenic *Helicobacter pylori* infection precedes gastric hypochlorhydria in cancer relatives, and *H. pylori* virulence evolves in these families. Clin. Cancer Res. *14*, 2227–2235.

Argent, R.H., Thomas, R.J., Letley, D.P., Rittig, M.G., Hardie, K.R., and Atherton, J.C. (2008b). Functional association between the *Helicobacter pylori* virulence factors VacA and CagA. J. Med. Microbiol. *57*, 145–150.

Asahi, M., Tanaka, Y., Izumi, T., Ito, Y., Naiki, H., Kersulyte, D., Tsujikawa, K., Saito, M., Sada, K., Yanagi, S., Fujikawa, A., Noda, M., and Itokawa, Y. (2003). *Helicobacter pylori* CagA containing ITAM-like sequences localized to lipid rafts negatively regulates VacA-induced signaling *in vivo*. Helicobacter *8*, 1–14.

Atherton, J.C. (2006). The pathogenesis of *Helicobacter pylori*-induced gastro-duodenal diseases. Annu. Rev. Pathol. *1*, 63–96.

Atherton, J.C., Cao, P., Peek, R.M., Jr., Tummuru, M.K., Blaser, M.J., and Cover, T.L. (1995). Mosaicism in vacuolating cytotoxin alleles of *Helicobacter pylori*. Association of specific vacA types with cytotoxin production and peptic ulceration. J. Biol. Chem. *270*, 17771–17777.

Atherton, J.C., Cover, T.L., Papini, E., and Telford, J.L. (2001). Vacuolating cytotoxin. In: *Helicobacter pylori*: physiology and genetics, eds. H.L.T. Mobley, G.L. Mendz, and S.L. Hazell, Washington, DC: ASM Press.

Atherton, J.C., Peek, R.M., Jr., Tham, K.T., Cover, T.L., and Blaser, M.J. (1997). Clinical and pathological importance of heterogeneity in vacA, the vacuolating cytotoxin gene of *Helicobacter pylori*. Gastroenterology *112*, 92–99.

Atherton, J.C., Sharp, P.M., Cover, T.L., Gonzalez-Valencia, G., Peek, R.M., Jr., Thompson, S.A., Hawkey, C.J., and Blaser, M.J. (1999). Vacuolating cytotoxin (vacA) alleles of *Helicobacter pylori* comprise two geographically widespread types, m1 and m2, and have evolved through limited recombination. Curr Microbiol *39*, 211–218.

Aviles-Jimenez, F., Letley, D.P., Gonzalez-Valencia, G., Salama, N., Torres, J., and Atherton, J.C. (2004). Evolution of the *Helicobacter pylori* vacuolating cytotoxin in a human stomach. J. Bacteriol. *186*, 5182–5185.

Ayala, G., Chihu, L., Perales, G., Fierros-Zarate, G., Hansen, L.M., Solnick, J.V., and Sanchez, J. (2004). Quantitation of *H. pylori* cytotoxin mRNA by real-time RT-PCR shows a wide expression range that does not correlate with promoter sequences. Microb Pathog *37*, 163–167.

Basso, D., Zambon, C.F., Letley, D.P., Stranges, A., Marchet, A., Rhead, J.L., Schiavon, S., Guariso, G., Ceroti, M., Nitti, D., Rugge, M., Plebani, M., and Atherton, J.C. (2008). Clinical relevance of *Helicobacter pylori* cagA and vacA gene polymorphisms. Gastroenterology *135*, 91–99.

Bedoya, A., Garay, J., Sanzon, F., Bravo, L.E., Bravo, J.C., Correa, H., Craver, R., Fontham, E., Du, J.X., and Correa, P. (2003). Histopathology of gastritis in *Helicobacter pylori*-infected children from populations at high and low gastric cancer risk. Hum Pathol *34*, 206–213.

Blaser, M.J., and Atherton, J.C. (2004). *Helicobacter pylori* persistence: biology and disease. J Clin Invest *113*, 321–333.

Bolek, B.K., Salih, B.A., and Sander, E. (2007). Genotyping of *Helicobacter pylori* strains from gastric biopsies by multiplex polymerase chain reaction. How advantageous is it? Diagn Microbiol Infect Dis *58*, 67–70.

Boncristiano, M., Paccani, S.R., Barone, S., Ulivieri, C., Patrussi, L., Ilver, D., Amedei, A., D'Elios, M.M., Telford, J.L., and Baldari, C.T. (2003). The *Helicobacter pylori* vacuolating toxin inhibits T-cell activation by two independent mechanisms. J Exp Med *198*, 1887–1897.

Bumann, D., Aksu, S., Wendland, M., Janek, K., Zimny-Arndt, U., Sabarth, N., Meyer, T.F., and Jungblut, P.R. (2002). Proteome analysis of secreted proteins of the gastric pathogen *Helicobacter pylori*. Infect. Immun. *70*, 3396–3403.

Con, S.A., Takeuchi, H., Valerin, A.L., Con-Wong, R., Con-Chin, G.R., Con-Chin, V.G., Nishioka, M., Mena, F., Brenes, F., Yasuda, N., Araki, K., and Sugiura, T. (2007). Diversity of *Helicobacter pylori* cagA and vacA genes in Costa Rica: its relationship with atrophic gastritis and gastric cancer. Helicobacter *12*, 547–552.

Correa, P. (1992). Human gastric carcinogenesis: a multistep and multifactorial process – First American Cancer Society Award Lecture on Cancer Epidemiology and Prevention. Cancer Res. *52*, 6735–6740.

Cover, T.L., and Blanke, S.R. (2005). *Helicobacter pylori* VacA, a paradigm for toxin multifunctionality. Nat Rev Microbiol *3*, 320–332.

Cover, T.L., and Blaser, M.J. (1992). Purification and characterization of the vacuolating toxin from *Helicobacter pylori*. J. Biol. Chem. *267*, 10570–10575.

Cover, T.L., Dooley, C.P., and Blaser, M.J. (1990). Characterization of and human serologic response to proteins in *Helicobacter pylori* broth culture supernatants with vacuolizing cytotoxin activity. Infect. Immun. *58*, 603–610.

Cover, T.L., Hanson, P.I., and Heuser, J.E. (1997). Acid-induced dissociation of VacA, the *Helicobacter pylori* vacuolating cytotoxin, reveals its pattern of assembly. J Cell Biol *138*, 759–769.

Cover, T.L., Tummuru, M.K.R., Cao, P., Thompson, S.A., and Blaser, M.J. (1994). Divergence of genetic sequences for the vacuolating cytotoxin among *Helicobacter pylori* strains. J. Biol. Chem. *269*, 10566–10573.

Czajkowsky, D.M., Iwamoto, H., Cover, T.L., and Shao, Z. (1999). The vacuolating toxin from *Helicobacter pylori* forms hexameric pores in lipid bilayers at low pH. Proc. Natl. Acad. Sci. USA *96*, 2001–2006.

Czajkowsky, D.M., Iwamoto, H., Szabo, G., Cover, T.L., and Shao, Z. (2005). Mimicry of a host anion channel by a *Helicobacter pylori* pore-forming toxin. Biophys J *89*, 3093–3101.

Dautin, N., and Bernstein, H. (2007). Protein secretion in Gram-negative bacteria via the autotransporter pathway. Annu. Rev. Microbiol. *61*, 89–112.

de Bernard, M., Arico, B., Papini, E., Rizzuto, R., Grandi, G., Rappuoli, R., and Montecucco, C. (1997). *Helicobacter pylori* toxin VacA induces vacuole formation by acting in the cell cytosol. Mol. Microbiol. *26*, 665–674.

de Bernard, M., Burroni, D., Papini, E., Rappuoli, R., Telford, J., and Montecucco, C. (1998). Identification of the *Helicobacter pylori* VacA toxin domain active in the cell cytosol. Infect. Immun. *66*, 6014–6016.

de Bernard, M., Cappon, A., Del Giudice, G., Rappuoli, R., and Montecucco, C. (2004). The multiple cellular activities of the VacA cytotoxin of *Helicobacter pylori*. Int. J. Med. Microbiol. *293*, 589–597.

de Bernard, M., Cappon, A., Pancotto, L., Ruggiero, P., Rivera, J., Del Giudice, G., and Montecucco, C. (2005). The *Helicobacter pylori* VacA cytotoxin activates RBL-2H3 cells by inducing cytosolic calcium oscillations. Cell. Microbiol. *7*, 191–198.

De Guzman, B.B., Hisatsune, J., Nakayama, M., Yahiro, K., Wada, A., Yamasaki, E., Nishi, Y., Yamazaki, S., Azuma, T., Ito, Y., Ohtani, M., van der Wijk, T., den Hertog, J., Moss, J., and Hirayama, T. (2005). Cytotoxicity and recognition of receptor-like protein tyrosine phosphatases, RPTPalpha and RPTPbeta, by *Helicobacter pylori* m2VacA. Cell. Microbiol. *7*, 1285–1293.

El-Bez, C., Adrian, M., Dubochet, J., and Cover, T.L. (2005). High resolution structural analysis of *Helicobacter pylori* VacA toxin oligomers by cryo-negative staining electron microscopy. J. Struct. Biol. *151*, 215–228.

Emsley, P., Charles, I.G., Fairweather, N.F., and Isaacs, N.W. (1996). Structure of *Bordetella pertussis* virulence factor P.69 pertactin. Nature *381*, 90–92.

Falush, D., Kraft, C., Taylor, N.S., Correa, P., Fox, J.G., Achtman, M., and Suerbaum, S. (2001). Recombination and mutation during long-term gastric colonization by *Helicobacter pylori*: estimates of clock rates, recombination size, and minimal age. Proc. Natl. Acad. Sci. USA *98*, 15056–15061.

Figura, N., Guglielmetti, P., Rossolini, A., Barberi, A., Cusi, G., Musmanno, R.A., Russi, M., and Quaranta, S. (1989). Cytotoxin production by Campylobacter pylori strains isolated from patients with peptic ulcers and from patients with chronic gastritis only. J. Clin. Microbiol. *27*, 225–226.

Fischer, W., Buhrdorf, R., Gerland, E., and Haas, R. (2001). Outer membrane targeting of passenger proteins by the vacuolating cytotoxin autotransporter of *Helicobacter pylori*. Infect. Immun. *69*, 6769–6775.

Forsyth, M.H., Atherton, J.C., Blaser, M.J., and Cover, T.L. (1998). Heterogeneity in levels of vacuolating cytotoxin gene (vacA) transcription among *Helicobacter pylori* strains. Infect. Immun. *66*, 3088–3094.

Forsyth, M.H., and Cover, T.L. (1999). Mutational analysis of the vacA promoter provides insight into gene transcription in *Helicobacter pylori*. J. Bacteriol. *181*, 2261–2266.

Fujikawa, A., Shirasaka, D., Yamamoto, S., Ota, H., Yahiro, K., Fukada, M., Shintani, T., Wada, A., Aoyama, N., Hirayama, T., Fukamachi, H., and Noda, M. (2003). Mice deficient in protein tyrosine phosphatase receptor type Z are resistant to gastric ulcer induction by VacA of *Helicobacter pylori*. Nat Genet *33*, 375–381.

Galmiche, A., Rassow, J., Doye, A., Cagnol, S., Chambard, J.C., Contamin, S., de Thillot, V., Just, I., Ricci, V., Solcia, E., Van Obberghen, E., and Boquet, P. (2000). The N-terminal 34 kDa fragment of *helicobacter pylori* vacuolating cytotoxin targets mitochondria and induces cytochrome c release. EMBO J. *19*, 6361–6370.

Gangwer, K.A., Mushrush, D.J., Stauff, D.L., Spiller, B., McClain, M.S., Cover, T.L., and Lacy, D.B. (2007). Crystal structure of the *Helicobacter pylori* vacuolating toxin p55 domain. Proc. Natl. Acad. Sci. USA *104*, 16293–16298.

Garner, J.A., and Cover, T.L. (1996). Binding and internalization of the *Helicobacter pylori* vacuolating cytotoxin by epithelial cells. Infect. Immun. *64*, 4197–4203.

Gauthier, N.C., Monzo, P., Gonzalez, T., Doye, A., Oldani, A., Gounon, P., Ricci, V., Cormont, M., and Boquet, P. (2007). Early endosomes associated with dynamic F-actin structures are required for late trafficking of *H. pylori* VacA toxin. J. Cell Biol. *177*, 343–354.

Gauthier, N.C., Monzo, P., Kaddai, V., Doye, A., Ricci, V., and Boquet, P. (2005). *Helicobacter pylori* VacA cytotoxin: a probe for a clathrin-independent and Cdc42-dependent pinocytic pathway routed to late endosomes. Mol. Biol. Cell *16*, 4852–4866.

Gauthier, N.C., Ricci, V., Gounon, P., Doye, A., Tauc, M., Poujeol, P., and Boquet, P. (2004). Glycosylphosphatidylinositol-anchored proteins and actin cytoskeleton modulate chloride transport by channels formed by the *Helicobacter pylori* vacuolating cytotoxin VacA in HeLa cells. J. Biol. Chem. *279*, 9481–9489.

Gauthier, N.C., Ricci, V., Landraud, L., and Boquet, P. (2006). *Helicobacter pylori* VacA toxin: a tool to study novel early endosomes. Trends Microbiol. *14*, 292–294.

Gebert, B., Fischer, W., and Haas, R. (2004). The *Helicobacter pylori* vacuolating cytotoxin: from cellular vacuolation to immunosuppressive activities. Rev. Physiol. Biochem. Pharmacol. *152*, 205–220.

Gebert, B., Fischer, W., Weiss, E., Hoffman, R., and Haas, R. (2003). *Helicobacter pylori* vacuolating cytotoxin inhibits T lymphocyte activation. Science *301*, 1099–1102.

Geisse, N.A., Cover, T.L., Henderson, R.M., and Edwardson, J.M. (2004). Targeting of *Helicobacter pylori* vacuolating toxin to lipid raft membrane domains analysed by atomic force microscopy. Biochem. J. *381*, 911–917.

Genisset, C., Puhar, A., Calore, F., de Bernard, M., Dell'Antone, P., and Montecucco, C. (2007). The concerted action of the *Helicobacter pylori* cytotoxin VacA and of the v-ATPase proton pump induces swelling of isolated endosomes. Cell. Microbiol. *9*, 1481–1490.

Gerhard, M., Lehn, N., Neumayer, N., Boren, T., Rad, R., Schepp, W., Miehlke, S., Classen, M., and Prinz, C. (1999). Clinical relevance of the *Helicobacter pylori* gene for blood-group antigen-binding adhesin. Proc. Natl. Acad. Sci. USA *96*, 12778–12783.

Ghose, C., Perez-Perez, G.I., Torres, V.J., Crosatti, M., Nomura, A., Peek, R.M., Jr., Cover, T.L., Francois, F., and Blaser, M.J. (2007). Serological assays for identification of human gastric colonization by *Helicobacter pylori* strains expressing VacA m1 or m2. Clin. Vaccine Immunol. *14*, 442–450.

Gonzalez-Valencia, G., Atherton, J.C., Munoz, O., Dehesa, M., la Garza, A.M., and Torres, J. (2000). *Helicobacter pylori* vacA and cagA genotypes in Mexican adults and children. J. Infect. Dis. *182*, 1450–1454.

Gunn, M.C., Stephens, J.C., Stewart, J.D., and Rathbone, B.J. (1998). Detection and typing of the virulence determinants cagA and vacA of *Helicobacter pylori* directly from biopsy DNA: are *in vitro* strains representative of *in vivo* strains? Eur J Gastroenterol Hepatol *10*, 683–687.

Gupta, V.R., Patel, H.K., Kostolansky, S.S., Ballivian, R.A., Eichberg, J., and Blanke, S.R. (2008). Sphingomyelin functions as a novel receptor for *Helicobacter pylori* VacA. PLoS Pathog *4*, e1000073.

Hogg, N., Laschinger, M., Giles, K., and McDowall, A. (2003). T-cell integrins: more than just sticking points. J. Cell Sci. *116*, 4695–4705.

Hussein, N.R., Mohammadi, M., Talebkhan, Y., Doraghi, M., Letley, D.P., Muhammad, M.K., Argent, R.H., and Atherton, J.C. (2008). Differences in virulence markers between *Helicobacter pylori* strains from Iraq and those from Iran: potential importance of regional differences in *H. pylori*-associated disease. J. Clin. Microbiol. *46*, 1774–1779.

Ilver, D., Barone, S., Mercati, D., Lupetti, P., and Telford, J.L. (2004). *Helicobacter pylori* toxin VacA is transferred to host cells via a novel contact-dependent mechanism. Cell. Microbiol. *6*, 167–174.

Ito, Y., Azuma, T., Ito, S., Suto, H., Miyaji, H., Yamazaki, Y., Kohli, Y., and Kuriyama, M. (1998). Full-length sequence analysis of the vacA gene from cytotoxic and noncytotoxic *Helicobacter pylori*. J. Infect. Dis. *178*, 1391–1398.

Ito, Y., Inagaki, T., Yamakawa, A., Satomi, S., Matsuda, H., *et al.* (2008). *Helicobacter pylori* vacA i region genotype is associated with an increased risk of gastric caner, vacuolating cytotoxin activity and CagA status in Okinawa, Japan. Gastroenterology *134* (Suppl. 1), A565 (abstract).

Ivie, S.E., McClain, M.S., Torres, V.J., Algood, H.M., Lacy, D.B., Yang, R., Blanke, S.R., and Cover, T.L. (2008). *Helicobacter pylori* VacA subdomain required for intracellular toxin activity and assembly of functional oligomeric complexes. Infect. Immun. *76*, 2843–2851.

Iwamoto, H., Czajkowsky, D.M., Cover, T.L., Szabo, G., and Shao, Z. (1999). VacA from *Helicobacter pylori*: a hexameric chloride channel. FEBS Lett. *450*, 101–104.

Janulaityte-Gunther, D., Kupcinskas, L., Pavilonis, A., Valuckas, K., Wadstrom, T., and Andersen, L.P. (2007). Combined serum IgG response to *Helicobacter pylori* VacA and CagA predicts gastric cancer. FEMS Immunol. Med. Microbiol. *50*, 220–225.

Ji, X., Fernandez, T., Burroni, D., Pagliaccia, C., Atherton, J.C., Reyrat, J.M., Rappuoli, R., and Telford, J.L. (2000). Cell specificity of *helicobacter pylori* cytotoxin is determined by a short region in the polymorphic midregion. Infect. Immun. *68*, 3754–3757.

Kim, J.M., Kim, J.S., Lee, J.Y., Kim, Y.J., Youn, H.J., Kim, I.Y., Chee, Y.J., Oh, Y.K., Kim, N., Jung, H.C., and Song, I.S. (2007). Vacuolating cytotoxin in *Helicobacter pylori* water-soluble proteins up-regulates chemokine expression in human eosinophils via

Ca2+ influx, mitochondrial reactive oxygen intermediates, and NF-kappaB activation. Infect. Immun. 75, 3373–3381.

Kim, S., Chamberlain, A.K., and Bowie, J.U. (2004). Membrane channel structure of *Helicobacter pylori* vacuolating toxin: role of multiple GXXXG motifs in cylindrical channels. Proc. Natl. Acad. Sci. USA 101, 5988–5991.

Lanzavecchia, S., Bellon, P.L., Lupetti, P., Dallai, R., Rappuoli, R., and Telford, J.L. (1998). Three-dimensional reconstruction of metal replicas of the *Helicobacter pylori* vacuolating cytotoxin. J. Struct. Biol. 121, 9–18.

Letley, D.P., and Atherton, J.C. (2000). Natural diversity in the N terminus of the mature vacuolating cytotoxin of *Helicobacter pylori* determines cytotoxin activity. J. Bacteriol. 182, 3278–3280.

Letley, D.P., Lastovica, A., Louw, J.A., Hawkey, C.J., and Atherton, J.C. (1999). Allelic diversity of the *Helicobacter pylori* vacuolating cytotoxin gene in South Africa: rarity of the vacA s1a genotype and natural occurrence of an s2/m1 allele. J. Clin. Microbiol. 37, 1203–1205.

Letley, D.P., Rhead, J.L., Bishop, K., and Atherton, J.C. (2006). Paired cysteine residues are required for high levels of the *Helicobacter pylori* autotransporter VacA. Microbiology 152, 1319–1325.

Letley, D.P., Rhead, J.L., Twells, R.J., Dove, B., and Atherton, J.C. (2003). Determinants of non-toxicity in the gastric pathogen *Helicobacter pylori*. J. Biol. Chem. 278, 26734–26741.

Leunk, R.D., P.T., J., David, B.C., Kraft, W.G., and Morgan, D.R. (1988). Cytotoxic activity in broth-culture filtrates of *Campylobacter pylori*. J. Med. Microbiol. 26, 93–99.

Lupetti, P., Heuser, J.E., Manetti, R., Massari, P., Lanzavecchia, S., Bellon, P.L., Dallai, R., Rappuoli, R., and Telford, J.L. (1996). Oligomeric and subunit structure of the *Helicobacter pylori* vacuolating cytotoxin. J. Cell Biol. 133, 801–807.

McClain, M.S., Cao, P., and Cover, T.L. (2001a). Amino-terminal hydrophobic region of *Helicobacter pylori* vacuolating cytotoxin (VacA) mediates transmembrane protein dimerization. Infect. Immun. 69, 1181–1184.

McClain, M.S., Cao, P., Iwamoto, H., Vinion-Dubiel, A.D., Szabo, G., Shao, Z., and Cover, T.L. (2001b). A 12-amino-acid segment, present in Type s2 but not Type s1 *Helicobacter pylori* VacA proteins, abolishes cytotoxin activity and alters membrane channel formation. J. Bacteriol. 183, 6499–6508.

McClain, M.S., Iwamoto, H., Cao, P., Vinion-Dubiel, A.D., Li, Y., Szabo, G., Shao, Z., and Cover, T.L. (2003). Essential role of a GXXXG motif for membrane channel formation by *Helicobacter pylori* vacuolating toxin. J. Biol. Chem. 278, 12101–12108.

Molinari, M., Galli, C., de Bernard, M., Norais, N., Ruysschaert, J.M., Rappuoli, R., and Montecucco, C. (1998a). The acid activation of *Helicobacter pylori* toxin VacA: structural and membrane binding studies. Biochem. Biophys. Res. Commun. 248, 334–340.

Molinari, M., Salio, M., Galli, C., Norais, N., Rappuoli, R., Lanzavecchia, A., and Montecucco, C. (1998b).

Selective inhibition of Ii-dependent antigen presentation by *Helicobacter pylori* toxin VacA. J. Exp. Med. 187, 135–140.

Moll, G., Papini, E., Colonna, R., Burroni, D., Telford, J., Rappuoli, R., and Montecucco, C. (1995). Lipid interaction of the 37-kDa and 58-kDa fragments of the *Helicobacter pylori* cytotoxin. Eur. J. Biochem. 234, 947–952.

Montecucco, C., and Rappuoli, R. (2001). Living dangerously: how *Helicobacter pylori* survives in the human stomach. Nat. Rev. Mol. Cell. Biol. 2, 457–466.

Morales-Espinosa, R., Castillo-Rojas, G., Gonzalez-Valencia, G., Ponce de Leon, S., Cravioto, A., Atherton, J.C., and Lopez-Vidal, Y. (1999). Colonization of Mexican patients by multiple *Helicobacter pylori* strains with different vacA and cagA genotypes. J. Clin. Microbiol. 37, 3001–3004.

Nguyen, V.Q., Caprioli, R.M., and Cover, T.L. (2001). Carboxy-terminal proteolytic processing of *Helicobacter pylori* vacuolating toxin. Infect. Immun. 69, 543–546.

Ogiwara, H., Graham, D.Y., and Yamaoka, Y. (2008). vacA i-region subtyping. Gastroenterology 134, 1267; author reply 1268.

Oswald-Richter, K., Torres, V.J., Sundrud, M.S., VanCompernolle, S.E., Cover, T.L., and Unutmaz, D. (2006). *Helicobacter pylori* VacA toxin inhibits human immunodeficiency virus infection of primary human T-cells. J. Virol. 80, 11767–11775.

Otto, B.R., Sijbrandi, R., Luirink, J., Oudega, B., Heddle, J.G., Mizutani, K., Park, S.Y., and Tame, J.R. (2005). Crystal structure of hemoglobin protease, a heme binding autotransporter protein from pathogenic Escherichia coli. J. Biol. Chem. 280, 17339–17345.

Owen, R.J., and Xerry, J. (2003). Tracing clonality of *Helicobacter pylori* infecting family members from analysis of DNA sequences of three housekeeping genes (ureI, atpA and ahpC), deduced amino acid sequences, and pathogenicity-associated markers (cagA and vacA). J. Med. Microbiol. 52, 515–524.

Owen, R.J., and Xerry, J. (2007). Geographical conservation of short inserts in the signal and middle regions of the *Helicobacter pylori* vacuolating cytotoxin gene. Microbiology 153, 1176–1186.

Padilla, P.I., Wada, A., Yahiro, K., Kimura, M., Niidome, T., Aoyagi, H., Kumatori, A., Anami, M., Hayashi, T., Fujisawa, J., Saito, H., Moss, J., and Hirayama, T. (2000). Morphologic differentiation of HL-60 cells is associated with appearance of RPTPbeta and induction of *helicobacter pylori* VacA sensitivity. J. Biol. Chem. 275, 15200–15206.

Pagliaccia, C., de Bernard, M., Lupetti, P., Ji, X., Burroni, D., Cover, T.L., Papini, E., Rappuoli, R., Telford, J.L., and Reyrat, J.M. (1998). The m2 form of the *Helicobacter pylori* cytotoxin has cell type-specific vacuolating activity. Proc. Natl. Acad. Sci. USA 95, 10212–10217.

Pan, Z.J., Berg, D.E., van der Hulst, R.W., Su, W.W., Raudonikiene, A., Xiao, S.D., Dankert, J., Tytgat, G.N., and van der Ende, A. (1998). Prevalence of vacuolating cytotoxin production and distribution of distinct vacA alleles in *Helicobacter pylori* from China. J. Infect. Dis. 178, 220–226.

Papini, E., Satin, B., Norais, N., de Bernard, M., Telford, J.L., Rappuoli, R., and Montecucco, C. (1998). Selective increase of the permeability of polarized epithelial cell monolayers by *Helicobacter pylori* vacuolating toxin. J. Clin. Invest. *102*, 813–820.

Patel, H.K., Willhite, D.C., Patel, R.M., Ye, D., Williams, C.L., Torres, E.M., Marty, K.B., MacDonald, R.A., and Blanke, S.R. (2002). Plasma membrane cholesterol modulates cellular vacuolation induced by the *Helicobacter pylori* vacuolating cytotoxin. Infect. Immun. *70*, 4112–4123.

Perez-Perez, G.I., Peek, R.M., Jr., Atherton, J.C., Blaser, M.J., and Cover, T.L. (1999). Detection of anti-VacA antibody responses in serum and gastric juice samples using type s1/m1 and s2/m2 *Helicobacter pylori* VacA antigens. Clin. Diagn. Lab. Immunol. 6, 489–493.

Prouzet-Mauleon, V., Hussain, M.A., Lamouliatte, H., Kauser, F., Megraud, F., and Ahmed, N. (2005). Pathogen evolution *in vivo*: genome dynamics of two isolates obtained 9 years apart from a duodenal ulcer patient infected with a single *Helicobacter pylori* strain. J. Clin. Microbiol. *43*, 4237–4241.

Reyrat, J.M., Lanzavecchia, S., Lupetti, P., de Bernard, M., Pagliaccia, C., Pelicic, V., Charrel, M., Ulivieri, C., Norais, N., Ji, X., Cabiaux, V., Papini, E., Rappuoli, R., and Telford, J.L. (1999). 3D imaging of the 58kDa cell binding subunit of the *Helicobacter pylori* cytotoxin. J. Mol. Biol. *290*, 459–470.

Rhead, J.L., Letley, D.P., Mohammadi, M., Hussein, N., Mohagheghi, M.A., Eshagh Hosseini, M., and Atherton, J.C. (2007). A new *Helicobacter pylori* vacuolating cytotoxin determinant, the intermediate region, is associated with gastric cancer. Gastroenterology *133*, 926–936.

Ricci, V., Galmiche, A., Doye, A., Necchi, V., Solcia, E., and Boquet, P. (2000). High cell sensitivity to *Helicobacter pylori* VacA toxin depends on a GPI-anchored protein and is not blocked by inhibition of the clathrin- mediated pathway of endocytosis. Mol Biol Cell *11*, 3897–3909.

Roche, N., Ilver, D., Angstrom, J., Barone, S., Telford, J.L., and Teneberg, S. (2007). Human gastric glycosphingolipids recognized by *Helicobacter pylori* vacuolating cytotoxin VacA. Microbes Infect. *9*, 605–614.

Rudi, J., Rudy, A., Maiwald, M., Kuck, D., Sieg, A., and Stremmel, W. (1999). Direct determination of *Helicobacter pylori* vacA genotypes and cagA gene in gastric biopsies and relationship to gastrointestinal diseases. Am. J. Gastroenterol. *94*, 1525–1531.

Schmitt, W., and Haas, R. (1994). Genetic analysis of the *Helicobacter pylori* vacuolating cytotoxin: structural similarities with the IgA protease type of exported protein. Mol. Microbiol. *12*, 307–319.

Schraw, W., Li, Y., McClain, M.S., van der Goot, F.G., and Cover, T.L. (2002). Association of *Helicobacter pylori* vacuolating toxin (VacA) with lipid rafts. J. Biol. Chem. *277*, 34642–34650.

Sewald, X., Gebert-Vogal, B., Prassl, S., Barwig, I., Weiss, E., Fabbri, M., Osicka, R., Schiemann, M., Busch, D.H., Semmrich, M., Holzmann, B., Sebo, P., and Haas, R. (2008). CD18 is the T-lymphocyte receptor of the *Helicobacter pylori* vacuolating cytotoxin. Cell Host Microbe 3, 20–29.

Sezikli, M., Guliter, S., Apan, T.Z., Aksoy, A., Keles, H., and Ozkurt, Z.N. (2006). Frequencies of serum antibodies to *Helicobacter pylori* CagA and VacA in a Turkish population with various gastroduodenal diseases. Int. J. Clin. Pract. *60*, 1239–1243.

Skibinski, D.A., Genisset, C., Barone, S., and Telford, J.L. (2006). The cell-specific phenotype of the polymorphic vacA midregion is independent of the appearance of the cell surface receptor protein tyrosine phosphatase beta. Infect. Immun. *74*, 49–55.

Soltermann, A., Koetzer, S., Eigenmann, F., and Komminoth, P. (2007). Correlation of *Helicobacter pylori* virulence genotypes vacA and cagA with histological parameters of gastritis and patient's age. Mod. Pathol. *20*, 878–883.

Strobel, S., Bereswill, S., Balig, P., Allgaier, P., Sonntag, H.G., and Kist, M. (1998). Identification and analysis of a new vacA genotype variant of *Helicobacter pylori* in different patient groups in Germany. J. Clin. Microbiol. *36*, 1285–1289.

Suerbaum, S., and Michetti, P. (2002). *Helicobacter pylori* infection. N. Engl. J. Med. *347*, 1175–1186.

Suerbaum, S., Smith, J.M., Bapumia, K., Morelli, G., Smith, N.H., Kunstmann, E., Dyrek, I., and Achtman, M. (1998). Free recombination within *Helicobacter pylori*. Proc. Natl. Acad. Sci. USA 95, 12619–12624.

Sundrud, M.S., Torres, V.J., Unutmaz, D., and Cover, T.L. (2004). Inhibition of primary human T-cell proliferation by *Helicobacter pylori* vacuolating toxin (VacA) is independent of VacA effects on IL-2 secretion. Proc. Natl. Acad. Sci. USA *101*, 7727–7732.

Supajatura, V., Ushio, H., Wada, A., Yahiro, K., Okumura, K., Ogawa, H., Hirayama, T., and Ra, C. (2002). Cutting edge: VacA, a vacuolating cytotoxin of *Helicobacter pylori*, directly activates masT-cells for migration and production of proinflammatory cytokines. J. Immunol. *168*, 2603–2607.

Szabo, I., Brutsche, S., Tombola, F., Moschioni, M., Satin, B., Telford, J.L., Rappuoli, R., Montecucco, C., Papini, E., and Zoratti, M. (1999). Formation of anion-selective channels in the cell plasma membrane by the toxin VacA of *Helicobacter pylori* is required for its biological activity. EMBO J. *18*, 5517–5527.

Telford, J.L., Ghiara, P., Dell'Orco, M., Comanducci, M., Burroni, D., Bugnoli, M., Tecce, M.F., Censini, S., Covacci, A., Xiang, Z., and et al. (1994). Gene structure of the *Helicobacter pylori* cytotoxin and evidence of its key role in gastric disease. J. Exp. Med. *179*, 1653–1658.

Tombola, F., Carlesso, C., Szabo, I., de Bernard, M., Reyrat, J.M., Telford, J.L., Rappuoli, R., Montecucco, C., Papini, E., and Zoratti, M. (1999a). *Helicobacter pylori* vacuolating toxin forms anion-selective channels in planar lipid bilayers: possible implications for the mechanism of cellular vacuolation. Biophys. J. *76*, 1401–1409.

Tombola, F., Oregna, F., Brutsche, S., Szabo, I., Del Giudice, G., Rappuoli, R., Montecucco, C., Papini, E., and Zoratti, M. (1999b). Inhibition of the vacuolating and anion channel activities of the VacA toxin of *Helicobacter pylori*. FEBS Lett. *460*, 221–225.

Tombola, F., Pagliaccia, C., Campello, S., Telford, J.L., Montecucco, C., Papini, E., and Zoratti, M. (2001).

How the loop and middle regions influence the properties of *Helicobacter pylori* VacA channels. Biophys. J. *81*, 3204–3215.

Torres, V.J., Ivie, S.E., McClain, M.S., and Cover, T.L. (2005). Functional properties of the p33 and p55 domains of the *Helicobacter pylori* vacuolating cytotoxin. J. Biol. Chem. *280*, 21107–21114.

Torres, V.J., McClain, M.S., and Cover, T.L. (2004). Interactions between p-33 and p-55 domains of the *Helicobacter pylori* vacuolating cytotoxin (VacA). J. Biol. Chem. *279*, 2324–2331.

Torres, V.J., VanCompernolle, S.E., Sundrud, M.S., Unutmaz, D., and Cover, T.L. (2007). *Helicobacter pylori* vacuolating cytotoxin inhibits activation-induced proliferation of human T and B lymphocyte subsets. J. Immunol. *179*, 5433–5440.

Van Doorn, L.J., Figueiredo, C., Megraud, F., Pena, S., Midolo, P., Queiroz, D.M., Carneiro, F., Vanderborght, B., Pegado, M.D., Sanna, R., De Boer, W., Schneeberger, P.M., Correa, P., Ng, E.K., Atherton, J., Blaser, M.J., and Quint, W.G. (1999). Geographic distribution of vacA allelic types of *Helicobacter pylori*. Gastroenterology *116*, 823–830.

van Doorn, L.J., Figueiredo, C., Rossau, R., Jannes, G., van Asbroek, M., Sousa, J.C., Carneiro, F., and Quint, W.G. (1998a). Typing of *Helicobacter pylori* vacA gene and detection of cagA gene by PCR and reverse hybridization. J Clin Microbiol 36, 1271–1276.

van Doorn, L.J., Figueiredo, C., Sanna, R., Pena, S., Midolo, P., Ng, E.K., Atherton, J.C., Blaser, M.J., and Quint, W.G. (1998b). Expanding allelic diversity of *Helicobacter pylori* vacA. J. Clin. Microbiol. 36, 2597–2603.

Vinion-Dubiel, A.D., McClain, M.S., Czajkowsky, D.M., Iwamoto, H., Ye, D., Cao, P., Schraw, W., Szabo, G., Blanke, S.R., Shao, Z., and Cover, T.L. (1999). A dominant negative mutant of *Helicobacter pylori* vacuolating toxin (VacA) inhibits VacA-induced cell vacuolation. J. Biol. Chem. 274, 37736–37742.

Wang, H.J., Kuo, C.H., Yeh, A.A., Chang, P.C., and Wang, W.C. (1998). Vacuolating toxin production in clinical isolates of *Helicobacter pylori* with different vacA genotypes. J. Infect. Dis. *178*, 207–212.

Wang, H.J., and Wang, W.C. (2000). Expression and binding analysis of GST-vacA fusions reveals that the C- terminal approximately 100-residue segment of exotoxin is crucial for binding in HeLa cells. Biochem. Biophys. Res. Commun. *278*, 449–454.

Wang, J., van Doorn, L.J., Robinson, P.A., Ji, X., Wang, D., Wang, Y., Ge, L., Telford, J.L., and Crabtree, J.E. (2003). Regional variation among vacA alleles of *Helicobacter pylori* in China. J. Clin. Microbiol. *41*, 1942–1945.

Wang, W.-C., Wang, H.-J., and Kuo, C.-H. (2001). Two distinctive cell binding patterns by vacuolating toxin fused with glutathione S-transferase: one high-affinity m1-specific binding and the other lower-affinity binding for variant m forms. Biochemistry *40*, 11887–11896.

Wang, X., Wattiez, R., Paggliacia, C., Telford, J.L., Ruysschaert, J., and Cabiaux, V. (2000). Membrane topology of VacA cytotoxin from *H. pylori*. FEBS Lett. *481*, 96–100.

Willhite, D.C., and Blanke, S.R. (2004). *Helicobacter pylori* vacuolating cytotoxin enters cells, localizes to the mitochondria, and induces mitochondrial membrane permeability changes correlated to toxin channel activity. Cell. Microbiol. 6, 143–154.

Willhite, D.C., Ye, D., and Blanke, S.R. (2002). Fluorescence resonance energy transfer microscopy of the *Helicobacter pylori* vacuolating cytotoxin within mammalian cells. Infect. Immun. 70, 3824–3832.

Yahiro, K., Niidome, T., Kimura, M., Hatakeyama, T., Aoyagi, H., Kurazono, H., Imagawa, K., Wada, A., Moss, J., and Hirayama, T. (1999). Activation of *Helicobacter pylori* VacA toxin by alkaline or acid conditions increases its binding to a 250-kDa receptor protein-tyrosine phosphatase beta. J. Biol. Chem. *274*, 36693–36699.

Yahiro, K., Wada, A., Nakayama, M., Kimura, T., Ogushi, K., Niidome, T., Aoyagi, H., Yoshino, K., Yonezawa, K., Moss, J., and Hirayama, T. (2003). Protein-tyrosine phosphatase alpha, RPTP alpha, is a *Helicobacter pylori* VacA receptor. J. Biol. Chem. *278*, 19183–19189.

Yamazaki, S., Yamakawa, A., Okuda, T., Ohtani, M., Suto, H., Ito, Y., Yamazaki, Y., Keida, Y., Higashi, H., Hatakeyama, M., and Azuma, T. (2005). Distinct diversity of vacA, cagA, and cagE genes of *Helicobacter pylori* associated with peptic ulcer in Japan. J. Clin. Microbiol. *43*, 3906–3916.

Ye, D., and Blanke, S.R. (2000). Mutational analysis of the *Helicobacter pylori* vacuolating toxin amino terminus: identification of amino acids essential for cellular vacuolation. Infect. Immun. 68, 4354–4357.

Ye, D., and Blanke, S.R. (2002). Functional complementation reveals the importance of intermolecular monomer interactions for *Helicobacter pylori* VacA vacuolating activity. Mol. Microbiol. 43, 1243–1253.

Ye, D., Willhite, D.C., and Blanke, S.R. (1999). Identification of the minimal intracellular vacuolating domain of the *Helicobacter pylori* vacuolating toxin. J. Biol. Chem. 274, 9277–9282.

Yokoyama, K., Higashi, H., Ishikawa, S., Fujii, Y., Kondo, S., Kato, H., Azuma, T., Wada, A., Hirayama, T., Aburatani, H., and Hatakeyama, M. (2005). Functional antagonism between *Helicobacter pylori* CagA and vacuolating toxin VacA in control of the NFAT signaling pathway in gastric epithelial cells. Proc. Natl. Acad. Sci. USA *102*, 9661–9666.

Zheng, P.Y., and Jones, N.L. (2003). *Helicobacter pylori* strains expressing the vacuolating cytotoxin interrupt phagosome maturation in macrophages by recruiting and retaining TACO (coronin 1) protein. Cell. Microbiol. 5, 25–40.

Staphylococcal Immune Evasion Toxins

8

Ries J. Langley, Thomas Proft and John D. Fraser

Abstract

With the advent of complete microbial genomes, the identification and characterization of novel immune evasion proteins from *Staphylococcus aureus* has increased significantly. Studies of these proteins have revealed significant conservation of protein structures and a range of activities that are all directed at the two key elements of host immunity, complement and neutrophils. This chapter focuses on some of these secreted virulence factors and the ways in which they assist the bacterium to survive in the face of a hostile immune response. In particular, the chapter discusses the structure and function of complement inhibiting molecules SSL7, CHIPS, Efb, Ehp, SCIN, and Sbi and the leucocyte-inhibiting SAgs, SSLs, CHIPS, and Eap.

Introduction

The genus *Staphylococcus* comprises over 30 species, of which ten are described as human commensals (Freney *et al.*, 1999). *Staphylococcus aureus*, the only coagulase-positive species, causes most of the diseases, ranging from minor skin infections to life-threatening conditions such as sepsis, endocarditis and toxic shock syndrome (Lowy, 1998). *S. aureus* is an opportunistic pathogen that mainly colonizes the anterior nares of humans without causing any symptoms, unless damages occur to mucosal and cutaneous membranes and the host's innate immunity is compromised. *Staphylococcus epidermidis*, another opportunistic human pathogen, causes a similar range of diseases, but is significantly less virulent than *S. aureus*. A major difference between these 2 species is a large set of additional species-specific virulence genes in *S. aureus* that are located on horizontally acquired genomic islands and are absent on the *S. epidermidis* genome (Gill *et al.*, 2005). Thus far, seven *S. aureus* pathogenicity islands (SaPIs) have been identified, which encode more than 50% of all virulence factors (Gill *et al.*, 2005). Many of these virulence factors are toxins and immune evasion proteins with specific activity against components of the human immune system (Foster, 2005; Rooijakkers *et al.*, 2005a) and it has been argued that these virulence factors play an important role for survival of the organism in an hostile environment, such as the epithelium of anterior nares, which is defended from microbial infections by nasal associated lymphoid tissue (NALT), a component of the host's immune system (Massey *et al.*, 2006).

In this chapter, we discuss recent findings in the field of *S. aureus* immune evasion toxins, including protein structures, biological functions and potential roles in staphylococcal disease.

Superantigens and the SSLs

The term 'superantigen' (SAg) describes a group of proteins with extremely high potency to stimulate human, and to a certain degree, other mammalian CD4 and CD8 T-cells (White *et al.*, 1989). At present 20 serologically distinct SAgs have been described in the literature, comprising toxic shock syndrome toxin-1 (TSST-1), the staphylococcal enterotoxins (SE) A-E, G-J and the staphylococcal enterotoxin-like toxins (SEl) K-R and U and the recently identified SEl-U2 and SEl-V (Thomas *et al.*, 2007; Thomas *et al.*,

2006). A new nomenclature was introduced in 2004 to distinguish between SAgs that also cause food poisoning (SEs) and SAgs without emetic activity or for which their potential role in staphylococcal food poisoning remains unconfirmed (SEls) (Lina *et al.*, 2004). SEs and SEls are part of a larger family of structurally related proteins that also includes the streptococcal superantigens (Proft and Fraser, 2007) and the staphylococcal superantigen-like toxins (SSLs, this chapter).

Superantigens are highly potent T-cell mitogens that simultaneously bind to the major histocompatibility complex (MHC) class II protein on antigen presenting cells and to T-cell receptor (TcR) molecules on T-cells. The interaction with the TcR involves predominantly the Vβ domains and results in a characteristic Vβ fingerprint of stimulated T-cells that depends on the Vβ-specificity of the SAg. Cross-linking of these two surface receptors results in massive release of pro-inflammatory cytokines, such as interleukin (IL)-1β, tumour necrosis factor (TNF)-α and interferon (IFN)-γ, and T-cell mediators, such as IL-2, which can cause toxic shock and death. The function of SAgs has been discussed extensively in the literature over recent years (reviewed in (Fraser *et al.*, 2000; McCormick *et al.*, 2001; Proft and Fraser, 2003, 2007) and in this chapter we will focus on the very recent findings.

Protein structure of staphylococcal superantigens

To date, 10 protein structures of staphylococcal SAgs have been solved by crystallography. These include, SEA (Schad *et al.*, 1995), SEB (Swaminathan *et al.*, 1992), SEC2 (Papageorgiou *et al.*, 1995), SEC3 (Fields *et al.*, 1996), SED (Sundstrom *et al.*, 1996), SEG (Fernandez *et al.*, 2007), SEH (Hakansson *et al.*, 2000), SEI (Fernandez *et al.*, 2006), SEK (Gunther *et al.*, 2007) and TSST-1 (Acharya *et al.*, 1994). All staphylococcal (and streptococcal) SAgs share a conserved protein fold, which is also found in the staphylococcal superantigen-like toxins (SSLs). Two structural domains are separated by a long central α-helix. The N-terminal domain is a mixed β-barrel with Greek-key topology and is called an oligonucleotide oligosaccharide-binding (OB)-fold (Zhang and Kim, 2000). The C-terminal β-grasp domain consists of a mixed β-sheet capped by the central

α-helix. Several SAgs possess a zinc-binding site in the C-terminal domain that is used for binding to the MHC II β-chain. A novel loop domain located between the α3 helix and the β8 strand was recently discovered in the SEK and SEI structures (Fernandez *et al.*, 2006; Gunther *et al.*, 2007) and is absent in all other staphylococcal SAgs. The α3-β8 loop plays an important role in TcR binding (see below).

Binding of SAgs to MHC class II and TcR

Based on crystallographic data of SAgs in complex with MHC II, the interaction of SAgs with MHC II has been classified into four distinct groups: (a) binding to MHC class II α-chain entirely peripheral to the bound antigen peptide (peptide-independent binding), e.g. HLA-DR1/SEB (Jardetzky *et al.*, 1994); (b) binding to MHC II α-chain and extension over the bound peptide (peptide-dependent binding), e.g. HLA-DR1/TSST-1 (Kim *et al.*, 1994); (c) zinc-mediated binding to MHC II β-chain and extension over the bound peptide (peptide-dependent binding), e.g. HLA-DR1/SEI (Fernandez *et al.*, 2006), and d) SAgs that combine binding modes (a) and (c), allowing them to cross-link MHC II molecules, e.g. SEA and SED. SAgs that bind to the generic MHC II α-chain possess an exposed hydrophobic loop within their N-terminal domain that interacts with a hydrophobic groove located in the distal region of the $DR\alpha_1$-domain. SAgs that bind to the MHC II β-chain possess a zinc-binding site in their C-terminal domain and the tetrahedrically complexed Zn^{2+} interacts with the highly conserved His 81 within the MHC II β-chain. It has been reported that zinc-dependent SAgs achieve promiscuous binding to the polymorphic MHC II β-chain by targeting conservatively substituted residues and circumvent peptide specificity by engaging MHC-bound peptides at their conformationally conserved N-terminal regions (Fernandez *et al.*, 2006).

SAgs interact with TcR molecules by binding to the variable region of the β-chain (Vβ-domain) resulting in oligoclonal stimulation of a defined T-cell repertoire potentially activation >20% of all T-cells. Crystal structures of SAgs complexed with Vβ domains have revealed structurally diverse interactions, although binding to the TcR Vβ CDR2 loop seems to be a requirement for all bacterial SAgs, whereas binding to other Vβ domain regions appears to result in Vβ domain specificity and cross-reactivity of SAgs (Sundberg et al., 2007). SEB and SEC engage the TcR Vβ chain through mostly conformationally dependent mechanisms that are mainly independent of specific Vβ amino acid site chains and are restricted to CDR2 and HV4 regions in the TcR (Sundberg et al., 2002). This results in a relatively low Vβ specificity, e.g. SEC-1 targets TcRs carrying human Vβ3.2, 6.4, 6.9, 12 and 15.1 domains. In contrast, TSST-1 is highly specific for human Vβ2 T-cells and this is caused by a unique interaction of TSST-1 with the TcR FR3 loop at the expense of binding to each of the hypervariable structures. Furthermore, TSST-1 targets a specific residue (Lys62) on the FR3 loop and this interaction is required for T-cell stimulation (Moza et al., 2007). The crystal structure of SEK in complex with human Vβ5.1 revealed similar characteristics (Gunther et al., 2007). SEK has an extended loop region (α3-β8 loop) with a specific residue that forms side chain-to-side chain hydrogen bonds with two relatively uncommon residues (63 and 75) on the TcR Vβ FR3 and FR4 regions. The α3-β8 loop region is also found in SEI (Fernandez et al., 2006), but its exact role in TcR binding is still elusive. Another variation in the TcR binding mechanism has been identified with SEH. In contrast to all other known staphylococcal SAgs, SEH recognizes the variable region of the α-chain (Vα 27), which was shown by Biacore real-time binding studies (Pumphrey et al., 2007).

Staphylococcal superantigens in disease

Toxic shock syndrome (TSS)

TSS is a major systemic disease characterized by high fever, desquamation of the skin, hypotension and systemic organ failure. TSS is a capillary leak syndrome that is primarily mediated by TNF-α. It is caused by the sudden release of superantigen into the blood where it triggers massive T-cell mediated cytokine release.

Menstrual associated Toxic Shock Syndrome (TSS) is caused by intravaginal growth of a phage group I strain of S. aureus producing large amounts of the superantigen TSST-1. An epidemic of TSS was observed in the late 1970s and early 1980s with the introduction of a high-absorbance tampon (McCormick et al., 2001). Non-menstrual cases of TSS are commonly associated with other SEs in particular SEA, SEB and SEC, whereas SAgs encoded on the enterotoxin gene cluster (egc), SEG, SEI, SElM, SElN and SElO, appear to be associated with suppurative infections (Dauwalder et al., 2006; Ferry et al., 2005).

Atopic dermatitis

It has recently been postulated that staphylococcal SAgs might play a role in atopic dermatitis (AD) (reviewed by Lin et al., 2007). A study by Skov et al. showed that application of SEB on intact skin induced local up-regulation of SEB-reactive T-cells and dermatitis in both AD patients and healthy subjects (Skov et al., 2000). When stimulated with SEB, peripheral blood mononuclear cells (PBMCs) isolated from AD patients produced more of the Th2-type cytokine IL-4 compared to PBMCs from healthy donors, which preferentially produced the Th1-type cytokine IFN-γ (Lin et al., 2003). Furthermore, it was shown in a murine model of AD that topical exposure of SEB provokes epidermal accumulation of CD8+ T-cells, a mixed Th1/Th2 type dermatitis and vigorous production of specific IgE and IgG2a antibodies, which can be related to the chronic phase of atopic skin inflammation (Savinko et al., 2005). Keratinocytes lack MHC class II molecules in healthy individuals, but can be activated by SAgs, thereby facilitating epithelial presentation of allergen to Th-2 cells. It has been shown that the SEB-induced T-cell response was mediated by two synergistic mechanisms; SEB-induced IFN-γ promoted expression of MHC class II and intercellular adhesion molecule-1 (ICAM-1) by presenting keratinocytes, and SEB-induced IL-4 directly amplified allergen-specific CD4+ T-cell production of many cytokines (Ardern-Jones et al., 2007).

Why do bacteria produce SAgs?

After more than a decade of intensive research, it is still uncertain why bacteria produce superantigens. A possible advantage might involve corrupting host immunity by interfering with the adaptive immune response. SAgs drive a profound Th1 type responses characterized by high levels of type 1 cytokines, such as IL-2, IFN-γ and TNF-α, and non-specific T-cell proliferation. One might argue that their prime function is to suppress a Th2 type response thereby preventing high affinity antibodies. However, recent reports have shown that SAgs are also able to stimulate a profound Th2-type response and might play a role in atopic dermatitis (AD) and other allergic diseases (Lin *et al.*, 2007). When stimulated with SEB, peripheral blood mononuclear cells (PBMCs) isolated from AD patients produced more of the Th2-type cytokine IL-4 compared to PBMCs from healthy donors, which preferentially produced the Th1-type cytokine IFN-γ (Lin *et al.*, 2003). This suggests that SAgs might be amplifiers rather than modifiers of the adaptive immune response.

While SAgs stimulate T-cells, the adaptive immune response is not considered critical in the immediate defence against *S. aureus*. What is important however is innate immunity mediated by complement, neutrophils and phagocytosis. Why then would the organism produce a toxin that so effectively stimulates so many T-cells? The reason may lie in the effect that T-cell cytokines have on the innate immune response such as neutrophil function and phagocytosis. One feature of SAgs is their extreme potency which suggests that they exert their crucial effects at very low concentrations at the early stage of bacterial colonization. At this level, the stimulation of T-cells producing INF-γ might suppress the local innate immunity thus increasing survival of the organism (Unnikrishnan *et al.*, 2002).

SSLs (staphylococcal superantigen-like proteins)

The staphylococcal superantigen-like proteins (SSLs) (Lina *et al.*, 2004), first termed the staphylococcal exotoxin-like proteins (SETs) by Williams *et al.* in 2000 (Williams *et al.*, 2000), were so named due to their sequence identity to the superantigen family of *S. aureus* and *S. pyogenes*. They were identified as SAg-like molecules based on a high degree of identity in the highly conserved α4 region of the SAg sequence as well as containing the SAg GGI/VT motif. The SSLs

have however failed to exhibit the Vβ-specific stimulation of T-cells that is a hallmark property of the superantigens (Langley *et al.*, 2005). They do not contain the FL motif of the MHC class II α-chain binding SAgs or the Zn^{2+} co-ordinating residues of the MHC class II β-chain binding SAgs. The SSLs possess low amino acid sequence identity to the SAgs at less than 20%. Greatest identity is shared with the TSST-1 sequence, creating an alignment sub-group with this previous superantigen outlier. Despite this the SSLs share structural identity with the SAgs with the common 2-domain OB-fold/β-grasp architecture.

Fourteen *ssl* genes have been discovered clustered in two regions of the genome. The main cluster contains *ssl1–ssl11*, while the second cluster, consisting of *ssl12*, *ssl13*, and *ssl14*, is located some 680 kb downstream. The SSLs possess N-terminal hydrophobic signal peptides, which after cleavage result in mature secreted exoproteins of approximately 200 amino acids.

Unlike the SAgs that exhibit a diversity of modes towards performing the common function of ligating MHC class II with the TCR, the SSLs do not show conservation in their interactions with the host organism. In fact the presence of so many *ssl* genes within a single genome is highly supportive of non-redundant roles for the various members of the family

SSL7

SSL7 is a superantigen-like protein that binds with high affinity to both IgA and complement C5 and blocks both the binding of IgA to FcαRI and complement-mediated haemolytic activity and serum bactericidal activity (Langley *et al.*, 2005). SSL7 is postulated to serve two distinct but perhaps complementary roles; the first is to inhibit IgA-mediated antistaphylococcal mechanisms at the mucosal surface where the organism binds to mucosal epithelial cells (Ramsland *et al.*, 2007) and the second is to inhibit end stage complement that results in the formation of the membrane attack complex (MAC). However it is uncertain how inhibition of complement C5 should benefit survival of *S. aureus* since the organism is essentially impervious to the lytic activity of MAC. Binding to the Fc region of IgA is through the smaller OB-fold domain while the C5 binding site has been mapped to the C-terminal β-grasp domain on the opposite end of the SSL7 molecule. Thus, the formation of IgA-SSL7-C5 trimers in serum is entirely feasible but the significance of this large complex remains to be determined.

SSL5/SSL11

SSL5 and SSL11 belong to a subgroup of SSLs that bind multiple cell surface receptors presenting the carbohydrate sialyllactosamine [Neu5Acα2–3Galβ1–4GlcNAc] (Baker *et al.*, 2007; Chung *et al.*, 2007). One important immune receptor expressing large amounts of the trisaccharide as part of the sialyl Lewis X (sLe^x) antigen is the P-selectin glycoprotein ligand 1 (PSGL-1) and Bestebroer *et al.* have shown that SSL5 binds to recombinant PSGL-1-Ig fusion protein and effectively inhibits the binding of neutrophils to sLe^x coated glass slides (Bestebroer *et al.*, 2007). Chung *et al.* have shown that SSL11 has very similar binding characteristics to SSL5 and effectively binds to multiple cell surface glycoproteins including PSGL-1 such as the FcαRI to block the binding of IgA (Chung *et al.*, 2007). SSL11 bound to cells of myeloid origin such as neutrophils and monocytes, and more weakly to lymphocytes. Binding was completely prevented by prior treatment of cells with the enzyme neuraminidase that removes the terminal sialic acid from the glycan moiety. Co-crystal structures of SSL11/sLe^x and SSL5/sLe^x show almost identical binding sites on the C-terminal β-grasp domain. SSL11 was shown to be internalized rapidly by human neutrophils, suggesting that binding to multiple glycoproteins induces a rapid energy-dependent internalization mechanism.

Sequence alignment with other members of the SSL family show that SSLs 2–6 and SSL11 all possess the residues required for binding sLacNac. One of these residues, Thr175 in SSL5 and Thr168 in SSL11, make two essential hydrogen bonds to sLex. Mutation to proline in SSL11 results in complete loss of binding to any of the ligands recognized by wild-type SSL and also to a lack of inhibition of neutrophil attachment to P-selectin.

The role of SSLs in staphylococcal infection has yet to be determined but under normal culture conditions they are not significantly expressed. However disruption of the staphylococcal *hrt* gene which codes for part of a two component system regulating iron, results in strong expression of the SSLs when an Δ*hrt* isogenic strain is incubated with haemin, the major form of iron used by the bacterium (Torres *et al.*, 2007). This Δ*hrt* isogenic mutant displays hypervirulence in a mouse systemic infection model with significant abscess formation in liver and kidney, suggesting that expression of the SSLs increases survival and the formation of abscesses.

Other staphylococcal immune evasion molecules

CHIPS (chemotaxis inhibitory protein)

Staphylococcus aureus produces a 14-kDa chemotaxis inhibitory protein called CHIPS that is structurally similar to the β-grasp domain of SAgs and SSLs. CHIPS binds to neutrophils and monocytes but not to lymphocytes and inhibits both neutrophil and monocyte chemotaxis. CHIPS was found to block fMLP and C5a activation of neutrophils and intracellular calcium mobilization by down-regulating their neutrophil receptors (de Haas *et al.*, 2004; Veldkamp *et al.*, 2000). CHIPS is a bacteriophage encoded 14.1-kDa secreted protein found in over 60% of clinical *S. aureus* isolates. Most of the chemotaxis inhibitory activity of *S. aureus* conditioned culture supernatant was shown to be caused by CHIPS using a *chips⁻* isogenic strain although these results did not account for the many virulence factors that are unlikely to be expressed under normal culture conditions.

Mice intravenously injected with recombinant CHIPS protein displayed reduced neutrophil infiltration in an intraperitoneal lavage when analysed 3 hours after injection of C5a.

Binding studies of CHIPS to a human monocytic cell line transfected with either human C5aR or FPR genes showed that CHIPS bound with high affinity to C5aR with an apparent K_d of 1.1 ± 0.2 nM and to FPR with an apparent K_d of 35.4 ± 7.7 nM (Postma *et al.*, 2004).

SCIN (staphylococcal complement inhibitor)

SCIN is a 9.8-kDa protein coded for by a gene located in the staphylococcal pathogenicity island 5 (SaPI5) and is located on the same cluster as the genes for CHIPS, staphylokinase, and SEA (Rooijakkers *et al.*, 2005b; van Wamel *et al.*, 2006). SCIN has been found to inhibit human complement by preventing the deposition of C3b onto the surface of *S. aureus* and thus SCIN blocks the classical, lectin and alternative pathways of complement activation, opsonization and phagocytosis of *S. aureus* by neutrophils. Surprisingly, direct binding of SCIN to C3 or C3b or to any other complement component including C1s, MASP-1, MASP-2, C2, C3, C3b, C4, factor D, factor B or properdin, has not been observed. Because SCIN inhibits C3b deposition but does not affect C4b deposition Roojakers *et al.* have proposed that SCIN acts between the deposition of these two opsonins to block formation of the C4b2a convertase. Indeed SCIN prevented the cleavage of both C2 and factor B in human serum when triggered by yeast zymosans, thus confirming that it blocks the formation of both C4b2a and C4bBb complexes (Rooijakkers *et al.*, 2005b). They found that 10 µg/ml SCIN prevented the production of C2 and factor B cleavage products thus affecting the formation of the 3C convertases C4b2a and C3bBb. The formation of C4b2a and C3bBb is dependent on C2 and factor B first binding their surface-bound cofactors C4b and C3b followed by C2 being cleaved by C1s or MASP-2 and factor B being cleaved by factor D. In the absence of C4b and C3b no inhibition of C2 or factor B cleavage by SCIN was observed suggesting that SCIN inhibits the generation of C3 convertases in the presence of an activator surface.

Map/Eap (extracellular adherence protein)

Map/Eap (major histocompatability complex class II analogous protein/extracellular adherence protein) is an extracellular staphylococcal protein. Initially isolated as a protein with broad binding specificity (McGavin *et al.*, 1993). Map was so named due to sequence homologies of its repeated subdomains with a region of the MHC class II β_1 domain which forms the peptide binding groove (Jonsson *et al.*, 1995). Map is composed of 689 amino acids, with a 30 amino acid signal sequence, and six repeats of a 110 amino acid domain. Map was later given the designation Eap owing to its affinity for at least

seven serum and matrix proteins (Palma *et al.*, 1999). Map/Eap binds fibrinogen, fibronectin, prothrombin, vitronectin, collagen, osteopontin, and bone sialoprotein. Palma *et al.* also showed that Map/Eap could form oligomers and rebind back to *S. aureus* although the binding was difficult to saturate suggesting that this was due to non-specific interactions.

Map/Eap promotes agglutination of staphylococcal cells and enhances the adherence of *S. aureus* to both epithelial cells and fibroblasts. Inactivation of the *eap* gene reduced the binding of the isogenic *S. aureus* to fibroblasts and binding could be complemented by transfecting in the *eap* gene (Hussain *et al.*, 2002). Chavakis *et al.* (2002) found that Eap also bound to ICAM-1 on endothelial cells and blocked the interaction of Mac-1 ($\alpha_M\beta_2$ integrin) binding with ICAM-1 or fibrinogen, as well as interfering with LFA-1 (lymphocyte function-associated antigen-1 or $\alpha_L\beta_2$ integrin) to ICAM-1. Eap also inhibited vitronectin binding to uPAR (the urokinase plasminogen activator receptor) that mediates leukocyte adhesion to vitronectin. Thus, Eap was described as a major anti-inflammatory factor that inhibits β2-integrin-dependent leukocyte recruitment. Mice pretreated either intravenously or intraperitoneally with Eap at concentrations comparable to those found in culture supernatant showed a significant reduction in thiglycollate induced neutrophil recruitment to the peritoneum. In addition 2–3 fold more neutrophils were found in the mouse peritoneum in response to injection of an *eap–* isogenic mutant of strain Newman compared to the parental strain. These authors did not, however, see any increased virulence associated with Eap in either a wound infection model, or an intravenous infection model in mice.

In contrast Lee *et al.* (2002) did observe Eap dependent affects on bacterial infection in a mouse model. In the initial stages of infection with either an Δ*eap* mutant or the parental Newman strain, there was little difference in the rate of infection. However by week 8, 86% of wt Newman infected mice had developed osteomyelitis, 57% had heart abscesses, and 75% had kidney abscesses. In comparison to only 6% of mice infected with the Eap deficient strain had osteomyelitis, none had developed heart disease, and only 3% had developed kidney abscesses. T-cell-deficient nude (*nu/nu*) mice and genotype control (*nu/+*) mice were infected with either wt Newman or Δ*eap*- Newman but showed similar results. In this case, *nu/nu* mice showed similar differences in osteoarthritis and osteomyelitis when infected with either the Δ*eap*- variant or parental strain Newman suggesting that the pathology was not T-cell mediated. However, lack of abscess formation in any of the *nu−/nu−* mice infected with the Δ*eap* strain suggested that abscess formation was T-cell dependent. Lee *et al.* (2002) showed that recombinant Eap inhibited the *in vitro* T-cell proliferation and presented evidence that Eap also induced apoptosis in T-cells.

A major difference between the infection model of Lee *et al.* and that of Chavakis *et al.* was that Lee *et al.* followed infection over eight weeks as opposed to the 5-day infection of mice by Chavakis *et al.* This suggested that Eap was more important for the development of chronic rather than acute infection. Additionally, the anti-inflammatory affect of Eap in the Chavakis *et al.* peritonitis model seemed at odds with the arthritis and abscess formation seen by Lee *et al.* Harraghy *et al.* (2003) suggest that while Eap binding to ICAM-1 on endothelial cells exhibits anti-inflammatory affects, Eap binding to ICAM-1 on antigen presenting cells (APC) may inhibit the interaction between APCs and T-cells, leading to immune suppression.

In 2004 Haggar *et al.* showed that *in vitro* Eap had the ability to inhibit the binding of neutrophils to endothelial cells under both static and flow conditions (Haggar *et al.*, 2004). This inhibition was to a similar degree as that obtained using anti-ICAM-1 antibodies suggesting that Eap was blocking neutrophil attachment via its ability to bind ICAM-1.

In a mouse wound model, healing and closure took significantly longer in the presence of wt *S aureus* when compared to an Δ*eap* strain (Athanasopoulos *et al.*, 2006). In addition, excision wounds took longer to heal when treated with recombinant Eap protein. Treated wounds showed significantly lower neutrophil and macrophage counts after 5 days, reduced tissue factor levels and inhibition of neovascularization as seen by a reduction in the number of CD31-positive blood vessels forming. Perfusion of these wounds with fluorescent microspheres showed a 40–50% reduction in blood flow.

Eap has also been found to impair angiogenesis by inhibiting vascular endothelial growth factor (VEGF) and basic fibroblast growth factor (bFGF) induced proliferation and capillary-like sprouting of endothelial cells (Athanasopoulos *et al.*, 2006; Sobke *et al.*, 2006). Sobke *et al.* showed that this anti-angiogenesis was caused by Eap inhibition of Ras-dependent signalling pathways in endothelial cells.

Eap blocked the binding of peripheral blood T-cells to immobilized ICAM-1 and partially inhibited T-cell binding to endothelial cells under static conditions but not under flow conditions (Xie *et al.*, 2006). However, greater inhibition of transendothelial migration of peripheral blood T-cells by Eap was observed under both static and flow conditions. Adhesion of peripheral blood T-cells to the endothelium is mainly mediated by VLA-4–VCAM-1 whereas transendothelial migration of peripheral blood T-cells is predominantly dependent on LFA-1–ICAM-1 interaction.

Finally, Lee *et al.* (2002) and Xie *et al.* (2006) demonstrate *in vivo* that delayed type hypersensitivity responses were significantly reduced in Eap treated mice primarily through inhibition of T-cell recruitment (Xie *et al.*, 2006).

Efb (extracellular fibrinogen binding protein)
Efb (extracellular fibrinogen binding protein) is a 19-kDa secreted protein formerly known as Fib (fibrinogen-binding protein). The N-terminal half of Efb possesses two 22 amino acid repeats with homology to the C-terminal repeat of coagulase and the N-terminal half of Efb possesses the fibrinogen binding site (Boden and Flock, 1994). Efb is expressed constitutively by all isolates of *S. aureus* examined (Boden Wastfelt and Flock, 1995; Smeltzer *et al.*, 1997). Palma *et al.* (1996) have generated an Δ*efb* mutant to test in an experimental wound infection model and an endocarditis model in rats. Wound infection was significantly reduced when the Δ*efb* strain was used to infect compared to the parental strain. However, no significant differences were observed in the experimental endocarditis model. The loss of Efb did not affect *S. aureus*

adherence to either fibrinogen or fibronectin and Palma *et al.* speculated that Efb interfered with fibrinogen-associated clot formation and wound healing (Palma *et al.*, 1998). Efb has also been shown to inhibit fibrinogen-dependent platelet aggregation (Palma *et al.*, 2001). Efb binds to a region on the A alpha-chain of the D fragment of fibrinogen close to the RGD sequence bound by the platelet integrin GPIIb/IIIa receptor complex.

Efb has also been shown to bind directly to activated platelets via a fibrinogen-independent mechanism and that fibrinogen bound by Efb on platelets cannot promote platelet cross-linking thus further preventing platelet activation and clumping (Shannon and Flock, 2004). Efb significantly prolonged bleeding time in mice injected with just the two N-terminal fibrinogen binding repeats of Efb into the tail vein prior to tail tip cutting (Shannon *et al.*, 2005). In a mouse model of thrombosis, pre-treatment of mice intravenously with the Efb N-terminal domain protected mice against thrombotic death. These *in vivo* results were attributed to the anti-platelet aggregation activity of Efb as no affect on coagulation cascade was evident (Shannon *et al.*, 2005).

Efb also binds to complement C3d and inhibits complement activation (Lee *et al.*, 2004a). The binding site for C3d was isolated to the non-fibrinogen-binding C-terminal region of Efb (Efb-C). It was proposed that Efb predominantly exerted its effect against the alternative pathway because it achieved half maximal inhibition of this pathway at equimolar concentrations to C3 whereas a 50-fold excess of Efb over C3 was required to reach half maximal inhibition of the classic pathway (Hammel *et al.*, 2007b). The authors proposed that the inhibition of the classical pathway was a consequence of Efb-mediated inhibition of the alternative pathway self-amplification loop, which contributes most of the C3b generated by the classical pathway (Harboe *et al.*, 2004). Isothermal titration calorimetry and surface plasmon resonance gave dissociation constants for the interaction between Efb-C and C3d containing fragments of C3 in the low nM range (Hammel *et al.*, 2007b).

The crystal structure of the C-terminal C3-binding domain of Efb (Efb-C) alone and in complex with C3d has been solved (Hammel

et al., 2007b). It revealed that Efb-C comprised three α-helices arranged in a canonical three-helix bundle. The co-crystal structure of Efb-C with C3d showed that basic residues from the α2 helix of Efb-C make extensive contacts with the conserved acidic pocket in the concave face of C3d. Homology modelling of this complex against the determined structure of C3 (Janssen *et al.*, 2005) indicated that residues from the Efb α3-helix could make contacts with the second α_2-macroglobulin domain of the C3 β-chain. Conversely, modelling of the complex against the structure of C3b showed that a steric clash between Efb-C and the first α_2-macroglobulin domain of the C3 β-chain would occur indicating that it was unlikely that Efb could bind C3b. Previous work had shown that this interaction could occur and prompted experimental confirmation that Efb-C causes C3 to undergo a conformational change that renders it incapable of being processed into C3b. Additionally, Efb may bind to already formed C3b and alter the conformation of this active form to prevent its function. Hammel *et al.* (2007b) explained that the weaker inhibition of the classical pathway by Efb may be due to Efb-mediated inhibition of the alternative pathway self-amplification loop which contributes most of the C3b generated by the classical pathway.

Genome scanning identified another molecule called Ehp (Efb homologous protein) which displays 44% sequence identity with the Efb-C domain, including conservation of residues important for binding C3 (Hammel *et al.*, 2007a). Ehp binds to C3d with a dissociation constant of 0.18 nM and a second lower-affinity binding of 125 nM. The C3b high-affinity site of Ehp has been identified and includes residues that are conserved with Efb while the second site was located in a tandem repeat region that contained one of these conserved residues. The ability of Ehp to inhibit the alternative pathway was two- to threefold greater than Efb and like Efb, its binding to C3d causes a conformational change in C3 that prevents the generation and deposition of C3b. Ehp was concurrently reported by a second group under another name Ecb (extracellular complement-binding protein) (Jongerius *et al.*, 2007).

FLIPr (formyl peptide receptor-like inhibitory protein)

Formyl peptide receptor-like 1 (FPRL1) inhibitory protein (FLIPr) was found by homology searches against the *S. aureus* genome sequence using CHIPS as a probe (Prat *et al.*, 2006). It was found to have 49% similarity to the gene encoding CHIPS (*chp*) and 29% homology to CHIPS. The gene for FLIPr (*flr*) was carried by 59% of clinical isolates screened (a similar proportion to those carrying *chp*). It codes for a protein possessing a signal peptide that is cleaved to give a 105 amino acid mature secreted protein. Like CHIPS, FLIPr was shown to inhibit fMLP-induced calcium mobilization of neutrophils although with lower potency to CHIPS.

To test the specificity of FLIPr binding, HEK293 cells were transfected with FPR, FPRL1, FPRL2, or C5aR. FLIPr-FITC bound primarily to FPRL1 transfected HEK293 cells but also displayed weak binding to FPR transfected cells. No specific binding of FLIPr to FPRL2 or C5aR was observed (Prat *et al.*, 2006).

Sbi

Protein A is a well-described protein which binds Ig from multiple species. A second IgG-binding protein has been identified from *S. aureus* (Zhang *et al.*, 1998a). This 436 amino acid protein, Sbi, contains a signal sequence but no cell wall anchoring LPXTG motif. Following the signal sequence are two repeat regions (Sbi-I, Sbi-II) that possess homology to the IgG-binding domains of SpA (Atkins *et al.*, 2008; Burman *et al.*, 2008; Zhang *et al.*, 1998a), two novel domains (Sbi-III, Sbi-IV), a proline-repeat predicted cell-wall spanning region, and a C-terminal tyrosine-rich region proposed to be involved in IgG-mediated signal transduction (Zhang *et al.*, 2000). Small angle X-ray scattering of the putative extracellular region of Sbi revealed that Sbi I-IV was most probably an elongated molecule consisting of four domains joined by short linker regions (Burman *et al.*, 2008). With a predicted pI of 9.8 it has been suggested that Sbi binds to the bacterial surface by electrostatic interactions (Zhang *et al.*, 1998a). However, more recently, Sbi was not found to be cell wall associated but was in fact secreted (Burman *et al.*, 2008). Expression of

Sbi was found to be induced by IgG (Zhang *et al.*, 2000).

The IgG-binding regions of Sbi are located at the N-terminus. Domains Sbi-I and Sbi-II both bind IgG Fc, but unlike SpA no interaction with Fab was observed (Atkins *et al.*, 2008). Homology modelling against the solution structures of SpA IgG-binding domains revealed that amino acids responsible for Fc-binding in helices 1 and 2 of SpA are conserved in Sbi-I and Sbi-II. FAb-binding residues in helices 2 and 3 of SpA are not conserved in Sbi though. Sbi forms large insoluble immune complexes with IgG via the binding of two Fcs per Sbi. Sbi was found to bind an additional serum protein. Expression of individual Sbi domains revealed that while Sbi-I and Sbi-II were responsible for binding IgG, the Sbi-III–IV and Sbi-IV domains bound to complement C3. Sbi-III alone did not possess the capacity to bind C3 (Burman *et al.*, 2008). The Sbi-IV domain bound C3 via interactions with both C3dg and C3a. Sbi-III-IV inhibited all three complement pathways, whereas the Sbi-IV domain alone only inhibited the alternative pathway. Sbi-III-IV was found to activate the alternative pathway leading to the rapid consumption of C3. When incubated together, a large fraction of the total C3 activated in serum became covalently bound to Sbi-III–IV presumably through the exposed C3b thioester. It was hypothesized that Sbi-III–IV caused the formation of a Sbi bound C3bBb that leads to the futile consumption of C3. The inhibition of alternative pathway by the Sbi-IV domain alone did not involve C3 consumption and its interaction with C3 is similar to that of Efb-C and Ehp/Ecb. Indeed, these protein domains are structurally homologous and possess conservation in their C3-binding residues (Upadhyay *et al.*, 2008).

Genetic location of immune evasion genes

Most of the genes for the staphylococcal superantigens are carried on mobile elements such as bacteriophage, plasmids, or pathogenicity islands. Pathogenicity islands are genomic regions of DNA with inferred exogenous origin that are present in pathogenic bacteria but not in non-pathogenic strains of the same or related species

(Dobrindt and Reidl, 2000). They differ in their GC content to the rest of the genome, are flanked by direct repeats, and contain genes encoding for genetic mobility.

The genes encoding SEA (Betley and Mekalanos, 1985), SEE (Borja et al., 1972), and SEP (Kuroda et al., 2001), were reported to be carried by bacteriophage while the genes for SED (Bayles and Iandolo, 1989), and SEJ (Zhang et al., 1998b) are located on the penicillinase plasmid pIB485. The gene encoding SEH is reported to be chromosomally encoded (Su and Wong, 1995). An enterotoxin gene cluster (egc), carried on a pathogenicity island, contains seg, sei, sem, sen, seo, and two enterotoxin pseudogenes ψent1 and ψent2 (Jarraud et al., 2001; Kuroda et al., 2001). The gene encoding SEK is carried by the pathogenicity island SaPI1 (Orwin et al., 2001). It is located at the 3′ end of the island downstream to a partial enterotoxin sequence ent and upstream of the integrase gene. The gene for TSST-1 is located at the 5′ end of this pathogenicity island (Lindsay et al., 1998). Recently a new pathogenicity island SaPI3 has been reported to carry the genes for SEB, SEQ, and SEK (Yarwood et al., 2002). Yarwood et al. found that SaPI1 and SaPI3 share large regions of high homology and that many of their coding regions show homology to phage sequences. The genes seq and sek are located in the same position as the partial ent and sek in SaPI1 while the gene for SEB replaces the gene for TSST-1. The att sites of SaPI1 and SaPI3 were found to be identical and the same integrase genes were located next the attR site. Islands with different integrases appear to be specific for different att sites, and as such are most likely the factor that determines the site of integration within the genome. With SaPI1 and SaPI3 having the same integration site one would exclude the other from integrating which may explain why SEB and TSST-1 have never been isolated from the same strain. Kuroda et al. describe a family of SaPI1-like islands from the strains N315, Mu50, and a bovine strain. The four islands were found to have high identity to one another though their integrases have only 34% amino acid identity and they are integrated into completely different positions in the genome. Interestingly the identical islands SaPIn1 and SaPIm1 (from N315 and

Mu50 respectively) carry the genes for SEL and SEC3 at their 5′ end, upstream to the gene for TSST-1.

The genes encoding SSL1 to SSL10 (ssl1–ssl10) are located clustered together in the genome on a pathogenicity island, SaPIn2 (N315) or SaPIm2 (mu50) (Baba et al., 2002; Kuroda et al., 2001). ssl11 is located 3.4 kb downstream of this cluster on the same pathogenicity island. The genes for SSL12, SSL13, and SSL14 are situated in reverse tandem order approximately 680 kb downstream in another immune evasion cluster. The ssls have so far been identified in every strain of S. aureus analysed. Comparison of the ssl gene clusters from different strains indicates that the order of genes is conserved and that while some strains; for example MW2 carry all 14 genes, others do not carry the full complement of ssls. Col for instance only has ten ssl genes (ssl1 – 4, ssl9 – 14). Homology within the SSL family ranges from approximately 19–70%. There is also a great degree of variation within alleles with some SSLs being highly conserved while other alleles display up to 20% difference in their sequences (R.J. Langley, unpublished).

The genes encoding CHIPS (chp), SCIN (scn), staphylokinase (sak), and SEA (sea) or SEP (sep) are located at the 3′ end of a β-hemolysin converting bacteriophage family (Rooijakkers et al., 2005b; van Wamel et al., 2006). Because of the presence of several immune modulating virulence factors on these bacteriophage, this 3′ 8kb region has been termed an Immune Evasion Cluster (IEC). Of 85 clinical and laboratory S. aureus strains analysed, 90% carry this IEC, with seven different types being identified. Though all carry sak and scn, chp is only present in four of the IEC, and two carry sea while two others carry sep.

A second Immune Evasion Cluster (IEC-2) has also been identified containing the genes for either FLIPr or its homologue FLIPr-like, one of the two SCIN homologues SCIN-B or SCIN-C, Efb, Ecb/Ehp, the SSLs 12 13 and 14, and α-haemolysin was identified (Jongerius et al., 2007). IEC-2 contains transposases and bacteriophage remnants. The genes encoding SCIN-B and SCIN-C are situated at the same genomic position in different strains and are carried by 47 and 32% of strains respectively.

Common structures – different functions

It is becoming increasingly apparent that many of staphylococcal immune evasion molecules are constructed from similar protein domains. This section reviews the general features of some of these structures and the way in which the domains have been adapted to provide different functions.

The SAgs and the SSLs and the most conserved and belong to a single structural superfamily. The N-terminal domain of the SAgs and SSLs is of the oligosaccharide/oligonucleotide binding fold or OB-fold motif and is formed by five β strands that form a compact and very stable concave β-barrel structure (Murzin, 1993). Other secreted bacterial toxins containing this motif include staphylococcal nuclease, and the B subunits of the AB_5 toxins, which include cholera toxin, heat-labile enterotoxins I and IIb, Verotoxin, and pertussis toxin (Mitchell *et al.*, 2000; Murzin, 1993). The OB-fold is used for DNA binding and carbohydrate binding respectively in these proteins and is thought to be a stable common fold able to accommodate a wide variety of sequences and functions. Orengo *et al.* (1994) have suggested that the OB-fold represents one of the superfolds which are extra-stable folds that allow a wide range of sequence variation and functions while retaining a similar topology.

The C-terminal domain of the SAgs and SSLs is a four stranded antiparallel β-sheet backing on to the central α-helix (the 'β-grasp' motif), plus 3 coplanar β-strands, and an additional α-helix. The β-grasp motif is another example of the highly stable superfolds (Orengo *et al.*, 1994).

Comparison of X-ray crystal structures of SSL5 and TSST (Arcus *et al.*, 2002) reveals that only 37 residues are conserved even though both proteins fold into very similar structures (Fig. 8.1A). The 37 residues are involved in maintaining structural integrity by contributing to the hydrophobic core, or have side chains forming buried charge-charge interactions or hydrogen bonding with main chain atoms. The β-grasp domain is utilized by the SSLs, SAgs, Eap and CHIPS (Arcus *et al.*, 2002; Geisbrecht *et al.*, 2005; Haas *et al.*, 2005; Proft and Fraser, 2003) (Fig. 8.1B).

SAgs display a high degree of heterogeneity in their interactions with the αβ T-cell receptor with some binding to a broad range of Vβs while others display a more restricted Vβ repertoire. TCR binding is predominantly through interactions involving the shallow cleft formed between the C-terminal and N-terminal domain of the SAg but there is no evidence of any conserved amino acid residues providing a framework for TcR binding across all the SAgs. The MHC class II binding of SAgs is slightly more conserved with two different sites on opposite sides of the molecule. The high affinity site is located in the β-grasp face of the C-terminal domain and employs a zinc atom to bind to a highly conserved histidine residue at position 81 located on the top of the β-chain of MHC class II (Hudson *et al.*, 1995; Li *et al.*, 1997; Li *et al.*, 2001; Petersson *et al.*, 2001, 2002). The second lower affinity binding site is on the SAg N-terminal domain and consists of a narrow hydrophobic ridge that binds into a shallow groove in the top of the conserved MHC class II α-chain (Jardetzky *et al.*, 1994; Kim *et al.*, 1994; Petersson *et al.*, 2002). Some SAgs such as SEB, SEC, SEG, TSST, SPEA and SSA use the low affinity site, while the majority (17/28) employ the C-terminal zinc site. A subset possesses both sites including SEA, SEE and SED and thus are able to cross-link MHC class II on the surface of the antigen presenting cell (Al-Daccak *et al.*, 1998; Mehindate *et al.*, 1995; Newton *et al.*, 1996; Petersson *et al.*, 2002; Tiedemann and Fraser, 1996).

The SSLs do not bind MHC class II or TCR and thus do not stimulate T-cells. The structures of SSL5 (Arcus *et al.*, 2002; Baker *et al.*, 2007), SSL7(Al-Shangiti *et al.*, 2004; Ramsland *et al.*, 2007) and SSL11 (Chung *et al.*, 2007) have been solved and confirm their structural similarity with SAgs even though they bind a wide range of host factors. X-ray crystallography and mutagenesis of various SSLs has revealed that different regions of individual SSLs are used for a widely diverse range of interactions with host factors.

SSL7 binding to the C_H2 -C_H3 region of IgA using the same residues that are used in IgA interaction with FcαRI (Ramsland *et al.*, 2007; Wines *et al.*, 2006). The region of SSL7 responsible for this binding is located on the outer face of the SSL7 N-terminal domain (Ramsland *et*

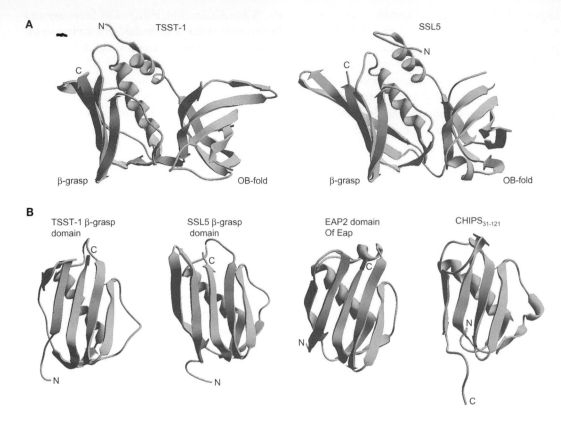

Figure 8.1 Structural comparison of the β-grasp domain-containing immune evasion molecules of *S. aureus*. A) Structures of TSST-1 (PDB 2tss) and SSL5 (1m4v) showing the OB-fold and β-grasp domains separated by the central α-helix. B) The β-grasp domains of TSST-1 (PDB 2tss) SSL5 (1m4v), the EAP2 domain of Eap (PDB 1yn3), and CHIPS (PDB 1xee).

al., 2007). Preliminary evidence indicates that the complement C5 binding site is located in the C-terminal domain (N. Willoughby and J.D. Fraser, unpublished data).

Crystal structures of the second domain of Eap (EAP2) and two other functionally uncharacterized Eap-like proteins have been determined (Geisbrecht *et al.*, 2005). The Eap domain consisted of an α-helix packed diagonally onto a five-stranded mixed β-sheet with high homology to the β-grasp fold C-terminal domains of the SAgs and SSLs. Small angle X-ray diffraction of recombinant Eap confirmed that the protein assumed an elongated conformation in solution. An averaged model based on the data collected from the small angle X-ray diffraction of Eap indicated that the protein consisted of four globular subunits connected by short stretched regions. An overall solution structure of Eap based on the EAP2 domain structure has been proposed in

which the first two domains (EAP1 and EAP2) are linearly aligned to each other and the EAP3 and EAP4 domains are similarly aligned with a distinct bend between EAP2 and EAP3. This proposed elongated conformation of Eap with its many solvent exposed β-grasp domains offers the potential for a multitude of binding surfaces both within each domain and across domains. This solution structure thus might help to explain the broad binding capacity and extended functionality displayed by Eap (Athanasopoulos *et al.*, 2006; Chavakis *et al.*, 2002; Lee *et al.*, 2002; McGavin *et al.*, 1993; Palma *et al.*, 1999).

The structure of a 90 amino acid portion of CHIPS (CHIPS$_{31-121}$) was determined by NMR and was found to also be a β-grasp fold protein consisting of an N-terminal α-helix placed diagonally across a four-stranded antiparallel β-sheet (Haas *et al.*, 2005). While the N-terminus of CHIPS has been found to be important for blocking the activity of the FPR, structural determination of the full length protein has not yet been possible. CHIPS$_{31-121}$ exhibits high structural homology to the SSLs with a lower degree of structural homology to the SAgs TSST-1 and SPE-C. A similar degree of similarity is shared with the EAP2 domain. The greatest region of sequence conservation between these protein families is located in the α-helix region and appears to be involved in maintaining conformational stability. The arrangement of the β-strands in CHIPS$_{31-121}$ differs to that seen in the SSLs, SAgs, and EAP2. Strands β1, β2 and β3 of CHIPS$_{31-121}$ correspond to β9, β10 and β12 of SSL5. However, the anti-parallel β4 strand of CHIPS$_{31-121}$ corresponds to the β6 strand of SSL5 which hydrogen bonds in parallel to the C-terminal β12 strand.

Residues R44 and K95 in CHIPs are important for C5aR binding and are located on the α-helix and the β3 strand respectively. This region is distinct from the sialic acid binding site of SSL5 and SSL11 and also the C5-binding residues that have been identified on

the β-grasp domain of SSL7. The flexibility and importance of the β-grasp fold as a domain used by multiple immune evasion molecules is clearly evident.

Comparison of the α-helix bundle IEMs

SCIN, EFB-C, Ehp, Sbi-IV are all involved in complement inhibition and the analysis of the structures of SCIN (Rooijakkers et al., 2007), EFB-C (Hammel et al., 2007b), Ehp (Hammel et al., 2007a), Sbi-IV (Upadhyay et al., 2008), SpA domains B (Gouda et al., 1992; Torigoe et al., 1990), D (Graille et al., 2000), and E (Starovasnik et al., 1996) reveal that they all comprise a triple α-helix bundle fold in which three helices linked by two short loops pack into a coiled coil with a core of hydrophobic residues and surface exposed charged residues (Fig. 8.2A). The arrangement of the helices differs between different members of this related family of complement controlling molecules as shown in Fig. 8.2A. In the case of SpA B and SCIN helix 2 sits on top and between helices 1 and 3 whereas Efb-C, Ehp, and Sbi-IV are arranged with helix 3 sitting atop

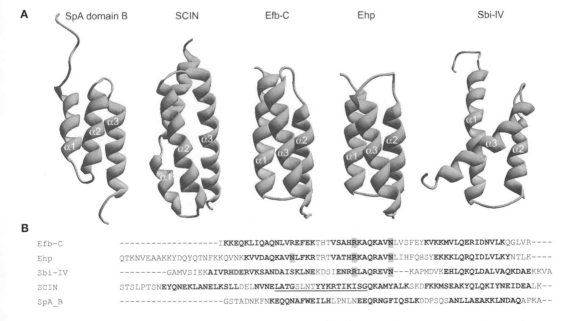

Figure 8.2 Comparison of the triple helix domain bundle-containing immune evasion molecules of S. aureus. A) Structural comparison of protein A domain B (PDB 1ss1), SCIN (PDB 2qff), Efb-C (PDB 2gom), Ehp (PDB 2noj), and Sbi-IV (PDB code 2jvh). B) Sequence alignment of the triple helix bundle domain immune evasion molecules. Alpha helices are indicated in bold type. Conserved C3-binding residues R and N from Efb-C, Ehp, and Sbi are highlighted in grey. The region implicated in C3-binding in SCIN is underlined.

and between helices 1 and 2 when viewed in the same orientation.

In the co-crystal structure of Efb-C bound to C3d (Hammel *et al.*, 2007b) the main region of contact was contributed by the α2 helix of Efb-C, notably residues H130, R131, K135, and N138, with additional contacts made by K106 and K110 on the α1 helix and α3 helix residue K148. Major contributions were made by the side chain of arginine-131, which protruded deeply into a solvent exposed pocket in C3d, and asparagine-138, which displayed extensive hydrogen bonding with residues that formed the H4–H5 loop of C3d. These two residues were shown by mutagenesis to be essential for the binding of C3 by Efb-C. The H4–H5 loop undergoes a conformational change during activation of C3 (Isenman and Cooper, 1981; Isenman *et al.*, 1981) and Hammel *et al.* proposed that Efb-C prevents the movement of this loop to the active state conformation.

The Efb-C homologous protein Ehp exhibits both similarities and differences in its interaction with C3 (Hammel *et al.*, 2007a). Arginine-75 and asparagine-82 of Ehp correspond to R131 and N138 of Efb-C (Fig. 8.2B). The co-crystal structure of Ehp bound to C3d supports the mutagenesis results that indicate the importance of these residues in C3 binding and inhibition. The structure of Ehp-C3d superimposes very well on that of Efb-C-C3d with the predominance of C3d contacting residues contributed by the α2 helix. Again, extensive contacts between the conserved arginine and asparagine with C3d are key features of the binding interface.

A solution structure of Sbi-IV revealed that this domain forms a three-helix bundle fold similar to that observed in Efb-C, Ehp, SCIN, and the SpA IgG-binding domains (Upadhyay *et al.*, 2008). Comparison of Sbi-IV with the structurally homologous Efb-C showed that only eight amino acids were conserved between the two domains. Five of these were located in the α2 helix. In particular were an Arg and an Asn that aligned with an Arg and Asn in both Efb-C and Ehp that are important in binding to C3d (Fig. 8.2B). Two other residues, His and Lys in Efb, which interact with C3d are conservatively substituted by arginines in Sbi-IV. Mutation of

the Arg and Asn to alanines rendered Sbi-IV incapable of binding C3d.

Although SCIN interacts with C3 to inhibit complement activation and is structurally similar to Efb-C, Ehp and Sbi-V, it does not possess identity with the α2 residues shown to be important for C3 binding by these other immune evasion molecules and the arrangement of its helices differs to that of the other three proteins. Unsurprisingly, its mode of action is quite distinct. SCIN functions by binding activated C3 convertases to stabilize and ultimately inactivate the convertase. The active site has been functionally mapped to a region that includes residues 31 to 48 of SCIN (Fig. 8.2B). These amino acids encompass the C-terminal end of α1 through to the N-terminal half of α2 (Rooijakkers *et al.*, 2007).

Inhibition of leucocyte migration
Migration of leucocytes to the site of an infection is an important aspect of the host immune response (reviewed in Seely *et al.*, 2003). It is not surprising then that several *S. aureus* virulence factors function to prevent this from occurring. Many of the steps involved in migration are targeted (Fig. 8.3). For instance, the initial rolling of neutrophils on the endothelial surface is mediated by the interaction of PSGL-1 on the neutrophil with P-selectin on the endothelial cell and is necessary for more stable adhesion to occur. This interaction is inhibited by SSL5 and SSL11 which bind PSGL-1 and block selectin binding. Stable adhesion follows the rolling event and is required for transmigration (or diapedesis) of the neutrophil. This step relies on the neutrophil intergrin LFA-1 binding to ICAM-1 on the endothelial cell and is blocked by Eap binding to ICAM-I thus inhibiting transendothelial migration of the neutrophil to the site of infection. CHIPs and FLIPr inhibit neutrophil activation by preventing the C5a, formyl peptide, and FPRL1 agonist-mediated stimulation of neutrophils that leads to the chemotaxis of these cells to the site of inflammation (de Haas *et al.*, 2004; Prat *et al.*, 2006).

Potential uses of staphylococcal immune evasion molecules
Some of the immune evasion molecules have shown suppression of inflammation *in vivo* and

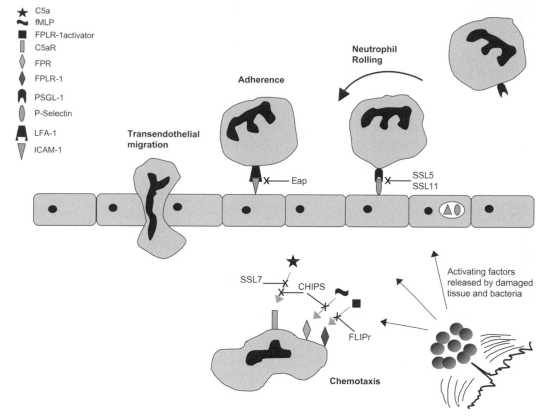

Figure 8.3 Inhibition of neutrophil migration by immune evasion molecules of *S. aureus*.

thus have been proposed as potential therapeutics. However, the multifunctional nature of these proteins means that there is a high risk that unexpected adverse effects might accompany the desired response. Sufficient care must be exercised to eliminate all potential adverse activities before use in humans.

Xie *et al.* (2006) have shown that purified Eap prevented the development and progression of Experimental Autoimmune Encephalitis (a model for multiple sclerosis) when injected into mice and these studies concluded that the activity was due to inhibition of T-cell infiltration into the brain.

Eap has been shown to inhibit bone metastases of breast cancer in nude mice by blocking α_v-integrin-dependent binding of the matrix associated osteopontin growth factor. Eap blocked the adhesion of a mammary carcinoma cell line to bone immobilized osteopontin and also inhibited osteopontin-induced transmigration of these cells in a transwell assay (Schneider *et al.*, 2007).

Efb has also been shown to have anti-platelet and anti-complement activities with each of these activities separated into the N-terminal or C-terminal domain proteins respectively (Lee *et al.*, 2004b; Shannon *et al.*, 2005).

Staphylococcal virulence factors use common structures as scaffolds to allow for a wide variety of interactions in its attempts to subvert and evade host immune responses. Future endeavours to investigate 'pattern recognition' of some of these conserved structures may be of potential therapeutic use for developing a broad-spectrum anti-staphylococcal target. However, allelic variation of virulence factors and low sequence homology between virulence factors with common structures means antibody therapy directed against one factor or allele may not be successful against others.

Concluding remarks

This chapter is not intended to be a comprehensive list of virulence factors from *S. aureus*.

Rather it presents some of the more recently characterized molecules to demonstrate that the organism targets both the innate (SSLs, CHIPS, FLIPr, SCIN, Sbi, Eap, Efb) and adaptive arms (SAgs, SpA, Sbi, Eap) of the immune system. What is apparent is that many of these compact molecules have multiple activities; the most common of which appears to be the inhibition of complement and/or leucocyte function. This clearly highlights the importance of these two immune components in the host defence against *S. aureus*.

There also appears to be significant redundancy in the staphylococcal armamentarium with several structurally different molecules performing the same task. The challenge will be to place their individual and collective activities in the context of staphylococcal growth and defence and to use this information to develop strategies that might reduce staphylococcal disease.

References

Acharya, K.R., Passalacqua, E.F., Jones, E.Y., Harlos, K., Stuart, D.I., Brehm, R.D., and Tranter, H.S. (1994). Structural basis of superantigen action inferred from crystal structure of toxic-shock syndrome toxin-1. Nature 367, 94–97.

Al-Daccak, R., Mehindate, K., Damdoumi, F., Etongue-Mayer, P., Nilsson, H., Antonsson, P., Sundstrom, M., Dohlsten, M., Sekaly, R.P., and Mourad, W. (1998). Staphylococcal enterotoxin D is a promiscuous superantigen offering multiple modes of interactions with the MHC class II receptors. J. Immunol. 160, 225–232.

Al-Shangiti, A.M., Naylor, C.E., Nair, S.P., Briggs, D.C., Henderson, B., and Chain, B.M. (2004). Structural relationships and cellular tropism of staphylococcal superantigen-like proteins. Infect. Immun. 72, 4261–4270.

Arcus, V.L., Langley, R., Proft, T., Fraser, J.D., and Baker, E.N. (2002). The Three-dimensional structure of a superantigen-like protein, SET3, from a pathogenicity island of the *Staphylococcus aureus* genome. J. Biol. Chem. 277, 32274–32281.

Ardern-Jones, M.R., Black, A.P., Bateman, E.A., and Ogg, G.S. (2007). Bacterial superantigen facilitates epithelial presentation of allergen to T helper 2 cells. Proc. Natl. Acad. Sci. USA 104, 5557–5562.

Athanasopoulos, A.N., Economopoulou, M., Orlova, V.V., Sobke, A., Schneider, D., Weber, H., Augustin, H.G., Eming, S.A., Schubert, U., Linn, T., et al. (2006). The extracellular adherence protein (Eap) of *Staphylococcus aureus* inhibits wound healing by interfering with host defense and repair mechanisms. Blood 107, 2720–2727.

Atkins, K.L., Burman, J.D., Chamberlain, E.S., Cooper, J.E., Poutrel, B., Bagby, S., Jenkins, A.T., Feil, E.J., and

van den Elsen, J.M. (2008). S. aureus IgG-binding proteins SpA and Sbi: host specificity and mechanisms of immune complex formation. Mol Immunol 45, 1600–1611.

Baba, T., Takeuchi, F., Kuroda, M., Yuzawa, H., Aoki, K., Oguchi, A., Nagai, Y., Iwama, N., Asano, K., Naimi, T., et al. (2002). Genome and virulence determinants of high virulence community-acquired MRSA. Lancet 359, 1819–1827.

Baker, H.M., Basu, I., Chung, M.C., Caradoc-Davies, T., Fraser, J.D., and Baker, E.N. (2007). Crystal structures of the staphylococcal toxin SSL5 in complex with sialyl Lewis X reveal a conserved binding site that shares common features with viral and bacterial sialic acid binding proteins. J Mol Biol 374, 1298–1308.

Bayles, K.W., and Iandolo, J.J. (1989). Genetic and molecular analyses of the gene encoding staphylococcal enterotoxin D. J. Bacteriol. 171, 4799–4806.

Bestebroer, J., Poppelier, M.J., Ulfman, L.H., Lenting, P.J., Denis, C.V., van Kessel, K.P., van Strijp, J.A., and de Haas, C.J. (2007). Staphylococcal superantigen-like 5 binds PSGL-1 and inhibits P-selectin-mediated neutrophil rolling. Blood 109, 2936–2943.

Betley, M.J., and Mekalanos, J.J. (1985). Staphylococcal enterotoxin A is encoded by phage. Science 229, 185–187.

Boden, M.K., and Flock, J.I. (1994). Cloning and characterization of a gene for a 19 kDa fibrinogen-binding protein from *Staphylococcus aureus*. Mol. Microbiol. 12, 599–606.

Boden Wastfelt, M.K., and Flock, J.I. (1995). Incidence of the highly conserved fib gene and expression of the fibrinogen-binding (Fib) protein among clinical isolates of *Staphylococcus aureus*. J Clin Microbiol 33, 2347–2352.

Borja, C.R., Fanning, E., Huang, I.Y., and Bergdoll, M.S. (1972). Purification and some physicochemical properties of staphylococcal enterotoxin Eur. J. Biol. Chem. 247, 2456–2463.

Burman, J.D., Leung, E., Atkins, K.L., O'Seaghdha, M.N., Lango, L., Bernado, P., Bagby, S., Svergun, D.I., Foster, T.J., Isenman, D.E., et al. (2008). Interaction of Human Complement with Sbi, a Staphylococcal Immunoglobulin-binding Protein: indications of a novel mechanism of complement evasion by *Staphylococcus aureus*. J. Biol. Chem. 283, 17579–17593.

Chavakis, T., Hussain, M., Kanse, S.M., Peters, G., Bretzel, R.G., Flock, J.I., Herrmann, M., and Preissner, K.T. (2002). *Staphylococcus aureus* extracellular adherence protein serves as anti-inflammatory factor by inhibiting the recruitment of host leukocytes. Nat Med 8, 687–693.

Chung, M.C., Wines, B.D., Baker, H., Langley, R.J., Baker, E.N., and Fraser, J.D. (2007). The crystal structure of staphylococcal superantigen-like protein 11 in complex with sialyl Lewis X reveals the mechanism for cell binding and immune inhibition. Mol. Microbiol. 66, 1342–1355.

Dauwalder, O., Thomas, D., Ferry, T., Debard, A.L., Badiou, C., Vandenesch, F., Etienne, J., Lina, G., and Monneret, G. (2006). Comparative inflammatory properties of staphylococcal superantigenic enterotoxins SEA and SEG: implications for septic shock. J Leukoc Biol 80, 753–758.

de Haas, C.J., Veldkamp, K.E., Peschel, A., Weerkamp, F., Van Wamel, W.J., Heezius, E.C., Poppelier, M.J., Van Kessel, K.P., and van Strijp, J.A. (2004). Chemotaxis inhibitory protein of Staphylococcus aureus, a bacterial antiinflammatory agent. J. Exp. Med. 199, 687–695.

Dobrindt, U., and Reidl, J. (2000). Pathogenicity islands and phage conversion: evolutionary aspects of bacterial pathogenesis. Int. J. Med. Microbiol. 290, 519–527.

Fernandez, M.M., Bhattacharya, S., De Marzi, M.C., Brown, P.H., Kerzic, M., Schuck, P., Mariuzza, R.A., and Malchiodi, E.L. (2007). Superantigen natural affinity maturation revealed by the crystal structure of staphylococcal enterotoxin G and its binding to T-cell receptor Vbeta8.2. Proteins 68, 389–402.

Fernandez, M.M., Guan, R., Swaminathan, C.P., Malchiodi, E.L., and Mariuzza, R.A. (2006). Crystal structure of staphylococcal enterotoxin I (SEI) in complex with a human major histocompatibility complex class II molecule. J. Biol. Chem. 281, 25356–25364.

Ferry, T., Thomas, D., Genestier, A.L., Bes, M., Lina, G., Vandenesch, F., and Etienne, J. (2005). Comparative prevalence of superantigen genes in Staphylococcus aureus isolates causing sepsis with and without septic shock. Clin. Infect. Dis. 41, 771–777.

Fields, B.A., Malchiodi, E.L., Li, H., Ysern, X., Stauffacher, C.V., Schlievert, P.M., Karjalainen, K., and Mariuzza, R.A. (1996). Crystal structure of a T-cell receptor beta-chain complexed with a superantigen. Nature 384, 188–192.

Foster, T.J. (2005). Immune evasion by staphylococci. Nat. Rev. Microbiol. 3, 948–958.

Fraser, J., Arcus, V., Kong, P., Baker, E., and Proft, T. (2000). Superantigens – powerful modifiers of the immune system. Mol. Med. Today 6, 125–132.

Freney, J., Kloos, W.E., Hajek, V., Webster, J.A., Bes, M., Brun, Y., and Vernozy-Rozand, C. (1999). Recommended minimal standards for description of new staphylococcal species. Subcommittee on the taxonomy of staphylococci and streptococci of the International Committee on Systematic Bacteriology. Int J Syst Bacteriol 49 Pt 2, 489–502.

Geisbrecht, B.V., Hamaoka, B.Y., Perman, B., Zemla, A., and Leahy, D.J. (2005). The crystal structures of EAP domains from Staphylococcus aureus reveal an unexpected homology to bacterial superantigens. J. Biol. Chem. 280, 17243–17250.

Gill, S.R., Fouts, D.E., Archer, G.L., Mongodin, E.F., Deboy, R.T., Ravel, J., Paulsen, I.T., Kolonay, J.F., Brinkac, L., Beanan, M., et al. (2005). Insights on evolution of virulence and resistance from the complete genome analysis of an early methicillin-resistant Staphylococcus aureus strain and a biofilm-producing methicillin-resistant Staphylococcus epidermidis strain. J. Bacteriol. 187, 2426–2438.

Gouda, H., Torigoe, H., Saito, A., Sato, M., Arata, Y., and Shimada, I. (1992). Three-dimensional solution structure of the B domain of staphylococcal protein A: comparisons of the solution and crystal structures. Biochemistry 31, 9665–9672.

Graille, M., Stura, E.A., Corper, A.L., Sutton, B.J., Taussig, M.J., Charbonnier, J.B., and Silverman, G.J. (2000). Crystal structure of a Staphylococcus aureus protein A domain complexed with the Fab fragment of a human IgM antibody: structural basis for recognition of B-cell receptors and superantigen activity. Proc. Natl. Acad. Sci. USA 97, 5399–5404.

Gunther, S., Varma, A.K., Moza, B., Kasper, K.J., Wyatt, A.W., Zhu, P., Rahman, A.K., Li, Y., Mariuzza, R.A., McCormick, J.K., et al. (2007). A novel loop domain in superantigens extends their T-cell receptor recognition site. J. Mol. Biol. 371, 210–221.

Haas, P.J., de Haas, C.J., Poppelier, M.J., van Kessel, K.P., van Strijp, J.A., Dijkstra, K., Scheek, R.M., Fan, H., Kruijtzer, J.A., Liskamp, R.M., et al. (2005). The structure of the C5a receptor-blocking domain of chemotaxis inhibitory protein of Staphylococcus aureus is related to a group of immune evasive molecules. J. Mol. Biol. 353, 859–872.

Haggar, A., Ehrnfelt, C., Holgersson, J., and Flock, J.I. (2004). The extracellular adherence protein from Staphylococcus aureus inhibits neutrophil binding to endothelial cells. Infect. Immun. 72, 6164–6167.

Hakansson, M., Petersson, K., Nilsson, H., Forsberg, G., Bjork, P., Antonsson, P., and Svensson, L.A. (2000). The crystal structure of staphylococcal enterotoxin H: implications for binding properties to MHC class II and TcR molecules. J Mol Biol 302, 527–537.

Hammel, M., Sfyroera, G., Pyrpassopoulos, S., Ricklin, D., Ramyar, K.X., Pop, M., Jin, Z., Lambris, J.D., and Geisbrecht, B.V. (2007a). Characterization of Ehp, a secreted complement inhibitory protein from Staphylococcus aureus. J. Biol. Chem. 282, 30051–30061.

Hammel, M., Sfyroera, G., Ricklin, D., Magotti, P., Lambris, J.D., and Geisbrecht, B.V. (2007b). A structural basis for complement inhibition by Staphylococcus aureus. Nat. Immunol. 8, 430–437.

Harboe, M., Ulvund, G., Vien, L., Fung, M., and Mollnes, T.E. (2004). The quantitative role of alternative pathway amplification in classical pathway induced terminal complement activation. Clin Exp Immunol 138, 439–446.

Harraghy, N., Hussain, M., Haggar, A., Chavakis, T., Sinha, B., Herrmann, M., and Flock, J.I. (2003). The adhesive and immunomodulating properties of the multifunctional Staphylococcus aureus protein Eap. Microbiology 149, 2701–2707.

Hudson, K.R., Tiedemann, R.E., Urban, R.G., Lowe, S.C., Strominger, J.L., and Fraser, J.D. (1995). Staphylococcal enterotoxin A has two cooperative binding sites on major histocompatibility complex class II. J. Exp. Med. 182, 711–720.

Hussain, M., Haggar, A., Heilmann, C., Peters, G., Flock, J.I., and Herrmann, M. (2002). Insertional inactivation of Eap in Staphylococcus aureus strain Newman confers reduced staphylococcal binding to fibroblasts. Infect. Immun. 70, 2933–2940.

Isenman, D.E., and Cooper, N.R. (1981). The structure and function of the third component of human complement – I. The nature and extent of conformational changes accompanying C3 activation. Mol. Immunol. 18, 331–339.

Isenman, D.E., Kells, D.I., Cooper, N.R., Muller-Eberhard, H.J., and Pangburn, M.K. (1981). Nucleophilic modification of human complement protein C3: correlation of conformational changes with acquisition of C3b-like functional properties. Biochemistry 20, 4458–4467.

Janssen, B.J., Huizinga, E.G., Raaijmakers, H.C., Roos, A., Daha, M.R., Nilsson-Ekdahl, K., Nilsson, B., and Gros, P. (2005). Structures of complement component C3 provide insights into the function and evolution of immunity. Nature 437, 505–511.

Jardetzky, T.S., Brown, J.H., Gorga, J.C., Stern, L.J., Urban, R.G., Chi, Y.I., Stauffacher, C., Strominger, J.L., and Wiley, D.C. (1994). Three-dimensional structure of a human class II histocompatibility molecule complexed with superantigen. Nature 368, 711–718.

Jarraud, S., Peyrat, M.A., Lim, A., Tristan, A., Bes, M., Mougel, C., Etienne, J., Vandenesch, F., Bonneville, M., and Lina, G. (2001). egc, a highly prevalent operon of enterotoxin gene, forms a putative nursery of superantigens in *Staphylococcus aureus*. J. Immunol. 166, 669–677.

Jongerius, I., Kohl, J., Pandey, M.K., Ruyken, M., van Kessel, K.P., van Strijp, J.A., and Rooijakkers, S.H. (2007). Staphylococcal complement evasion by various convertase-blocking molecules. J. Exp. Med. 204, 2461–2471.

Jonsson, K., McDevitt, D., McGavin, M.H., Patti, J.M., and Hook, M. (1995). *Staphylococcus aureus* expresses a major histocompatibility complex class II analog. J. Biol. Chem. 270, 21457–21460.

Kim, J., Urban, R.G., Strominger, J.L., and Wiley, D.C. (1994). Toxic shock syndrome toxin-1 complexed with a class II major histocompatibility molecule HLA-DR1. Science 266, 1870–1874.

Kuroda, M., Ohta, T., Uchiyama, I., Baba, T., Yuzawa, H., Kobayashi, I., Cui, L., Oguchi, A., Aoki, K., Nagai, Y., *et al.* (2001). Whole genome sequencing of meticillin-resistant *Staphylococcus aureus*. Lancet 357, 1225–1240.

Langley, R., Wines, B., Willoughby, N., Basu, I., Proft, T., and Fraser, J.D. (2005). The staphylococcal superantigen-like protein 7 binds IgA and complement C5 and inhibits IgA-Fc alpha RI binding and serum killing of bacteria. J. Immunol. 174, 2926–2933.

Lee, L.Y., Hook, M., Haviland, D., Wetsel, R.A., Yonter, E.O., Syribeys, P., Vernachio, J., and Brown, E.L. (2004a). Inhibition of complement activation by a secreted *Staphylococcus aureus* protein. J. Infect. Dis. 190, 571–579.

Lee, L.Y., Liang, X., Hook, M., and Brown, E.L. (2004b). Identification and characterization of the C3 binding domain of the *Staphylococcus aureus* extracellular fibrinogen-binding protein (Efb). J. Biol. Chem. 279, 50710–50716.

Lee, L.Y., Miyamoto, Y.J., McIntyre, B.W., Hook, M., McCrea, K.W., McDevitt, D., and Brown, E.L. (2002). The *Staphylococcus aureus* Map protein is an immunomodulator that interferes with T-cell-mediated responses. J. Clin. Invest. 110, 1461–1471.

Li, P.L., Tiedemann, R.E., Moffat, S.L., and Fraser, J.D. (1997). The superantigen streptococcal pyrogenic exotoxin C (SPE-C) exhibits a novel mode of action. J Exp Med 186, 375–383.

Li, Y., Li, H., Dimasi, N., McCormick, J.K., Martin, R., Schnuck, P., Schlievert, P.M., and Mariuzza, R.A. (2001). Crystal Structure of a superantigen bound to the high-affinity, zinc-dependent site on MHC class II. Immunity 14, 93–104.

Lin, Y.T., Wang, C.T., and Chiang, B.L. (2007). Role of bacterial pathogens in atopic dermatitis. Clin. Rev. Allergy Immunol. 33, 167–177.

Lin, Y.T., Wang, C.T., Hsu, C.T., Wang, L.F., Shau, W.Y., Yang, Y.H., and Chiang, B.L. (2003). Differential susceptibility to staphylococcal superantigen (SsAg)-induced apoptosis of CD4+ T-cells from atopic dermatitis patients and healthy subjects: the inhibitory effect of IL-4 on SsAg-induced apoptosis. J. Immunol. 171, 1102–1108.

Lina, G., Bohach, G.A., Nair, S.P., Hiramatsu, K., Jouvin-Marche, E., and Mariuzza, R. (2004). Standard nomenclature for the superantigens expressed by Staphylococcus. J. Infect. Dis. 189, 2334–2336.

Lindsay, J.A., Ruzin, A., Ross, H.F., Kurepina, N., and Novick, R.P. (1998). The gene for toxic shock toxin is carried by a family of mobile pathogenicity islands in *Staphylococcus aureus*. Mol. Microbiol. 29, 527–543.

Lowy, F.D. (1998). *Staphylococcus aureus* infections. N. Engl. J. Med. 339, 520–532.

Massey, R.C., Horsburgh, M.J., Lina, G., Hook, M., and Recker, M. (2006). The evolution and maintenance of virulence in *Staphylococcus aureus*: a role for host-to-host transmission? Nat. Rev. Microbiol. 4, 953–958.

McCormick, J.K., Yarwood, J.M., and Schlievert, P.M. (2001). Toxic shock syndrome and bacterial superantigens: an update. Annu. Rev. Microbiol 55, 77–104.

McGavin, M.H., Krajewska-Pietrasik, D., Ryden, C., and Hook, M. (1993). Identification of a *Staphylococcus aureus* extracellular matrix-binding protein with broad specificity. Infect. Immun. 61, 2479–2485.

Mehindate, K., Thibodeau, J., Dohlsten, M., Kalland, T., Sekaly, R.P., and Mourad, W. (1995). Cross-linking of major histocompatibility complex class II molecules by staphylococcal enterotoxin A superantigen is a requirement for inflammatory cytokine gene expression. J. Exp. Med. 182, 1573–1577.

Mitchell, D.T., Levitt, D.G., Schlievert, P.M., and Ohlendorf, D.H. (2000). Structural evidence for the evolution of pyrogenic toxin superantigens. J. Mol. Evol. 51, 520–531.

Moza, B., Varma, A.K., Buonpane, R.A., Zhu, P., Herfst, C.A., Nicholson, M.J., Wilbuer, A.K., Seth, N.P., Wucherpfennig, K.W., McCormick, J.K., *et al.* (2007). Structural basis of T-cell specificity and activation by the bacterial superantigen TSST-1. EMBO J. 26, 1187–1197.

Murzin, A.G. (1993). OB(oligonucleotide/oligosaccharide binding)-fold: common structural and functional solution for non-homologous sequences. Embo J 12, 861–867.

Newton, D.W., Dohlsten, M., Olsson, C., Segren, S., Lundin, K.E., Lando, P.A., Kalland, T., and Kotb, M. (1996). Mutations in the MHC class II binding domains of staphylococcal enterotoxin A differentially affect T-cell receptor Vbeta specificity. J. Immunol. *157*, 3988–3994.

Orengo, C.A., Jones, D.T., and Thornton, J.M. (1994). Protein superfamilies and domain superfolds. Nature *372*, 631–634.

Orwin, P.M., Leung, D.Y., Donahue, H.L., Novick, R.P., and Schlievert, P.M. (2001). Biochemical and biological properties of Staphylococcal enterotoxin K. Infect. Immun. *69*, 360–366.

Palma, M., Haggar, A., and Flock, J.I. (1999). Adherence of *Staphylococcus aureus* is enhanced by an endogenous secreted protein with broad binding activity. J. Bacteriol. *181*, 2840–2845.

Palma, M., Nozohoor, S., Schennings, T., Heimdahl, A., and Flock, J.I. (1996). Lack of the extracellular 19-kilodalton fibrinogen-binding protein from *Staphylococcus aureus* decreases virulence in experimental wound infection. Infect. Immun. *64*, 5284–5289.

Palma, M., Shannon, O., Quezada, H.C., Berg, A., and Flock, J.I. (2001). Extracellular fibrinogen-binding protein, Efb, from *Staphylococcus aureus* blocks platelet aggregation due to its binding to the alpha-chain. J. Biol. Chem. *276*, 31691–31697.

Palma, M., Wade, D., Flock, M., and Flock, J.I. (1998). Multiple binding sites in the interaction between an extracellular fibrinogen-binding protein from *Staphylococcus aureus* and fibrinogen. J. Biol. Chem. *273*, 13177–13181.

Papageorgiou, A.C., Acharya, K.R., Shapiro, R., Passalacqua, E.F., Brehm, R.D., and Tranter, H.S. (1995). Crystal structure of the superantigen enterotoxin C2 from *Staphylococcus aureus* reveals a zinc-binding site. Structure *3*, 769–779.

Petersson, K., Hakansson, M., Nilsson, H., Forsberg, G., Svensson, L.A., Liljas, A., and Walse, B. (2001). Crystal structure of a superantigen bound to MHC class II displays zinc and peptide dependence. EMBO J. *20*, 3306–3312.

Petersson, K., Thunnissen, M., Forsberg, G., and Walse, B. (2002). Crystal Structure of a SEA variant in complex with MHC class II reveals the ability of SEA to crosslink MHC molecules. Structure (Camb.) *10*, 1619–1626.

Postma, B., Poppelier, M.J., van Galen, J.C., Prossnitz, E.R., van Strijp, J.A., de Haas, C.J., and van Kessel, K.P. (2004). Chemotaxis inhibitory protein of *Staphylococcus aureus* binds specifically to the C5a and formylated peptide receptor. J. Immunol. *172*, 6994–7001.

Prat, C., Bestebroer, J., de Haas, C.J., van Strijp, J.A., and van Kessel, K.P. (2006). A new staphylococcal anti-inflammatory protein that antagonizes the formyl peptide receptor-like 1. J. Immunol. *177*, 8017–8026.

Proft, T., and Fraser, J.D. (2003). Bacterial superantigens. Clin. Exp. Immunol. *133*, 299–306.

Proft, T., and Fraser, J.D. (2007). Streptococcal superantigens. Chem. Immunol. Allergy *93*, 1–23.

Pumphrey, N., Vuidepot, A., Jakobsen, B., Forsberg, G., Walse, B., and Lindkvist-Petersson, K. (2007). Cutting edge: Evidence of direct TCR alpha-chain interaction with superantigen. J. Immunol. *179*, 2700–2704.

Ramsland, P.A., Willoughby, N., Trist, H.M., Farrugia, W., Hogarth, P.M., Fraser, J.D., and Wines, B.D. (2007). Structural basis for evasion of IgA immunity by *Staphylococcus aureus* revealed in the complex of SSL7 with Fc of human IgA1. Proc. Natl. Acad. Sci. USA *104*, 15051–15056.

Rooijakkers, S.H., Milder, F.J., Bardoel, B.W., Ruyken, M., van Strijp, J.A., and Gros, P. (2007). Staphylococcal complement inhibitor: structure and active sites. J. Immunol. *179*, 2989–2998.

Rooijakkers, S.H., van Kessel, K.P., and van Strijp, J.A. (2005a). Staphylococcal innate immune evasion. Trends Microbiol. *13*, 596–601.

Rooijakkers, S.H., Ruyken, M., Roos, A., Daha, M.R., Presanis, J.S., Sim, R.B., van Wamel, W.J., van Kessel, K.P., and van Strijp, J.A. (2005b). Immune evasion by a staphylococcal complement inhibitor that acts on C3 convertases. Nat. Immunol. *6*, 920–927.

Savinko, T., Lauerma, A., Lehtimaki, S., Gombert, M., Majuri, M.L., Fyhrquist-Vanni, N., Dieu-Nosjean, M.C., Kemeny, L., Wolff, H., Homey, B., et al. (2005). Topical superantigen exposure induces epidermal accumulation of CD8+ T-cells, a mixed Th1/Th2-type dermatitis and vigorous production of IgE antibodies in the murine model of atopic dermatitis. J. Immunol. *175*, 8320–8326.

Schad, E.M., Zaitseva, I., Zaitsev, V.N., Dohlsten, M., Kalland, T., Schlievert, P.M., Ohlendorf, D.H., and Svensson, L.A. (1995). Crystal structure of the superantigen staphylococcal enterotoxin type A. EMBO J. *14*, 3292–3301.

Schneider, D., Liaw, L., Daniel, C., Athanasopoulos, A.N., Herrmann, M., Preissner, K.T., Nawroth, P.P., and Chavakis, T. (2007). Inhibition of breast cancer cell adhesion and bone metastasis by the extracellular adherence protein of *Staphylococcus aureus*. Biochem. Biophys. Res. Commun. *357*, 282–288.

Seely, A.J., Pascual, J.L., and Christou, N.V. (2003). Science review: Cell membrane expression (connectivity) regulates neutrophil delivery, function and clearance. Crit. Care *7*, 291–307.

Shannon, O., and Flock, J.I. (2004). Extracellular fibrinogen binding protein, Efb, from *Staphylococcus aureus* binds to platelets and inhibits platelet aggregation. Thromb. Haemost. *91*, 779–789.

Shannon, O., Uekotter, A., and Flock, J.I. (2005). Extracellular fibrinogen binding protein, Efb, from *Staphylococcus aureus* as an antiplatelet agent *in vivo*. Thromb. Haemost. *93*, 927–931.

Skov, L., Olsen, J.V., Giorno, R., Schlievert, P.M., Baadsgaard, O., and Leung, D.Y. (2000). Application of Staphylococcal enterotoxin B on normal and atopic skin induces up-regulation of T-cells by a superantigen-mediated mechanism. J. Allergy Clin. Immunol. *105*, 820–826.

Smeltzer, M.S., Gillaspy, A.F., Pratt, F.L., Jr., Thames, M.D., and Iandolo, J.J. (1997). Prevalence and chromosomal map location of *Staphylococcus aureus* adhesin genes. Gene *196*, 249–259.

Sobke, A.C., Selimovic, D., Orlova, V., Hassan, M., Chavakis, T., Athanasopoulos, A.N., Schubert, U., Hussain, M., Thiel, G., Preissner, K.T., et al. (2006). The extracellular adherence protein from

Staphylococcus aureus abrogates angiogenic responses of endothelial cells by blocking Ras activation. FASEB J. 20, 2621–2623.

Starovasnik, M.A., Skelton, N.J., O'Connell, M.P., Kelley, R.F., Reilly, D., and Fairbrother, W.J. (1996). Solution structure of the E-domain of staphylococcal protein A. Biochemistry 35, 15558–15569.

Su, Y.C., and Wong, A.C. (1995). Identification and purification of a new staphylococcal enterotoxin, H. Appl. Environ. Microbiol. 61, 1438–1443.

Sundberg, E.J., Deng, L., and Mariuzza, R.A. (2007). TCR recognition of peptide/MHC class II complexes and superantigens. Semin. Immunol. 19, 262–271.

Sundberg, E.J., Li, H., Llera, A.S., McCormick, J.K., Tormo, J., Schlievert, P.M., Karjalainen, K., and Mariuzza, R.A. (2002). Structures of two streptococcal superantigens bound to TCR beta chains reveal diversity in the architecture of T-cell signaling complexes. Structure 10, 687–699.

Sundstrom, M., Abrahmsen, L., Antonsson, P., Mehindate, K., Mourad, W., and Dohlsten, M. (1996). The crystal structure of staphylococcal enterotoxin type D reveals Zn2+-mediated homodimerization. EMBO J. 15, 6832–6840.

Swaminathan, S., Furey, W., Pletcher, J., and Sax, M. (1992). Crystal structure of staphylococcal enterotoxin B, a superantigen. Nature 359, 801–806.

Thomas, D., Chou, S., Dauwalder, O., and Lina, G. (2007). Diversity in *Staphylococcus aureus* enterotoxins. Chem. Immunol. Allergy 93, 24–41.

Thomas, D.Y., Jarraud, S., Lemercier, B., Cozon, G., Echasserieau, K., Etienne, J., Gougeon, M.L., Lina, G., and Vandenesch, F. (2006). Staphylococcal enterotoxin-like toxins U2 and V, two new staphylococcal superantigens arising from recombination within the enterotoxin gene cluster. Infect. Immun. 74, 4724–4734.

Tiedemann, R.E., and Fraser, J.D. (1996). Cross-linking of MHC class II molecules by staphylococcal enterotoxin A is essential for antigen-presenting cell and T-cell activation. J. Immunol. 157, 3958–3966.

Torigoe, H., Shimada, I., Saito, A., Sato, M., and Arata, Y. (1990). Sequential 1H NMR assignments and secondary structure of the B domain of staphylococcal protein A: structural changes between the free B domain in solution and the Fc-bound B domain in crystal. Biochemistry 29, 8787–8793.

Torres, V.J., Stauff, D.L., Pishchany, G., Bezbradica, J.S., Gordy, L.E., Iturregui, J., Anderson, K.L., Dunman, P.M., Joyce, S., and Skaar, E.P. (2007). A *Staphylococcus aureus* regulatory system that responds to host heme and modulates virulence. Cell Host Microbe 1, 109–119.

Unnikrishnan, M., Altmann, D.M., Proft, T., Wahid, F., Cohen, J., Fraser, J.D., and Sriskandan, S. (2002). The bacterial superantigen streptococcal mitogenic exotoxin Z is the major immunoactive agent of Streptococcus pyogenes. J. Immunol. 169, 2561–2569.

Upadhyay, A., Burman, J.D., Clark, E.A., Leung, E., Isenman, D.E., van den Elsen, J.M., and Bagby, S. (2008). Structure–function analysis of the C3 bind-

ing region of *Staphylococcus aureus* immune subversion protein Sbi. J. Biol. Chem. 283, 22113–22120.

van Wamel, W.J., Rooijakkers, S.H., Ruyken, M., van Kessel, K.P., and van Strijp, J.A. (2006). The innate immune modulators staphylococcal complement inhibitor and chemotaxis inhibitory protein of *Staphylococcus aureus* are located on beta-hemolysin-converting bacteriophages. J. Bacteriol. 188, 1310–1315.

Veldkamp, K.E., Heezius, H.C., Verhoef, J., van Strijp, J.A., and van Kessel, K.P. (2000). Modulation of neutrophil chemokine receptors by *Staphylococcus aureus* supernate. Infect. Immun. 68, 5908–5913.

White, J., Herman, A., Pullen, A.M., Kubo, R., Kappler, J.W., and Marrack, P. (1989). The V beta-specific superantigen staphylococcal enterotoxin B: stimulation of mature T-cells and clonal deletion in neonatal mice. Cell 56, 27–35.

Williams, R.J., Ward, J.M., Henderson, B., Poole, S., O'Hara, B.P., Wilson, M., and Nair, S.P. (2000). Identification of a novel gene cluster encoding staphylococcal exotoxin-like proteins: characterization of the prototypic gene and its protein product, SET1. Infect. Immun. 68, 4407–4415.

Wines, B.D., Willoughby, N., Fraser, J.D., and Hogarth, P.M. (2006). A competitive mechanism for staphylococcal toxin SSL7 inhibiting the leukocyte IgA receptor, Fc alphaRI, is revealed by SSL7 binding at the C alpha2/C alpha3 interface of IgA. J. Biol. Chem. 281, 1389–1393.

Xie, C., Alcaide, P., Geisbrecht, B.V., Schneider, D., Herrmann, M., Preissner, K.T., Luscinskas, F.W., and Chavakis, T. (2006). Suppression of experimental autoimmune encephalomyelitis by extracellular adherence protein of *Staphylococcus aureus*. J. Exp. Med. 203, 985–994.

Yarwood, J.M., McCormick, J.K., Paustian, M.L., Orwin, P.M., Kapur, V., and Schlievert, P.M. (2002). Characterization and expression analysis of *Staphylococcus aureus* pathogenicity island 3. Implications for the evolution of staphylococcal pathogenicity islands. J. Biol. Chem. 277, 13138–13147.

Zhang, C., and Kim, S.H. (2000). The anatomy of protein beta-sheet topology. J. Mol. Biol. 299, 1075–1089.

Zhang, L., Jacobsson, K., Vasi, J., Lindberg, M., and Frykberg, L. (1998a). A second IgG-binding protein in *Staphylococcus aureus*. Microbiology 144 (Pt 4), 985–991.

Zhang, L., Rosander, A., Jacobsson, K., Lindberg, M., and Frykberg, L. (2000). Expression of staphylococcal protein Sbi is induced by human IgG. FEMS Immunol. Med. Microbiol. 28, 211–218.

Zhang, S., Iandolo, J.J., and Stewart, G.C. (1998b). The enterotoxin D plasmid of *Staphylococcus aureus* encodes a second enterotoxin determinant (sej). FEMS Microbio.l Lett. 168, 227–233.

Fungal Ribotoxins: Structure, Function and Evolution

Elías Herrero-Galán, Elisa Álvarez-García, Nelson Carreras-Sangrà, Javier Lacadena, Jorge Alegre-Cebollada, Álvaro Martínez del Pozo, Mercedes Oñaderra and José G. Gavilanes

Abstract

Ribotoxins are a family of fungal extracellular ribonucleases which inactivate ribosomes by specifically cleaving a single phosphodiester bond located at the universally conserved sarcin/ricin loop of the large rRNA. The subsequent inhibition of protein biosynthesis is followed by cell death *via* apoptosis. Ribotoxins are also able to interact with membranes containing acid phospholipids, their cytotoxicity being preferentially directed towards cells showing altered membrane permeability, e.g. transformed or virus-infected cells. Many features of their cytotoxic action and their ribonucleolytic mechanism have been elucidated by comparison with other extracellular non-toxic fungal RNases, best represented by RNase T1. The study of structure–function relationships in ribotoxins is of particular interest, since they are postulated as potential therapeutic agents against different human pathologies. The production of hypoallergenic variants with application in several *Aspergillus*-related allergic syndromes and the construction of immunotoxins against different carcinomas are promising examples of such potential therapeutic utilization.

Introduction

In 1963, during a screening program started seven years earlier by the Michigan Department of Health searching for new antibiotics and antitumour agents, the culture filtrates of a mould isolated from a sample of a Michigan farm soil were found to contain a substance inhibitory to both *sarcoma 180* and *carcinoma 755* induced in mice (Jennings *et al.*, 1965; Olson and Goerner, 1965). The mould was identified as a strain of *Aspergillus giganteus* and the protein responsible for these effects was named α-sarcin (Olson and Goerner, 1965). Two other antitumour proteins with similar activities, restrictocin and mitogillin, both produced by *Aspergillus restrictus*, were later found, and came to be part of a new family of proteins, today called ribotoxins, which would be joined much later by Asp f 1, identified three decades after as one of the major allergens of *Aspergillus fumigatus* (Arruda *et al.*, 1992). The unspecificity exhibited by these proteins in their cytotoxic action (Roga *et al.*, 1971) caused the abandon of their study until the mid-1970s, when it was demonstrated that they were capable of inhibiting protein biosynthesis by specifically cleaving a unique phosphodiester bond in the large ribosomal RNA subunit (Schindler and Davies, 1977; Endo and Wool, 1982). This bond is of particular interest because it is located at a universally conserved site with important roles in ribosome function (Wool *et al.*, 1992). This site is known as the sarcin–ricin loop (SRL), because it is the target for the toxins α-sarcin and ricin, best representatives of ribotoxins and ribosome inhibiting proteins (RIPs), respectively. Such similarity in their action is the reason why ribotoxins were suggested to be included in the RIPs family, but some authors (Nielsen and Boston, 2001; Peumans *et al.*, 2001) claim that this name should be restricted to plant N-glycosidases that depurinate a single nucleotide contiguous to the phosphodiester bond cleaved by ribotoxins.

Seven proteins produced by *Aspergillus* have been described as ribotoxins to date, together with six from *Penicillium* and two more from *Neosartorya glabra* and *Hirsutella thompsonii*, and genes have been characterized in five additional strains of the first *genus* (Jennings *et al.*, 1965; Olson and Goerner, 1965; Arruda *et al.*, 1992; Lin *et al.*, 1995; Huang *et al.*, 1997; Wirth *et al.*, 1997; Martínez-Ruiz *et al.*, 1999a; Kao *et al.*, 2001, Herrero-Galán *et al.*, 2008). α-Sarcin, restrictocin and Asp f 1 are the most exhaustively characterized members of the family. These proteins show a high degree of conservation, displaying amino acid sequence similarities above 85% (Fig. 9.1). However, hirsutellin A (HtA), an extracellular protein produced by the invertebrate fungal pathogen *Hirsutella thompsonii*, has been recently demonstrated to be a ribotoxin (Herrero-Galán *et al.*, 2008), though it displays only about 25% sequence identity with previ-

ously known members of the same family (Fig. 9.1) (Boucias *et al.*, 1998; Martínez-Ruiz *et al.*, 1999b; Herrero-Galán *et al.*, 2008). This suggests that ribotoxins are more widely distributed among fungi than previously believed (Martínez-Ruiz *et al.*, 1999a).

Ribotoxins belong to a larger group of fungal extracellular unspecific RNases that show a high degree of sequence and structural similarity to those of the α-sarcin family but are not cytotoxic (Figs. 9.1 and 9.2). RNase T1 is the best known representative of this family, as well as one of the most exhaustively characterized enzymes (Heinemann and Hahn, 1989; Gohda *et al.*, 1994; Zegers *et al.*, 1994, 1998; Steyeaert, 1997; Arni *et al.*, 1999; Loverix and Steyaert, 2001). RNase U2 from *Ustilago sphaerogena* also stands out as the unspecific fungal extracellular RNase

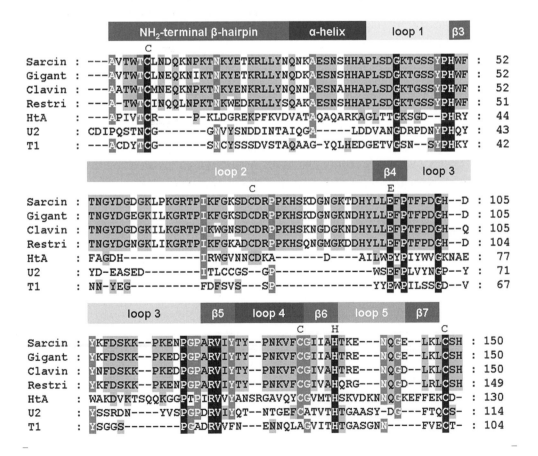

Figure 9.1 Sequence alignment of several ribotoxins (α-sarcin, gigantin, clavin, restrictocin and hirsutellin A) and RNases T1 and U2. Elements of secondary structure in α-sarcin are delimited at the top of the alignment, as well as the essential catalytic residues and the cysteines involved in disulphide bridges formation in ribotoxins. Positions conserved in the seven or at least four sequences are highlighted in black or grey, respectively.

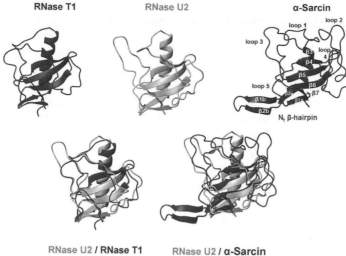

RNase T1   **RNase U2**   **α-Sarcin**

RNase U2 / RNase T1   RNase U2 / **α-Sarcin**

Figure 9.2 Representation of the three-dimensional structures of ribonucleases T1 (Martínez-Oyanedel *et al.*, 1991; PDB entry: 9RNT), U2 (Noguchi *et al.*, 1995; PDB entry: 1RTU) and α-sarcin (Pérez-Cañadillas *et al.*, 2000; PDB entry: 1DE3). Superpositions of the RNase U2 structure with those of RNase T1 and α-sarcin fitted to the active site residues are shown. The diagrams were generated with the MOLMOL program (Koradi *et al.*, 1996).

most closely related to ribotoxins (Sacco *et al.*, 1983; Martínez del Pozo *et al.*, 1988; Martínez-Ruiz *et al.*, 1999a). Ribotoxins share with RNases of the T1 family their main structural core but contain longer and positively charged loops (Fig. 9.2). So, these loops are supposed to be essential for their specific toxicity (Martínez del Pozo *et al.*, 1988), which gives rise to the hypothesis that a T1-like RNase could have acquired ribosome specificity and membrane interacting ability by the insertion of short recognition domains (Lamy *et al.*, 1992; Kao and Davies, 1995). Thus, the study of the evolution and mechanism of action of ribotoxins is of particular interest, as it could lead to the identification of the structural determinants that have allowed these proteins to become such efficient toxins, which would be a major step towards their biomedical utilization as weapons against different human pathologies. Comparative structural and functional studies between ribotoxins and RNases from the T1 family are crucial in the course for achieving that goal.

Structure

Ribotoxins are basic proteins of 149–150 amino acids with a high degree of identity, including two disulphide bridges conserved along the

whole family (Fig. 9.1) (Rodríguez *et al.*, 1982; Sacco *et al.*, 1983; López-Otín *et al.*, 1984; Fernández-Luna *et al.*, 1985; Arruda *et al.*, 1990; Wirth *et al.*, 1997; Martínez-Ruiz *et al.*, 1999b). This observation includes HtA, although it is 20 residues shorter than the other known ribotoxins (Martínez-Ruiz *et al.*, 1999b). Sequence differences are mainly concentrated at the loops, where ribotoxins also differ from RNases of the T1 family (Fig. 9.2) (Martínez-Ruiz *et al.*, 1999a).

The three dimensional structures of restrictocin (Yang and Moffat, 1996; Yang *et al.*, 2001) and α-sarcin (Pérez-Cañadillas *et al.*, 2000, 2002; García-Mayoral *et al.*, 2005a) have been elucidated. For α-sarcin, nuclear magnetic resonance and other techniques have been used to make a very detailed map of its structural and dynamic properties (Campos-Olivas *et al.*, 1996a, 1996b; Pérez-Cañadillas *et al.*, 2000, 2002; García-Mayoral *et al.*, 2005a, b). This protein folds into an α+β structure with a central five-stranded antiparallel β-sheet and an α-helix of almost three turns (Fig. 9.2). The sheet is composed of strands β3, β4, β5, β6 and β7, arranged in a −1, −1, −1, −1 topology. It is highly twisted in a right-handed sense, defining a convex face against which the α-helix is orthogonally packed, and a concave surface that holds the active site residues:

His50, Glu96, Arg121 and His137, all of them with their side chains projecting outwards from the cleft. This main structural core, including the active site, is conserved among all fungal extracellular RNases, the α-helix being longer in the T1 family. Major differences are concentrated in the loops of non periodic structure and the amino terminal region. In ribotoxins, residues 1–26 form a long β-hairpin that can be considered as two consecutive minor β-hairpins connected by a hinge region. The first one is closer to the open end of the hairpin, whereas the second sub-β-hairpin is formed by two short strands, β1b and β2b, connected by a type I β-turn. This last part of the N-terminal hairpin is exposed to the solvent and shows a high mobility (Pérez-Cañadillas *et al.*, 2002). The secondary structure elements are connected by large loops of non periodic structure but with very well defined conformations, maintained by a complex network of intraloop and interloop interactions, including hydrogen bonds, hydrophobic interactions and salt bridges (Yang and Moffat, 1996; Pérez-Cañadillas *et al.*, 2000). From a dynamic point of view, NMR studies have shown that these loops undergo fast internal motions, ranging from picoseconds to nanoseconds (Pérez-Cañadillas *et al.*, 2002).

Loop 2 of α-sarcin, one of the regions exhibiting more differences with RNases from the T1 family, deserves special attention because of its functional implications. It is rich in Gly and Lys residues, largely solvent exposed and highly mobile, though very well defined. In this loop, the stretch comprising residues 52–54 is essentially frozen within the molecular framework and includes Asn54, a conserved residue among fungal extracellular RNases (Mancheño *et al.*, 1995a) that establishes a hydrogen bond between its amide side chain proton and the carbonyl group of Ile69. This interaction is also conserved in RNases from the T1 family (Sevcik *et al.*, 1991; Pfeiffer *et al.*, 1997; Hebert *et al.*, 1998) and it has been suggested that this region could form the substrate recognition pocket in restrictocin (Yang and Moffat, 1996), a hypothesis that was later confirmed by the results obtained with α-sarcin Asn54 mutants, located in the equivalent region of this protein (53–56).

Thermal denaturation studies, by both differential scanning calorimetry (DSC) and circular dichroism (CD) measurements, have certified the high stability of ribotoxins (Gasset *et al.*, 1995a). The presence of eight Tyr and two Trp residues has allowed a great variety of spectroscopic studies, some of them leading to the determination of the pK_a values corresponding to pH-induced conformational transitions (Martínez del Pozo *et al.*, 1988; De Antonio *et al.*, 2000). These results were completed later by the assignment, thanks to NMR measurements and predictions, of the pK_a values of all titratable residues in the molecule (Pérez-Cañadillas *et al.*, 1998; García-Mayoral *et al.*, 2003). Such a detailed structural characterization has culminated with the determination of the different tautomeric states of every histidine residue in the protein (Pérez-Cañadillas *et al.*, 2003).

Finally, as far as the active site is concerned, it is important to mention that it is composed of, at least, Tyr48, His50, Glu96, His137, Arg121 and Leu145, although only three of them (His50, Glu96 and His137) are directly involved in proton transfer during the catalysis (Lacadena *et al.*, 1999; Martínez-Ruiz *et al.*, 2001). These three residues present unusual pK_a values, and the two histidines adopt unusual tautomeric forms, which is a common feature of microbial RNases. In addition, His137 establishes an important hydrogen bond with a backbone oxygen in loop 5. This loop is in part responsible for the low surface accessibility of all titratable atoms, which translates into important restrictions for the substrate, as will be discussed below (Pérez-Cañadillas *et al.*, 1998, 2000, 2003).

Function

Enzymatic activity

Both ribotoxins and T1-like RNases act as acid cyclizing RNases, following a two-step mechanism (Fig. 9.3) (Lacadena *et al.*, 1998). This mechanism, as well as the roles of most of the residues forming its active site, have been clearly established for RNase T1 (Steyaert, 1997; Loverix and Steyaert, 2001; Yoshida, 2001). In the first step, a transphosphorylation reaction occurs to form a 2',3'-cyclic phosphate intermediate.

A

R-**G/A**pN-R'   R-**2',3'-G/A**MP + N-R'   R-**G/A**MP

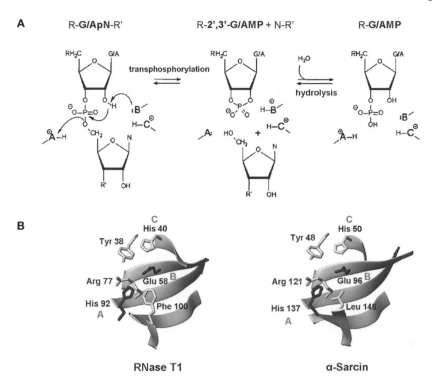

Figure 9.3 A Catalytic mechanism proposed for cyclizing ribonucleases. The type of substrate (dinucleotide, homo or heteropolynucleotide) is determined by R and R'. The catalytic residues A, B and C are represented within the active site structures of RNase T1 and α-sarcin in **B**. Essential residues are darkened. The diagrams were generated with MOLMOL (Koradi *et al.*, 1996).

Secondly, this intermediate is hydrolysed to the corresponding 3'-phosphate (Fig. 9.3).

This mechanism is shared by most RNases and is based on the general acid-base type endonucleolytic cleavage of RNA. However, the type of substrate (single- or double-stranded RNA), the specificity, the catalytic residues and the parameters defining the enzyme vary depending on the family. RNases from the T1 family can hydrolyse single stranded RNA, acting specifically on 3'-GpN-5' sequences. RNase U2 is an exception, cleaving additionally 3'-ApN-5' bonds (Egami *et al.*, 1980). RNase T1 optimum activity occurs at pH values around neutrality, whereas RNase U2 reaches its highest efficiency at acid pH (Arima *et al.*, 1968a, b; Uchida and Egami, 1971). Analysis of the cleavage reactions performed by α-sarcin against different dinucleoside monophosphates proved that this protein is also a cyclizing RNase with an optimum pH of 5.0 (Lacadena *et al.*, 1998, 1999; Pérez-Cañadillas *et al.*, 1998). However, ribotoxins are much

more exquisite enzymes, their specificity going further than a single preference for a type of nucleotide. As it has been already mentioned, ribotoxins exert their ribonucleolytic action on a single phosphodiester bond at the SRL, the one between G4325 and A4326 in 28S rRNA (rat ribosome numbering; G2661-A2662 in *E. coli*), releasing the so-called α fragment (488 bp in the rat ribosome) (Schindler and Davies, 1977; Endo and Wool, 1982; Endo *et al.*, 1983, 1988). This single cut is enough to inhibit protein biosynthesis, as it interferes with elongation factors function (Brigotti *et al.*, 1989; Wool *et al.*, 1992). Finally, cell death by apoptosis occurs (Olmo *et al.*, 2001).

There is evidence that all ribotoxins isolated to homogeneity conserve their specificity in the nanomolar range. However, at micromolar concentrations they are capable of hydrolysing RNA exhaustively, exhibiting preference for guanine at the 3'-end (Endo *et al.*, 1983). This loss of specificity is shown by the hydrolysis of different substrate analogues, as it is the case of homopolymers such as poly(A), poly(G) or poly(I) (Endo *et al.*, 1983). Activity has been detected even on dinucleoside monophosphates, which can be considered as their minimum substrate, though the way they are recognized by the enzyme seem to differ from the natural substrate. Despite this, the advantage in this case is that the products, substrates and intermediates of the reaction can be separated and quantified by HPLC, providing information about the different steps of the catalysis (Lacadena *et al.*, 1998). This kind of studies have allowed the establishment of the mechanism followed by ribotoxins when performing their ribonucleolytic action, the non-cytotoxic microbial RNases T1 and U2 being of great help as reference models. Accordingly, α-sarcin behaves as a cyclizing RNase, following the same general reaction scheme as the other members of the RNase T1 family (Fig. 9.3).

The production and characterization of many site-directed and randomly produced mutants have allowed the determination of not only the ribotoxin residues involved in the catalytic reaction, but also their different roles during the cleavage of a phosphosiester bond. All the identified α-sarcin's active site residues have their corresponding counterparts in RNase T1 (Figs. 9.2 and 9.3). Residues His137 and Glu96 are essential for the catalytic reaction, acting as a general acid and a general base, respectively, during the first step and reversing their roles during the subsequent hydrolysis of the cyclic derivative (Brandhorst *et al.*, 1994; Kao and Davies, 1995, 1999; Lacadena *et al.*, 1995, 1999; Sylvester *et al.*, 1997; Kao *et al.*, 1998). His50 would also contribute to the stabilization of the transition state but, in this case, would not be able to substitute for Glu96 as the general base in the E96Q mutant (Lacadena *et al.*, 1999), as is the case for the equivalent RNase T1 His residue (Steyaert *et al.*, 1990, 1997). Mutagenesis studies with residues Arg121, Leu145 and Tyr48 have confirmed their participation in the protein function (Masip *et al.*, 2001, 2003; Álvarez-García *et al.*, 2006). Arg121 is involved in the correct orientation of large substrates in the active site and its positive charge is essential for the interaction of the protein with membranes (Masip *et al.*, 2001). Leu145 interacts with His137, contributing to

the low pK$_a$ of this histidine residue, and keeps relation to the particular orientation of loop 5, which diminishes the accessibility of the active site to the solvent (Masip *et al.*, 2003). Finally, mutation of residue Tyr48 to Phe gives rise to a variant that is inactive against ribosomes but keeps the ability to degrade ApA, which reveals the essential role of the OH group in the phenolic ring in degradation of polymeric RNA (Álvarez-García *et al.*, 2006). Thus, Tyr48, Arg121 and Leu145 appear to be determinants of the ribotoxin activity of α-sarcin. In the crystal geometric complex of RNase T1 with the minimal substrate 3'-GMP (Loverix and Steyaert, 2001), their counterpart residues Tyr38, Arg77 and Phe100 also appear to be part of the catalytic site of the enzyme. There has been speculation that these three residues, together with His40, would form a prearranged structural and dielectric microenvironment that is complementary in shape, charge and hydrogen-bonding formation to the equatorial oxygens of the transition state, contributing to its optimal solvation/desolvation during the catalysis (Loverix and Steyaert, 2001). Additionally, studies on the crystal structures of complexes of restrictocin with inhibitors led to the proposal that ribotoxins may use base flipping to enable cleavage at the correct site of the SRL (Yang *et al.*, 2001). All studies so far suggest that residues Tyr48, Arg121 and Leu145 would enable the base flipping performed by the catalytic triad that permits RNase cleavage at a unique phosphodiester bond (Yang *et al.*, 2001).

Despite all these conclusions being obtained throughout dinucleoside phosphates assays, the catalytic efficiency of ribotoxins against these substrates is several orders of magnitude lower than that of T1-like RNases (Lacadena *et al.*, 1998). One of the main reasons for this behaviour is the solvent restriction imposed on the active site by the surrounding loops, especially loop 5. Length and interactions of this loop with the amino terminal hairpin affect the environment of the principal catalytic histidine (His137 in α-sarcin). This observation would explain the differences in catalytic efficiency, but not in specificity, leading to the suggestion that there might be interactions of the substrate with other loops.

In summary, the main advantage of ribotoxins over T1-like RNases relies on their specificity, which makes them capable of inhibiting protein biosynthesis by cleaving a single bond out of more than 7000 present in the ribosome.

Interaction with the ribosome

This specificity, which makes these proteins extraordinarily efficient toxins, depends on the recognition of a very specific region within the ribosome. The SRL is located at domain VI of 28S rRNA (23S in prokaryotes) and is composed of 30–35 nucleotides of double stranded RNA with a universally conserved sequence and a compact structure that contains several purine–purine base pairs, a GAGA tetraloop, and a bulged guanosine adjacent to a reverse Hoogsteen AU pair. This structure is stabilized by an unusual set of cross-strand base-stacking interactions and imino proton to phosphate oxygen hydrogen bonds (Fig. 9.4) (Szewczak et al., 1993; Szewczak and Moore, 1995; Seggerson

and Moore, 1998). Together with the ribosomal protein L11-binding region, the L7/L12 stalk, and the ribosomal proteins L6 and L14 (Fig. 9.4), the SRL constitutes a binding site for elongation factors that is required for correct ribosome function (Endo and Wool, 1982; Cameron et al., 2002; Van Dyke et al., 2002). The L11-binding domain sequence is also universally conserved, in good agreement with its essential role (Mears et al., 2002). Interestingly, the spatial orientation in the ribosome of both the SRL and the L11-binding domain varies not only among the different *phyla* (Ramakrishnan and Moore, 2001; Mears et al., 2002; Uchiumi et al., 2002), but also during the different steps of peptide bond formation (Gabashvili et al., 2000). These variations might explain why different toxins display different affinities when assayed against distinct ribosomal substrates (Schindler and Davies, 1977; Endo and Wool, 1982; Wool et al., 1992; Uchiumi

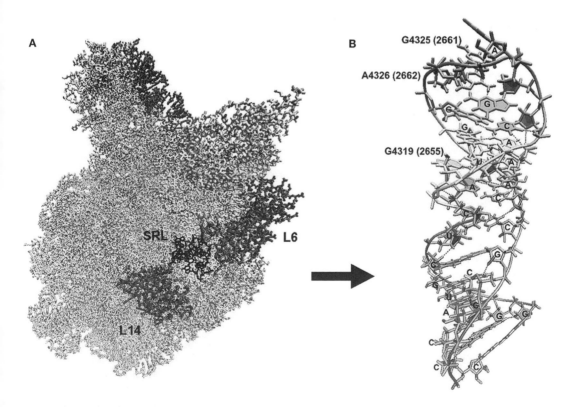

Figure 9.4 A Diagram showing the position of the SRL (black) in the structure of the *Haloarcula marismortui* large ribosomal subunit (PDB entry: 1JJ2). Ribosomal proteins L6 and L14 are also shown (grey). **B** Structure of the SRL. Bases are filled in the 5′ half and empty in the 3′ half of the sequence. Numbers correspond to rat or *E. coli* (in brackets) nucleotide positions within the 28S (23S) rRNA. Ribotoxins cleave the bond between G4325 (2661) and A4326 (2662). The bulged G4319 (2655) is also indicated. The diagrams were generated with VMD (Humphrey et al., 1996).

et al., 2002). Mutations affecting the sequence contained in the SRL result in defective binding of elongation factors and aminoacyl-tRNA, as well as a decreased translational fidelity (Liu and Liebman, 1996). Some of those mutations are lethal, reinforcing the importance of this region for the translational machinery (Leonov et al., 2003). Studies on the dynamics and kinetics of the ribosome show considerable mobility of this region, known as the GTPase centre, and its potential involvement in conformational changes essential for the correct performance of translation (Nilsson and Nissen, 2005).

Recognition of the SRL by ribotoxins mainly depends on interactions between the protein and both the GAGA tetraloop and the bulged G (Moazed et al., 1988; Glück and Wool, 1996; Munishkin and Wool, 1997; Pérez-Cañadillas et al., 2000). G2655 is the most critical site for binding of elongation factors and it seems to be the only really essential nucleotide for the specific ribonucleolytic activity of ribotoxins

(Macbeth and Wool, 1999). However, the primary determinant of recognition does not seem to be the type of nucleotide, but rather the SRL conformation (Munishkin and Wool, 1997; Correll et al., 1999; Correll and Swinger, 2003). As can be deduced from the crystal structure of the restrictocin-SRL analogue complex, loops 1 and 3 of the ribotoxin interact with the bulged G and the S-turn, which includes this nucleotide (Fig. 9.5) (Macbeth and Wool, 1999; Yang et al., 2001). Some residues of loop 5 and loop 2, the latter comprising an abundance of positive charge residues, would be involved in interactions with the GAGA tetraloop including the target bond. Once the ribotoxin is anchored, it exerts its action on the GA bond 12 Å distant from the bulged G, though participation of a G in the cleaved bond does not seem to be strictly necessary (Glück and Wool, 1996).

Nevertheless, all these interactions with the SRL do not explain by themselves the exquisite specificity of ribotoxins against their target ribo-

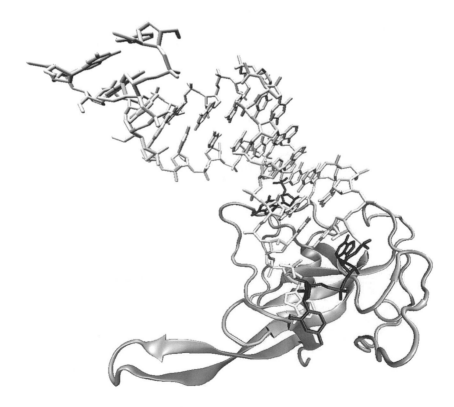

Figure 9.5 Structure of a restrictocin-SRL analogue complex (Yang et al., 2001; PDB entry: 1JBS). The bulged G, interacting with loops 1 and 3 of the protein, and the nucleotides linked by the bond cleaved by ribotoxins are highlighted in black. The GAGA tetraloop is distorted in this substrate analogue and thus is not degraded. The diagrams were generated with VMD (Humphrey et al., 1996).

somes. Accordingly, it has been recently shown that the ribosomal context enhances the reaction rate several orders of magnitude, probably due to favourable electrostatic interactions (Korennykh et al., 2006, 2007). Ribotoxins are basic proteins with a high density of charged and polar side chains exposed to the solvent (Pérez-Cañadillas et al., 2000), which would agree with an electrostatic localization of ribotoxins to the ribosome and subsequent diffusion within the ribosomal electrostatic field to the SRL. Additionally, the high internal mobility exhibited by some regions of the ribotoxins structure would enable accessibility of these proteins to other potential recognition sites, which would increase the probability of successful binding. In this regard, one of the regions with the highest conformational flexibility in α-sarcin is the N terminal β-hairpin (amino acids 1–26) (Pérez-Cañadillas et al., 2000; García-Mayoral et al., 2005a). Obtention of the deletion mutants α-sarcin Δ(7–22) and Asp f 1 Δ(7–22) made it possible to assess that these variants maintained the ribonucleolytic activity against dinucleoside phosphates, as well as the ability to specifically degrade oligonucleotides mimicking the sequence and structure of the SRL, but could not act against intact ribosomes, resulting to be much less cytotoxic

proteins (García-Ortega et al., 2002, 2005). The three-dimensional structure of the mentioned α-sarcin deletion mutant showed that the general folding of the wild-type protein was preserved, including the spatial conformation of the loops of non periodic structure (García-Mayoral et al., 2004). Modelling the ribotoxin recognition of ribosomes by both wild-type α-sarcin and its Δ(7–22) mutant, two additional interacting regions were identified (García-Mayoral et al., 2005b). One of them would involve a sequence stretch of loop 2 and ribosomal protein L6, whereas the other would depend on the contact between the N terminal β-hairpin and ribosomal protein L14 (Fig. 9.6). This latter interaction would not be possible for the Δ(7–22) mutant, thus being crucial for the specific recognition of the ribosomes, as proved by the results obtained with this variant. This conclusion is supported by the observation that a sequence homologous to the 11–16 region of α-sarcin can be found in EF-2 from *Saccharomyces cerevisiae* (Kao and Davies, 1999; García-Mayoral et al., 2005b). Additionally, sequence variability in this region among proteins of the L14 family would explain the different specificity exhibited by ribotoxins depending on the species from which the ribosomes assayed are obtained (Schindler and

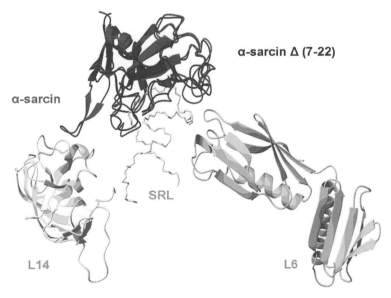

Figure 9.6 Minimized docking model showing the interaction of wild-type α-sarcin (dark grey) and its Δ(7–22) deletion mutant (black) (García-Mayoral *et al.*, 2004; PDB entry: 430D) with the SRL and the *Haloarcula marismortui* ribosomal proteins L6 and L14. The diagrams were generated with MOLMOL (Koradi *et al.*, 1996).

Davies, 1977; Endo and Wool, 1982; Endo *et al.*, 1983; Uchiumi *et al.*, 2002; García-Mayoral *et al.*, 2005b).

In summary, all these interactions contribute to the extraordinary specificity of ribotoxins against their substrate, inactivating it with a second order rate constant ($k_{cat}/K_M = 1.7 \times 10^{10}$ M^{-1} s^{-1}) that matches the catalytic efficiency of the fastest known enzymes (Korennykh *et al.*, 2006). This specific action is so effective that a single molecule of α-sarcin is enough to kill a cell (Lamy *et al.*, 1992).

Interaction with membranes and cytotoxicity

In order to completely explain the cytotoxic character of ribotoxins, mention must be made of their ability to interact with cell membranes. This is the main difference between ribotoxins and RNases from the T1 family, as well as the limiting factor for cytotoxicity. Although knowledge about the mechanism of cell entry followed by ribotoxins is very scarce, the most relevant data concerning this topic have been obtained for α-sarcin.

Studies with vesicle-model systems proved that α-sarcin interacts specifically with acid phospholipid vesicles, such as phosphatidylserine (PS) or phosphatidylglycerol (PG), at neutral or slightly acid pH (Gasset *et al.*, 1989, 1991a). This fact would be in agreement with the preference exhibited by ribotoxins for tumour or virus-infected target cells, where the loss of symmetry in the plasma membrane induces a higher exposure of PS or other acid phospholipids to the extracellular medium (Bergelson *et al.*, 1970; Turnay *et al.*, 1993; Orntoft and Vestergaard, 1999; Ran *et al.*, 2002; Papo and Shai, 2005). The recent discovery of the involvement in malignant transformation of the enzymes responsible for phosphatidic acid synthesis (diacylglycerol kinases) seems to further support this hypothesis (Filigheddu *et al.*, 2007; Griner and Kazanietz, 2007; Mérida *et al.*, 2008). Binding experiments allowed determination of a $K_d = 60.0$ nM for lipid–protein complexes that caused vesicle aggregation followed by fusion, but whose formation was abolished at basic pH (Gasset *et al.*, 1990). In the initial step of this interaction, α-sarcin acts as a bridge to dimerize vesicles

(Mancheño *et al.*, 1994a) and then fusion is triggered by the destabilizing effect of the protein, which simultaneously suffers conformational changes upon binding to the vesicles (Gasset *et al.*, 1991b; Mancheño *et al.*, 1994a). α-Sarcin is capable of translocating across the lipid bilayer of the vesicles thanks to a hydrophobic interaction involving region 116–139, as demonstrated by a synthetic peptide with that sequence (Oñaderra *et al.*, 1993; Gasset *et al.*, 1994, 1995b; Mancheño *et al.*, 1995b). Even a peptide comprising only residues 131–139 mimics the effects of the whole protein in this respect (Mancheño *et al.*, 1998). Passage across membranes is accompanied by structural changes and a decrease in protein stability, but the protein is ribonucleolytically active once inside the vesicle (Gasset *et al.*, 1991b, 1995b; Oñaderra *et al.*, 1993; Mancheño *et al.*, 1994b).

α-Sarcin Δ(7–22) deletion mutant and some other variants affecting residues in this region of the protein have shown that the N terminal β-hairpin is also involved in the interaction with cell membranes (García-Ortega *et al.*, 2001, 2002) as they display a different pattern of interaction with lipid vesicles, compatible with the absence of one vesicle-interacting protein region (García-Ortega *et al.*, 2002). Restrictocin also behaves differently from α-sarcin as far as lipid-interacting abilities are concerned (García-Ortega *et al.*, 2001), being noteworthy the fact that six residues out of the only 20 differences between both proteins are located in the N terminal β-hairpin. Loop 2 has been proposed by some authors to be involved as well in the interaction with membranes (Martínez del Pozo *et al.*, 1988; Yang and Moffat, 1996; Kao and Davies, 1999; Pérez-Cañadillas *et al.*, 2000). The two tryptophan residues of α-sarcin have not been proven necessary for the interaction (De Antonio *et al.*, 2000), but studies with α-sarcin's R121Q mutant have demonstrated the important role played by Arg121 during this process (Masip *et al.*, 2001). This interesting result led to the proposal that proteins that had evolved to interact with RNA, such as ribotoxins, would have developed structural and chemical determinants to recognize polyphosphate lattices that might as well allow recognition of a phospholipid bilayer (Masip *et al.*, 2001). Interestingly, when the crystalline structure of restrictocin was elucidated, Arg120, the counterpart to α-sarcin's Arg121, was found to be hydrogen bonded to a cocrystallized phosphate molecule at the active site (Yang and Moffat, 1996).

In general, the basic character of ribotoxins seems to be one of the key factors for cytotoxicity, as has been shown for other RNases (Di Donato et al., 1994; Vatzaki et al., 1999; Ilinskaya et al., 2002). Passage across the cell membrane is the rate-limiting step for α-sarcin's cytotoxic activity (Turnay et al., 1993), endocytosis being the internalization mechanism (Olmo et al., 2001). As no protein receptor has been found so far, the toxic specificity must be related to a differential interaction with the lipid components of the membranes. α-Sarcin has been reported to be a powerful inhibitor of protein synthesis in picornavirus-infected cells and several transformed cell lines (Fernández-Puentes and Carrasco, 1980; Carrasco and Esteban, 1982; Turnay et al., 1993; Olmo et al., 2001; Stuart and Brown, 2006). Besides, ionophores, external ATP or phospholipase C treatment made mammalian cells more sensitive to α-sarcin entry (Alonso and Carrasco, 1981, 1982; Otero and Carrasco, 1986, 1988). All these observations were interpreted in terms of the existence of altered membrane permeability. The toxin reaches the cytosol after clathrin-independent transport by acid endosomes and the Golgi (Olmo et al., 2001). In this regard, it has been recently shown that polycationic proteins tend to associate with phosphatidylserine enriched compartments such as endosomes (Yeung et al., 2008). The abundance of anionic phospholipids in the cytosolic leaflet of these organelles might be directly related to ribotoxins' cytotoxicity.

Studies with rhabdomyosarcoma cells proved that apoptosis is the mechanism of cell death, although it does not seem to be a general direct consequence of protein biosynthesis inhibition, as deduced from a comparative analysis of the effects of α-sarcin and cycloheximide (Olmo et al., 2001). However, variants with mutations affecting the enzymatic specificity of the protein showed diminished cytotoxic effects on these cells, revealing the relationship between ribonucleolytic activity and cytotoxicity (Lacadena et al., 1995; García-Ortega et al., 2002).

Evolution

Ribotoxins are an intriguing group of proteins regarding structure–function relationships. Their high degree of sequence and structural similarity with non-toxic fungal RNases of the T1 family has led to the suggestion that both families could have a common ancestor (Lamy et al., 1992; Kao and Davies, 1995). Comparing ribotoxins and T1-like RNases, 25% sequence homology can be found (Sacco et al., 1983), as well as the conservation of the main structural core including the active site responsible for the phosphodiesterase activity of these enzymes (Pérez-Cañadillas et al., 2000) (Figs. 9.1 and 9.2). RNase U2 stands out as the unspecific fungal extracellular RNase most closely related to ribotoxins (Sacco et al., 1983; Martínez del Pozo et al., 1988; Martínez-Ruiz et al., 1999a). It is 10 residues longer than the rest of the proteins of the T1 family (114 versus 101–106) and displays 34% sequence identity with ribotoxins. These are the reasons why both families are considered as members of the same group of proteins (Aravind and Koonin, 2001).

However, ribotoxins present a number of characteristics that make them unique within this superfamily. They are around 40 residues longer, basic, and show a high specificity for their natural substrate, resulting besides cytotoxic due to their ability to interact with cell membranes (Lacadena et al., 2007). From a structural point of view, the main differences with T1-type RNases lie in length and arrangement of the loops of non-periodic structure and the N terminal β-hairpin, these elements being thus considered as the determinants of the extra activities of ribotoxins. Consequently, the study of the evolution of these proteins is of particular interest, as they appear to be naturally engineered target toxins that could have evolved from a non-toxic microbial RNase (Lamy et al., 1992; Kao and Davies, 1995), maybe from a guanine- or purine-specific one. If so, it would be reasonable to consider the existence of evolutionary intermediates that could have acquired only some of the extra regions conferring additional activities to these RNases.

After its discovery, hirsutellin A (HtA), an insecticidal protein from the mite fungal pathogen *Hirsutella thompsonii*, appeared as a feasible candidate to be such intermediate (Martínez-

Ruiz *et al.*, 1999b). Some of its biological properties resembled those of the ribotoxin family (Liu *et al.*, 1996) and the alignment of the primary structure deduced from the HtA cDNA sequence (Boucias *et al.*, 1998) with those of ribotoxins revealed a significant similarity (Martínez-Ruiz *et al.*, 1999b). Sequence identity between HtA and ribotoxins was instead of only about 25%, a value much lower than that among known ribotoxins (always above 60%), but conservation of the catalytic residues and the four cysteines presumably involved in disulphide bridge formation was observed (Fig. 9.1). Interestingly, HtA is 20 residues shorter than α-sarcin-type proteins, the deletion of amino acids being presumably located at the protein loops and the N terminal hairpin, where HtA also differs from RNases of the T1 family. Characterization of its enzymatic properties has shown that HtA specifically inactivates ribosomes releasing the α-fragment characteristic of ribotoxin activity on rRNA (Herrero-Galán *et al.*, 2008). In addition, HtA specifically cleaves oligonucleotides that mimic the SRL, as well as selected polynucleotides and dinucleosides, behaving as a cyclizing ribonuclease too (Herrero-Galán *et al.*, 2008). Finally, it interacts with phospholipid membranes and exhibits cytotoxic activity on human tumour cells as the other ribotoxins (Herrero-Galán *et al.*, 2008). Based on all these results, HtA has been considered as a new ribotoxin, resulting to be the smallest member so far described in this family.

The characterization performed (Herrero-Galán *et al.*, 2008) proves that the abilities of ribotoxins can be accommodated into a shorter amino acid sequence of intermediate size between those of T1-type RNases and previously known ribotoxins. Therefore, comparative studies with HtA can shed light on the structure–function relationships of this family of proteins, maybe revealing unknown roles of the longer loops of the other ribotoxins. In this regard, the current determination of its three-dimensional structure in solution by NMR (A. Viegas *et al.*, unpublished) will be of great help. Additionally, the insecticidal character of HtA opens a new way for exploration of the biological function of ribotoxins, unknown to date. In this sense, it has been suggested that they could avoid destruction of the ribotoxin-producing fungi deterring

insect-feeding on their phialides (Brandhorst *et al.*, 1996).

Current trends

Despite the fact that the potential use of ribotoxins as antitumour agents was abandoned early due to high toxicity (Roga *et al.*, 1971), the present accumulation of data about their mechanism of action allows an optimistic view regarding the therapeutic utilization of these proteins. In relation to this, studies on the allergenic character of some ribotoxins and the production of hypoallergenic mutants, together with the development of immunotoxins based on these fungal ribonucleases, stand out as the most feasible alternatives in the mid-term future. On the other hand, the design of chimaeric ribotoxins with convenient activities must also be considered.

Allergenicity

Allergens are usually identified as substances recognized by IgE antibodies contained in the sera of allergic patients. In this sense, ribotoxins have been related to allergies caused by *Aspergillus*, the main ribotoxin-producing genus and the most important pathogen involved in human allergic syndromes provoked by fungi (Kurup *et al.*, 2002; Kurup, 2003). In fact, ribotoxins were found in the urine of patients with disseminated aspergillosis (Arruda *et al.*, 1990; Lamy *et al.*, 1991) and antibodies have been used to prove that they accumulate in the vicinity of the nodes of fungal infection (Lamy *et al.*, 1991), Asp f 1 from *Aspergillus fumigatus* being the ribotoxin most deeply studied as an allergen. This protein is involved in the pathogenicity of allergic bronchopulmonary aspergillosis (ABPA), the most severe form of allergic inhalant diseases, as high levels of Asp f 1-specific IgE are found in the sera of patients affected by this syndrome (Kurup *et al.*, 1994; García-Ortega *et al.*, 2005). Asp f 1 was, besides, the first recombinant allergen tested *in vivo* (Moser *et al.*, 1992), showing complete concordance with serologic determinations (Moser *et al.*, 1992; Crameri *et al.*, 1998; Hemmann *et al.*, 1999). Unfortunately, the recombinant protein is not devoid of cytotoxic activity and can trigger anaphylaxis.

Attempts to improve diagnosis of allergic diseases are focusing on the employment of homogeneous preparations of recombinantly produced allergens, much easier to standardize than complex fungal extracts (Piechura *et al.*, 1983; Crameri *et al.*, 1998; Kurup *et al.*, 2006). Recent studies with the Asp f 1 Δ(7–22) deletion mutant have shown that one of the major allergenic determinants of this protein is located at the N terminal β-hairpin (García-Ortega *et al.*, 2005). This region displays the highest sequence variability among ribotoxins (Fig. 9.1) (Martínez-Ruiz *et al.*, 1999a, b, 2001), and is highly flexible and solvent exposed (Pérez-Cañadillas *et al.*, 2000; García-Mayoral *et al.*, 2004). Asp f 1 differs from α-sarcin in only 19 residues, but five of these differences are located at this N terminal β-hairpin. Responses of α-sarcin and its deletion mutant against Asp f 1-containing sera were even lower than that of Asp f 1 Δ(7–22) mutant, indicating that the deleted portion, although important, is not the only allergenic epitope within the molecule and that the essential residues for the other determinants of the immunoreactivity are changed in wild-type α-sarcin (García-Ortega *et al.*, 2002, 2005). Despite this decrease in IgE reactivity, the prevalence of the three Asp f 1 variants remained essentially unaffected, and they retained most of the IgG epitopes (García-Ortega *et al.*, 2005). This fact, together with the absence of cytotoxic activity in these ribotoxins deletion variants, states them as promising molecules for use in immunomodulating therapy and diagnosis of *Aspergillus* hypersensitivity, though this possibility should still be corroborated by *in vivo* assays. With this purpose, development of an allergic murine system sensitized against Asp f 1 is currently under way (E. Álvarez-García *et al.*, unpublished).

Immunotoxins

Immunotoxins have emerged as a powerful alternative for the treatment of a variety of human pathologies because of their ability to specifically direct their action to certain cell types. Immunotoxin design is based on the 'magic bullet' concept, introduced by Ehrlich in 1906, according to which these molecules would consist of a tissue-specific carrier that would deliver toxic agents to neoplastic tissues (Ehrlich, 1956; Sandvig and van Deurs, 2000). The discovery of monoclonal antibodies in 1975

allowed the preparation of new toxins specifically directed against particular tumour cells, thanks to their conjugation to immunoglobulins specific for cancer cell antigens. The targeting moiety of these first-generation immunotoxins was the whole antibody molecule (Kreitman, 2000). As the recognition sites for antigens are on the variable regions of immunoglobulins, further studies were performed to verify that Fab fragments, obtained after IgG papain digestion, retained the ability to interact with the epitopes (Ward *et al.*, 1989; Worn and Pluckthun, 2001), leading to the so called Fab or Fv immunotoxins, which were more easily internalized because of their smaller size (Brinkmann, 2000). The development of new technologies allowed the production of recombinant immunotoxins, stabilized by a flexible peptide (scFv) or by a disulphide bridge between the variable domains (dsFv). These domains can be easily modified by genetic engineering, are more stable, and can be expressed in several organisms (Kreitman, 2003; Li *et al.*, 2004).

Regarding the toxin moiety, different toxins from bacteria and several ribosome inhibiting proteins (RIPs) from plants or fungi, mainly ricin, have been employed for immunotoxin design (Ghetie *et al.*, 1993; Engert *et al.*, 1997; Schnell *et al.*, 1998). The cytotoxic character of ribotoxins against carcinomas, together with their high thermostability, low immunogenicity and resistance to proteases, states them as ideal candidates for the construction of immunotoxins. Initial attempts were performed by chemical conjugation for restrictocin (Orlandi *et al.*, 1988; Conde *et al.*, 1989; Rathore and Batra, 1997a,b), mitogillin (Better *et al.*, 1992) and α-sarcin (Wawrzynczak *et al.*, 1991; Rathore *et al.*, 1997). Immunotoxins based on this latter protein offered promising results in preliminary assays (Wawrzynczak *et al.*, 1991) but did not proceed to *in vivo* studies probably because of their large size, that could hinder correct internalization, or because of low structural stability of the immunoconjugates.

Second generation immunotoxins attempt to solve these problems by fusing the toxin to a single chain containing only the variable domains, needed for antigen recognition. Recombinant immunotoxins of this kind have been already obtained based on restrictocin (Rathore and Batra,

1997a,b). These single-chain immunotoxins (scFv-IMTX) can be easily modified by genetic engineering to improve their cytotoxic activity or to diminish immunogenicity or unspecific toxicity *in vivo*. In relation to this, a single-chain immunotoxin composed of the variable domains of the B5 monoclonal antibody bound to α-sarcin through a peptide containing a furine cleavage site (scFv-IMTXαS) has been recently produced in the methylotrophic yeast *Pichia pastoris* (Lacadena *et al.*, 2005). The monoclonal antibody (mAb) B5 belongs to a family of mAbs directed against a Lewis^Y-related carbohydrate antigen that is overexpressed on the surface of many carcinomas, including breast and colon solid tumours (Pastan and FitzGerald, 1991). Different members of the family have been used as the targeting moiety in many immunotoxins, and three of them have been evaluated in phase I trials in cancer patients, with promising results (Pai *et al.*, 1996; Brinkmann, 2000; Woo *et al.*, 2008). As far as the expression system is concerned, *Pichia pastoris* has emerged as a robust heterologous host in which several immunotoxins have already been successfully produced extracellularly (Woo *et al.*, 2002, 2004, 2006; Liu *et al.*, 2005).

In this sense, scFv-IMTXαS produced in *P. pastoris* displays the characteristic ribonucleolytic activity and specific cytotoxicity of α-sarcin against targeted cells containing the Lewis^Y antigen (Lacadena *et al.*, 2005; N. Carreras-Sangrà, unpublished). Studies on genetically engineered variants of this immunotoxin with increased stability and affinity are currently being performed.

Biotechnology

In recent years, oral vaccination using Gram-positive bacteria as probiotics is being developed as a promising approach for treatment of several human pathologies, allergies included, thanks to the 'generally regarded as safe' (GRAS) status of some of these microorganisms (Pouwels *et al.*, 1996; Robinson *et al.*, 1997; Kirjavainen *et al.*, 1999; Maassen, 1999; Steidler *et al.*, 2000). One of these Gram-positive bacteria is *Lactococcus lactis*, a non-pathogenic, non-invasive, non-colonizing microorganism mainly used to produce fermented foods, but also proven useful, for example, in producing IL-10 for the treatment of inflammatory bowel disease in mice (Steidler *et al.*, 2000). Although *L. lactis* passes rapidly through the gastrointestinal tract without colonization (Gruzza *et al.*, 1994; Klijn *et al.*, 1995), genetically modified versions of this bacterium are still effective in delivering antigens to the mucosal immune system and capable of inducing a local immune response, which seems to happen because *L. lactis* lacks the ability to multiply *in vivo* but can readily be sampled by dendritic cells (Robinson *et al.*, 1997; Maassen, 1999; Adel-Patient *et al.*, 2005; Perez *et al.*, 2005). This process seems to be involved in the development of efficient immune responses (Xin *et al.*, 2003), including the selective induction of IgA (Macpherson and Uhr, 2004). In addition, antigens within lactococci are protected against direct contact with gastric acid and proteolytic enzymes.

Following this idea, the *Lactococcus lactis* MG1363 strain has been recently engineered to produce and secrete wild-type Asp f 1 and α-sarcin, as well as three different mutants with reduced cytotoxicity and/or IgE-binding affinity, such as the above mentioned Δ(7–22) variants of both proteins and H137Q active site mutant of α-sarcin (Álvarez-García *et al.*, 2008). The proteins were secreted in native and active form when the extracellular medium was buffered at pH values around 8.0. Intragastric administration of either the bacterial strain alone or the transformed producing wild-type α-sarcin did not induce any deleterious effect on mice intestinal tract, indicating that even the highly toxic protein could be safely delivered using this vehicle for oral administration (Álvarez-García *et al.*, 2008). *Lactococcus lactis* utilization as a potential delivery system for hypoallergenic variants of Asp f 1 and for ribotoxins in general as antitumoral agents against gastrointestinal tumours must thus be considered, though this possibility needs further evaluation.

Acknowledgements
This work was supported by Grant BFU2006/04404 from the Ministerio de Educación y Ciencia (Spain). E. Herrero-Galán, E. Álvarez-García, N. Carreras-Sangrà and J. Alegre-Cebollada are recipients of fellowships from the Ministerio de Educación y Ciencia (Spain).

Web resources

http://ccvweb.csres.utexas.edu/ccv/gallery/gallery.php?cat0ID=1&cat1ID=2&cat2ID=0&softwareID=4

Excellent movie on protein synthesis by ribosomes.

http://www.fgsc.net

One of the most popular fungal genetics web sites. A resource available for the fungal genetics research community with links to different Aspergillus genome sites (http://www.fgsc.net/aspergenome.htm) and the Aspergillus information web site (http://www.fgsc.net/Aspergillus/asperghome.html).

http://www. aspergillus.org.uk

The Aspergillus web site. A worldwide comprehensive resource providing information about these fungi and the diseases they can cause.

http://www.cbs.know.nl

The Centraal Bureau Voor Schimmelcultures, an institute of the Royal Netherlands Academy of Science.

http://rna.ucsc.edu/rnacenter/ribosome_movies.html

Different movies on ribosome structure and protein biosynthesis.

http://biochem4.okstate.edu/~biocukm/N1/N14212.html

Definition and description of different GNRA tetraloops such as that one cleaved by ribotoxins.

http://www.ehime-u.ac.jp/~cellfree/english/english_research.html

Link to a robust wheat germ cell-free protein synthesis system.

http://rmn.iqfr.csic.es/

Connection to the RMN group where many of the ribotoxins' three-dimensional structures have been solved.

http://bmb.bsd.uchicago.edu/Faculty_and_Research/01_Faculty/01_Faculty_Alphabetically.php?faculty_id=52

Connection to Ira Wool Home Page, where many of the initial discoveries regarding ribotoxins were made.

References

Adel-Patient, K., Ah-Leung, S., Creminon, C., Nouaille, S., Chatel, J.M., Langella, P., and Wal, J.M. (2005). Oral administration of recombinant *Lactococcus lactis* expressing bovine beta-lactoglobulin partially prevents mice from sensitization. Clin. Exp. Allergy 35, 539–546.

Alonso, M.A., and Carrasco, L. (1981). Permeabilization of mammalian cells to proteins by the ionophore nigericin. FEBS Lett. *127*, 112–114.

Alonso, M.A., and Carrasco, L. (1982). Molecular basis of the permeabilization of mammalian cells by ionophores. Eur. J. Biochem. *127*, 567–569.

Álvarez-García, E., García-Ortega, L., Verdún, Y., Bruix, M., Martínez del Pozo, A., and Gavilanes, J.G. (2006). Tyr-48, a conserved residue in ribotoxins, is involved in the RNA-degrading activity of α-sarcin. Biol. Chem. *387*, 535–541.

Álvarez-García, E., Alegre-Cebollada, J., Batanero, E., Monedero, V., Pérez-Martínez, G., García-Fernández, R., Gavilanes, J.G., and Martínez del Pozo, A. (2008). *Lactococcus lactis* as a vehicle for the heterologous expression of fungal ribotoxin variants with reduced IgE-binding affinity. J. Biotechnol. *134*, 1–8.

Aravind, L., and Koonin, E.V. (2001). A natural classification of ribonucleases. Meth. Enzymol. *341*, 3–28.

Arima, T., Uchida, T., and Egami, F. (1968a). Studies on extracellular ribonucleases of *Ustilago sphaerogena*. Purification and properties. Biochem. J. *106*, 601–607.

Arima, T., Uchida, T., and Egami, F. (1968b). Studies on extracellular ribonucleases of *Ustilago sphaerogena*. Characterization of substrate specificity with special reference to purine-specific ribonucleases. Biochem. J. *106*, 609–613.

Arni, R.K., Watanabe, L., Ward, R.J., Kreitman, R.J., Kumar, K., and Walz, F.G., Jr. (1999). Three-dimensional structure of ribonuclease T1 complexed with and isosteric phosphonate substrate analogue of GpU: alternate substrate binding modes and catalysis. Biochemistry 38, 2452–2461.

Arruda, L.K., Platts-Mills, T.A., Fox, J.W., and Chapman, M.D. (1990). *Aspergillus fumigatus* allergen I, a major IgE-binding protein, is a member of the mitogillin family of cytotoxins. J. Exp. Med. *172*, 1529–1532.

Arruda, L.K., Mann, B.J., and Chapman, M.D. (1992). Selective expression of a major allergen and cytotoxin, Asp f 1, in *Aspergillus fumigatus*. Implications for the immunopathogenesis of *Aspergillus*-related diseases. J. Immunol. *149*, 3354–3359.

Bergelson, L.D., Dyatlovitskaya, E.V., Torkhovskaya, T.I., Sorokina, I.B., and Gorkova, N.P. (1970). Phospholipid composition of membranes in the tumor cell. Biochim. Biophys. Acta *210*, 287–298.

Better, M., Bernhard, S.L., Lei, S.P., Fishwild, D.M., and Carroll, S.F. (1992). Activity of recombinant mitogillin and mitogillin immunoconjugates. J. Biol. Chem. 267, 16712–16718.

Boucias, D.G., Farmerie, W.G., and Pendland, J.C. (1998). Cloning and sequencing of cDNA of the insecticidal toxin Hirsutellin A. J. Invertebr. Pathol. 72, 258–261.

Brandhorst, T., Yang, R., and Kenealy, W.R. (1994). Heterologous expression of the cytotoxin restrictocin in *Aspergillus nidulans* and *Aspergillus niger*. Protein Expr. Purif. 5, 486–497.

Brandhorst, T., Dowd, P.F., and Kenealy, W.R. (1996). The ribosome-inactivating protein restrictocin deters insect feeding on *Aspergillus restrictus*. Microbiology 142, 1551–1556.

Brigotti, M., Rambelli, F., Zamboni, M., Montanaro, L., and Sperti, S. (1989). Effect of α–sarcin and ribosome-inactivating proteins on the interaction of elongation factors with ribosomes. Biochem. J. *257*, 723–727.

Brinkmann, U. (2000). Recombinant antibody fragments and immunotoxin fusions for cancer therapy. In vivo *14*, 21–27.

Cameron, D.M., Thompson, J., March, P.E., and Dahlberg, A.E. (2002). Initiation factor IF2, thiostrepton and micrococcin prevent the binding of elongation factor G to the *Escherichia coli* ribosome. J. Mol. Biol. *319*, 27–35.

Campos-Olivas, R., Bruix, M., Santoro, J., Martínez del Pozo, A., Lacadena, J., Gavilanes, J.G., and Rico, M. (1996a). H-1 and N-15 nuclear magnetic resonance assignment and secondary structure of the cytotoxic ribonuclease α–sarcin. Protein Sc. 5, 969–972.

Campos-Olivas, R., Bruix, M., Santoro, J., Martínez del Pozo, A., Lacadena, J., Gavilanes, J.G., and Rico, M. (1996b). Structural basis for the catalytic mechanism and substrate specificity of the ribonuclease α–sarcin. FEBS Lett. *399*, 163–165.

Carrasco, L., and Esteban, M. (1982). Modification of membrane permeability in vaccinia virus-infected cells. Virology *117*, 62–69.

Conde, F.P., Orlandi, R., Canevari, S., Mezzanzanica, D., Ripamonti, M., Muñoz, S.M., Jorge, P., and Colnaghi, M.I. (1989). The *Aspergillus* toxin restriction is a suitable cytotoxic agent for generation of immunoconjugates with monoclonal antibodies directed against human carcinoma cells. Eur. J. Biochem. *178*, 795–802.

Correll, C.C., and Swinger, K. (2003). Common and distinctive features of GNRA tetraloops based on a GUAA tetraloop structure at 1.4 Å resolution. RNA *9*, 355–363.

Correll, C.C., Wool, I.G., and Munishkin, A. (1999). The two faces of the *Escherichia coli* 23 S rRNA sarcin/ricin domain: The structure at 1.11 Å resolution. J. Mol. Biol. *292*, 275–287.

Crameri, R., Hemmann, S., Ismail, C., Menz, G., and Blaser, K. (1998). Disease-specific recombinant allergens for the diagnosis of allergic bronchopulmonary aspergillosis. Int. Immunol. *10*, 1211–1216.

De Antonio, C., Martínez del Pozo, A., Mancheño, J.M., Oñaderra, M., Lacadena, J., Martínez-Ruiz, A., Pérez-Cañadillas, J.M., Bruix, M., and Gavilanes, J.G. (2000). Assignment of the contribution of the tryptophan residues to the spectroscopic and functional properties of the ribotoxin α–sarcin. Proteins *41*, 350–361.

Di Donato, A., Cafaro, V., and Dalessio, G. (1994). Ribonuclease A can be transformed into a dimeric ribonuclease with antitumor activity. J. Biol. Chem. *269*, 17394–17396.

Egami, F., Oshima, T., and Uchida, T. (1980). Specific interaction of base-specific nucleases with nucleosides and nucleotides. F. Chapeville, and A.-L. Haenni, eds. (Berlin, Germany: Springer-Verlag), pp. 250–277.

Ehrlich, P. (1956). The relationship between chemical constitution, distribution and pharmacological action. In The collected papers of Paul Ehrlich, F. Himmelweit, M. Marquardt, and H. Dale, eds. (New York, Pergamon Press), p. 596.

Endo, Y., and Wool, I.G. (1982). The site of action of α–sarcin on eukaryotic ribosomes. The sequence at the α–sarcin cleavage site in 28 S ribosomal ribonucleic acid. J. Biol. Chem. *257*, 9054–9060.

Endo, Y., Huber, P.W., and Wool, I.G. (1983). The ribonuclease activity of the cytotoxin α–sarcin. The characteristics of the enzymatic activity of α–sarcin with ribosomes and ribonucleic acids as substrates. J. Biol. Chem. *258*, 2662–2667.

Endo, Y., Chan, Y.-L., Lin, A., Tsurugi, K., and Wool, I. (1988). The cytotoxins alpha sarcin and ricin retain their specificity when tested on a synthetic oligoribonucleotide (35-mer) that mimics a region of 28 S ribosomal ribonucleic acid. J. Biol. Chem. *263*, 7917–7920.

Engert, A., Diehl, V., Schnell, R., Radszuhn, A., Hatwig, M.T., Drillich, S., Schon, G., Bohlen, H., Tesch, H., Hansmann, M.L., Barth, S., Schindler, J., Ghetie, V., Uhr, J., and Vitetta, E. (1997). A phase-I study of an anti-CD25 ricin A-chain immunotoxin (RFT5-SMPT-dgA) in patients with refractory Hodgkin's lymphoma. Blood *89*, 403–410.

Fernández-Luna, J.L., López-Otín, C., Soriano, F., and Méndez, E. (1985). Complete amino acid sequence of the *Aspergillus* cytotoxin mitogillin. Biochemistry *24*, 861–867.

Fernández-Puentes, C., and Carrasco, L. (1980). Viral infection permeabilizes mammalian cells to protein toxins. Cell *20*, 769–775.

Filigheddu, N., Cutrupi, S., Porporato, P.E., Riboni, F., Baldanzi, G., Chianale, F., Fortina, E., Piantanida, P., De Bortoli, M., Vacca, G., Graziani, A., and Surico, N. (2007). Diacylglycerol kinase is required for HGF-induced invasiveness and anchorage-independent growth of MDA-MB-231 breast cancer cells. Anticancer Res. *27*, 1489–1492.

Gabashvili, I.S., Agrawal, R.K., Spahn, C.M., Grassucci, R.A., Svergun, D.I., Frank, J., and Penczek, P. (2000). Solution structure of the *E. coli* 70S ribosome at 11.5 Å resolution. Cell *100*, 537–549.

García-Mayoral, M.F., Pérez-Cañadillas, J.M., Santoro, J., Ibarra-Molero, B., Sánchez-Ruiz, J.M., Lacadena, J., Martínez del Pozo, A., Gavilanes, J.G., Rico, M., and Bruix, M. (2003). Dissecting structural and electrostatic interactions of charged groups in α-sarcin. An NMR study of some sutants involving the catalytic residues. Biochemistry *42*, 13122–13133.

García-Mayoral, M.F., García-Ortega, L., Lillo, M.P., Santoro, J., Martínez del Pozo, A., Gavilanes, J.G., Rico, M., and Bruix, M. (2004). NMR structure of the noncytotoxic α-sarcin mutant Δ(7–22): the importance of the native conformation of peripheral loops for activity. Protein Sci. *13*, 1000–1011.

García-Mayoral, M.F., Pantoja-Uceda, D., Santoro, J., Martínez del Pozo, A., Gavilanes, J.G., Rico, M., and Bruix, M. (2005a). Refined NMR structure of α-sarcin by ¹⁵N-¹H residual dipolar couplings. Eur. Biophys. J *34*, 1057–1065.

García-Mayoral, M.F., García-Ortega, L., Álvarez-García, E., Bruix, M., Gavilanes, J.G., and Martínez del Pozo, A. (2005b). Modeling the highly specific ribotoxin recognition of ribosomes. FEBS Lett. *579*, 6859–6864.

García-Ortega, L., Lacadena, J., Mancheño, J.M., Oñaderra, M., Kao, R., Davies, J., Olmo, N., Pozo, A.M., and Gavilanes, J.G. (2001). Involvement of the amino-terminal β-hairpin of the *Aspergillus* ribotoxins on the interaction with membranes and nonspecific ribonuclease activity. Protein Sci. *10*, 1658–1668.

García-Ortega, L., Masip, M., Mancheño, J.M., Oñaderra, M., Lizarbe, M.A., García-Mayoral, M.F., Bruix, M., Martínez del Pozo, A., and Gavilanes, J.G. (2002). Deletion of the NH₂-terminal β-hairpin of the ribotoxin α-sarcin produces a nontoxic but active ribonuclease. J. Biol. Chem. *277*, 18632–18639.

García-Ortega, L., Lacadena, J., Villalba, M., Rodríguez, R., Crespo, J.F., Rodríguez, J., Pascual, C., Olmo, N., Oñaderra, M., Martínez del Pozo, A., and Gavilanes, J.G. (2005). Production and characterization of a noncytotoxic deletion variant of the *Aspergillus fumigatus* allergen Asp f 1 displaying reduced IgE binding. FEBS J. *272*, 2536–2544.

Gasset, M., Martínez del Pozo, A., Oñaderra, M., and Gavilanes, J.G. (1989). Study of the interaction between the antitumour protein α–sarcin and phospholipid vesicles. Biochem. J. *258*, 569–575.

Gasset, M., Oñaderra, M., Thomas, P.G., and Gavilanes, J.G. (1990). Fusion of phospholipid vesicles produced by the anti-tumour protein α–sarcin. Biochem. J. *265*, 815–822.

Gasset, M., Oñaderra, M., Martínez del Pozo, A., Schiavo, G.-P., Laynez, J., Usobiaga, P., and Gavilanes, J.G. (1991a). Effect of the antitumour protein α–sarcin on the thermotropic behaviour of acid phospholipid vesicles. Biochim. Biophys. Acta *1068*, 9–16.

Gasset, M., Oñaderra, M., Goormaghtigh, E., and Gavilanes, J.G. (1991b). Acid phospholipid vesicles produce conformational changes on the antitumour protein α–sarcin. Biochim. Biophys. Acta *1080*, 51–58.

Gasset, M., Mancheño, J.M., Lacadena, J., Turnay, J., Olmo, N., Lizarbe, M.A., Martínez del Pozo, A., Oñaderra, M., and Gavilanes, J.G. (1994). α–sarcin, a ribosome-inactivating protein that translocates across the membranes of phospholipid vesicles. Curr. Topics Pept. Membr. Res. *1*, 99–104.

Gasset, M., Mancheño, J.M., Laynez, J., Lacadena, J., Fernández-Ballester, G., Martínez del Pozo, A., Oñaderra, M., and Gavilanes, J.G. (1995a). Thermal unfolding of the cytotoxin α-sarcin: Phospholipid binding induces destabilization of the protein structure. Biochim. Biophys. Acta *1252*, 126–134.

Gasset, M., Mancheño, J.M., Lacadena, J., Martínez del Pozo, A., Oñaderra, M., and Gavilanes, J.G. (1995b). Spectroscopic characterization of the alkylated α–sarcin cytotoxin: Analysis of the structural requirements for the protein-lipid bilayer hydrophobic interaction. Biochim. Biophys. Acta *1252*, 43–52.

Ghetie, V., Swindell, E., Uhr, J.W., and Vitetta, E.S. (1993). Purification and properties of immunotoxins containing one vs. two deglycosylated ricin A chains. J. Immunol. Meth. *166*, 117–122.

Glück, A., and Wool, I.G. (1996). Determination of the 28 S ribosomal RNA identity element (G4319)

for α–sarcin and the relationship of recognition to the selection of the catalytic site. J. Mol. Biol. *256*, 838–848.

Gohda, K., Oka, K., Tomita, K., and Hakoshima, T. (1994). Crystal structure of RNase T1 complexed with the product nucleotide 3'-GMP. Structural evidence for direct interaction of histidine 40 and glutamic acid 58 with the 2'-hydroxyl group of the ribose. J. Biol. Chem. *269*, 17531–17536.

Griner, E.M. and Kazanietz, M.G. (2007). Protein kinase C and other diacylglycerol effectors in cancer. Nat. Rev. Cancer *7*, 281–294.

Gruzza, M., Fons, M., Ouriet, M.F., Duval-Iflah, Y., and Ducluzeau, R. (1994). Study of gene transfer *in vitro* and in the digestive tract of gnotobiotic mice from *Lactococcus lactis* strains to various strains belonging to human intestinal flora. Microb. Releases *2*, 183–189.

Hebert, E.J., Giletto, A., Sevcik, J., Urbanikova, L., Wilson, K.S., Dauter, Z., and Pace, C.N. (1998). Contribution of a conserved asparagine to the conformational stability of ribonucleases Sa, Ba, and T1. Biochemistry *38*, 16192–16200.

Heinemann, U. and Hahn, U. (1989). Structural and functional studies of ribonuclease T1. In Protein-nucleic acid interaction, W. Saenger, and U. Heinemann, eds. (London, Macmillan), pp. 111–141.

Hemmann, S., Menz, G., Ismail, C., Blaser, K., and Crameri, R. (1999). Skin test reactivity to 2 recombinant *Aspergillus fumigatus* allergens in *A. fumigatus*-sensitized asthmatic subjects allows diagnostic separation of allergic bronchopulmonary aspergillosis from fungal sensitization. J. Allergy Clin. Immunol. *104*, 601–607.

Herrero-Galán, E., Lacadena, J., Martínez del Pozo, A., Boucias, D.G., Olmo, N., Oñaderra, M., and Gavilanes, J.G. (2008). The insecticidal protein hirsutellin A from the mite fungal pathogen *Hirsutella thompsonii* is a ribotoxin. Proteins *72*, 217–228.

Huang, K.-C., Hwang, Y.-Y., Hwang, L., and Lin, A. (1997). Characterization of a new ribotoxin gene (c-sar) from *Aspergillus clavatus*. Toxicon *35*, 383–392.

Humphrey, W., Dalke, A., and Schulten, K., (1996). VMD: visual molecular dynamics. J. Mol. Graph. *14*, 33–38, 27–28

Ilinskaya, O.N., Dreyer, F., Mitkevich, V.A., Shaw, K.L., Pace, C.N., and Makarov, A.A. (2002). Changing the net charge from negative to positive makes ribonuclease Sa cytotoxic. Protein Sci. *11*, 2522–2525.

Jennings, J.C., Olson, B.H., Roga, V., Junek, A.J., and Schuurmans, D.M. (1965). α-Sarcin, a new antitumor agent. II. Fermentation and Antitumor Spectrum. Appl. Microbiol. *13*, 322–326.

Kao, R., and Davies, J. (1995). Fungal ribotoxins: A family of naturally engineered targeted toxins? Biochem. Cell Biol. *73*, 1151–1159.

Kao, R., and Davies, J. (1999). Molecular dissection of mitogillin reveals that the fungal ribotoxins are a family of natural genetically engineered ribonucleases. J. Biol. Chem. *274*, 12576–12582.

Kao, R., Shea, J.E., Davies, J., and Holden, D.W. (1998). Probing the active site of mitogillin, a fungal ribotoxin. Mol. Microbiol. *29*, 1019–1027.

Kao, R., Martínez-Ruiz, A., Martínez del Pozo, A., Crameri, R., and Davies, J. (2001). Mitogillin and related fungal ribotoxins. Meth. Enzymol. *341*, 324–335.

Kirjavainen, P.V., Apostolou, E., Salminen, S.J., and Isolauri, E. (1999). New aspects of probiotics: a novel approach in the management of food allergy. Allergy *54*, 909–915.

Klijn, N., Weerkamp, A.H., and de Vos, W.M. (1995). Genetic marking of *Lactococcus lactis* shows its survival in the human gastrointestinal tract. Appl. Environ. Microbiol. *61*, 2771–2774.

Koradi, R., Billeter, M., and Wüthrich, K. (1996). MOLMOL: a program for display and analysis of macromolecular structures. J. Mol. Graph. *14*, 51–55, 29–32.

Korennykh, A.V., Piccirilli, J.A., and Correll, C.C. (2006). The electrostatic character of the ribosomal surface enables extraordinarily rapid target location by ribotoxins. Nat. Struct. Mol. Biol. *13*, 436–443.

Korennykh, A.V., Correll, C.C., and Piccirilli, J.A. (2007). Evidence for the importance of electrostatics in the function of two distinct families of ribosome inactivating toxins. RNA *13* 1391–1396.

Kreitman, R.J. (2000). Immunotoxins. Expert Opin. Pharmacother. *1*, 1117–1129.

Kreitman, R.J. (2003). Recombinant toxins for the treatment of cancer. Curr. Opin. Mol. Ther. *5*, 44–51.

Kurup, V.P. (2003). Fungal allergens. Curr. Allergy Asthma Rep. *3*, 416–423.

Kurup, V.P., Kumar, A., Kenealy, W.R., and Greenberger, P.A. (1994). *Aspergillus ribotoxins* react with IgE and IgG antibodies of patients with allergic bronchopulmonary aspergillosis. J. Lab. Clin. Med. *123*, 749–756.

Kurup, V.P., Shen, H.D., and Vijay, H. (2002). Immunobiology of fungal allergens. Int. Arch. Allergy Immunol. *129*, 181–188.

Kurup, V.P., Knutsen, A.P., Moss, R.B., and Bansal, N.K. (2006). Specific antibodies to recombinant allergens of *Aspergillus fumigatus* in cystic fibrosis patients with ABPA. Clin. Mol. Allergy *4*, 11.

Lacadena, J., Mancheño, J.M., Martínez Ruiz, A., Martínez del Pozo, A., Gasset, M., Oñaderra, M., and Gavilanes, J.G. (1995). Substitution of histidine-137 by glutamine abolishes the catalytic activity of the ribosome-inactivating protein α-sarcin. Biochem. J. *309*, 581–586.

Lacadena, J., Martínez del Pozo, A., Lacadena, V., Martínez-Ruiz, A., Mancheño, J.M., Oñaderra, M., and Gavilanes, J.G. (1998). The cytotoxin α–sarcin behaves as a cyclizing ribonuclease. FEBS Lett. *424*, 46–48.

Lacadena, J., Martínez del Pozo, A., Martínez-Ruiz, A., Pérez-Cañadillas, J.M., Bruix, M., Mancheño, J.M., Oñaderra, M., and Gavilanes, J.G. (1999). Role of histidine-50, glutamic acid-96, and histidine-137 in the ribonucleolytic mechanism of the ribotoxin α–sarcin. Proteins *37*, 474–484.

Lacadena, J., Carreras-Sangrà, N., Oñaderra, M., Martínez del Pozo, A., and Gavilanes, J.G. (2005). Production and purification of an immunotoxin based on the ribotoxin α-sarcin. In: 7th International

Meeting on Ribonucleases. Urbániková, ed. (Bratislava, Stará Lesná, Slovak Republic: ASCO Art & Science), abstract number P10, pp. 65.

Lacadena, J., Álvarez-García, E., Carreras-Sangrà, N., Herrero-Galán, E., Alegre-Cebollada, J., García-Ortega, L., Oñaderra, M., Gavilanes, J.G., and Martínez del Pozo, A. (2007). Fungal ribotoxins: molecular dissection of a family of natural killers. FEMS Microbiol. Rev. *31*, 212–237.

Lamy, B., Moutaouakil, M., Latge, J.P., and Davies, J. (1991). Secretion of a potential virulence factor, a fungal ribonucleotoxin, during human aspergillosis infections. Mol. Microbiol. 5, 1811–1815.

Lamy, B., Davies, J. and Schindler, D. (1992). The *Aspergillus* ribonucleolytic toxins (ribotoxins). Genetically engineered toxins (Ed: AE Frankel; Publ: Marcel Dekker, Inc) pp 237–258.

Leonov, A.A., Sergiev, P.V., Bogdanov, A.A., Brimacombe, R., and Dontsova, O.A. (2003). Affinity purification of ribosomes with a lethal G2655C mutation in 23 S rRNA that affects the translocation. J. Biol. Chem. *278*, 25664–25670.

Li, Q., Verschraegen, C.F., Mendoza, J., and Hassan, R. (2004). Cytotoxic activity of the recombinant anti-mesothelin immunotoxin, SS1(dsFv)PE38, towards tumor cell lines established from ascites of patients with peritoneal mesotheliomas. Anticancer Res. *24*, 1327–1335.

Lin, A.H., Huang, K.C., Hwu, L., and Tzean, S.S. (1995). Production of type II ribotoxins by *Aspergillus* species and related fungi in Taiwan. Toxicon *33*, 105–110.

Liu, J.-C., Boucias, D.G., Pendland, J.C., Liu, W.-Z., and Maruniak, J. (1996). The mode of action of hirsutellin A on eukaryotic cells. J. Invertebr. Pathol. *67*, 224–228.

Liu, R. and Liebman, S.W. (1996). A translational fidelity mutation in the universally conserved sarcin/ricin domain of 25S yeast ribosomal RNA. RNA 2, 254–263.

Liu, Y.Y., Woo, J.H., and Neville, D.M., Jr. (2005). Overexpression of an anti-CD3 immunotoxin increases expression and secretion of molecular chaperone BiP/Kar2p by Pichia pastoris. Appl. Environ. Microbiol. *71*, 5332–5340.

López-Otín, C., Barber, D., Fernández-Luna, J.L., Soriano, F., and Méndez, E. (1984). The primary structure of the cytotoxin restrictocin. Eur. J. Biochem. *143*, 621–634.

Loverix, S. and Steyaert, J. (2001). Deciphering the mechanism of RNase T1. Meth. Enzymol. *341*, 305–323.

Maassen, C.B. (1999). A rapid and safe plasmid isolation method for efficient engineering of recombinant lactobacilli expressing immunogenic or tolerogenic epitopes for oral administration. J. Immunol. Meth. *223*, 131–136.

Macbeth, M.R. and Wool, I.G. (1999). The phenotype of mutations of G2655 in the sarcin/ricin domain of 23 S ribosomal RNA. J. Mol. Biol. *285*, 965–975.

Macpherson, A.J., and Uhr, T. (2004). Induction of protective IgA by intestinal dendritic cells carrying commensal bacteria. Science *303*, 1662–1665.

Mancheño, J.M., Gasset, M., Lacadena, J., Ramon, F., Martínez del Pozo, A., Oñaderra, M., and Gavilanes, J.G. (1994a). Kinetic study of the aggregation and lipid mixing produced by α−sarcin on phosphatidylglycerol and phosphatidylserine vesicles: Stopped-flow light scattering and fluorescence energy transfer measurements. Biophys. J. 67, 1117–1125.

Mancheño, J.M., Gasset, M., Lacadena, J., Martínez del Pozo, A., Oñaderra, M., and Gavilanes, J.G. (1994b). Molecular interactions involved in the passage of the cytotoxic protein α-sarcin across membranes. In Structure, Biogenesis and Dynamics of Biological membranes, NATO ASI Series J. A. F. Op de Kamp, ed. (Berlin, Germany: Springer Verlag), pp. 269–276.

Mancheño, J.M., Gasset, M., Lacadena, J., Martínez del Pozo, A., Oñaderra, M., and Gavilanes, J.G. (1995a). Predictive study of the conformation of the cytotoxic protein α-sarcin: A structural model to explain α-sarcin-membrane interaction. J. Theor. Biol. 172, 259–267.

Mancheño, J.M., Gasset, M., Albar, J.P., Lacadena, J., Martínez del Pozo, A., Oñaderra, M., and Gavilanes, J.G. (1995b). Membrane interaction of a β-structure-forming synthetic peptide comprising the 116–139th sequence region of the cytotoxic protein α−sarcin. Biophys. J. 68, 2387–2395.

Mancheño, J.M., Martínez del Pozo, A., Albar, J.P., Oñaderra, M., and Gavilanes, J.G. (1998). A peptide of nine amino acid residues from α−sarcin cytotoxin is a membrane-perturbing structure. J. Pept. Res. 51, 142–148.

Martínez-Oyanedel, J., Choe, H.W., Heinemann, U., and Saenger, W. (1991). Ribonuclease T1 with free recognition and catalytic site: crystal structure analysis at 1.5 Å resolution. J. Mol. Biol. 222, 335–352.

Martínez-Ruiz, A., Kao, R., Davies, J., and Martínez del Pozo, A. (1999a). Ribotoxins are a more widespread group of proteins within the filamentous fungi than previously believed. Toxicon 37, 1549–1563.

Martínez-Ruiz, A., Martínez del Pozo, A., Lacadena, J., Oñaderra, M., and Gavilanes, J.G. (1999b). Hirsutellin A displays significant homology to microbial extracellular ribonucleases. J. Invertebr. Pathol. 74, 96–97.

Martínez-Ruiz, A., García-Ortega, L., Kao, R., Lacadena, J., Oñaderra, M., Mancheño, J.M., Davies, J., Martínez del Pozo, A., and Gavilanes, J.G. (2001). RNase U2 and α−sarcin: a study of relationships. Meth. Enzymol. 341, 335–351.

Martínez del Pozo, A., Gasset, M., Oñaderra, M., and Gavilanes, J.G. (1988). Conformational study of the antitumor protein α−sarcin. Biochim. Biophys. Acta 953, 280–288.

Masip, M., Lacadena, J., Mancheño, J.M., Oñaderra, M., Martínez-Ruiz, A., Martínez del Pozo, A., and Gavilanes, J.G. (2001). Arginine 121 is a crucial residue for the specific cytotoxic activity of the ribotoxin α-sarcin. Eur. J. Biochem. 268, 6190–6196.

Masip, M., García-Ortega, L., Olmo, N., García-Mayoral, M.F., Pérez-Cañadillas, J.M., Bruix, M., Oñaderra, M., Martínez del Pozo, A., and Gavilanes, J.G. (2003). Leucine 145 of the ribotoxin α-sarcin plays a key role for determining the specificity of the ribosome-inactivating activity of the protein. Protein Sci. 12, 161–169.

Mears, J.A., Cannone, J.J., Stagg, S.M., Gutell, R.R., Agrawal, R.K., and Harvey, S.C. (2002). Modeling a minimal ribosome based on comparative sequence analysis. J. Mol. Biol. 321, 215–234.

Mérida, I., Ávila-Flores, A., Merino, E. (2008). Diacylglycerol kinases: at the hub of cell signalling. Biochem. J. 409, 1–18.

Moazed, D., Robertson, J.M., and Noller, H.F. (1988). Interaction of elongation factors EF-G and EF-Tu with a conserved loop in 23S RNA. Nature 334, 362–364.

Moser, M., Crameri, R., Menz, G., Schneider, T., Dudler, T., Virchow, C., Gmachl, M., Blaser, K., and Suter, M. (1992). Cloning and expression of recombinant Aspergillus fumigatus allergen I/a (rAsp f I/a) with IgE binding and type I skin test activity. J. Immunol. 149, 454–460.

Munishkin, A. and Wool, I.G. (1997). The ribosome-in-pieces: Binding of elongation factor EF-G to oligoribonucleotides that mimic the sarcin/ricin and thiostrepton domains of 23S ribosomal RNA. Proc. Nat. Acad. Sci. USA 94, 12280–12284.

Nielsen, K., and Boston, R.S. (2001). Ribosome-inactivating proteins: a plant perspective. Annu. Rev. Plant. Physiol. Plant. Mol. Biol. 52, 785–816.

Nilsson, J. and Nissen, P. (2005). Elongation factors on the ribosome. Curr. Opin. Struct. Biol. 15, 349–354.

Noguchi, S., Satow, Y., Uchida, T., Sasaki, C., and Matsuzaki, T. (1995). Crystal structure of Ustilago sphaerogena ribonuclease U2 at 1.8 angstrom resolution. Biochemistry 34, 15583–15591.

Olmo, N., Turnay, J., González de Buitrago, G., López de Silanes, I., Gavilanes, J.G., and Lizarbe, M.A. (2001). Cytotoxic mechanism of the ribotoxin α-sarcin. Induction of cell death via apoptosis. Eur. J. Biochem. 268, 2113–2123.

Olson, B.H., and Goerner, G.L. (1965). α-Sarcin, a new antitumor agent. I. Isolation, purification, chemical composition, and the identity of a new amino acid. Appl. Microbiol. 13, 314–321.

Oñaderra, M., Mancheño, J.M., Gasset, M., Lacadena, J., Schiavo, G., Martínez del Pozo, A., and Gavilanes, J.G. (1993). Translocation of α−sarcin across the lipid bilayer of asolectin vesicles. Biochem. J. 295, 221–225.

Orlandi, R., Canevari, S., Conde, F.P., Leoni, F., Mezzanzanica, D., Ripamonti, M., and Colnaghi, M.I. (1988). Immunoconjugate generation between the ribosome inactivating protein restrictocin and an anti-human breast carcinoma MAB. Cancer Immunol. Immunother. 26, 114–120.

Orntoft, T.F. and Vestergaard, E.M. (1999). Clinical aspects of altered glycosylation of glycoproteins in cancer. Electrophoresis 20, 362–371.

Otero, M.J. and Carrasco, L. (1986). External ATP permeabilizes transformed cells to macromolecules. Biochem. Biophys. Res. Commun. 134, 453–460.

Otero, M.J. and Carrasco, L. (1988). Exogenous phospholipase C permeabilizes mammalian cells to proteins. Exp. Cell Res. 177, 154–161.

Pai, L.H., Wittes, R., Setser, A., Willingham, M.C., and Pastan, I. (1996). Treatment of advanced solid tumors with immunotoxin LMB-1: an antibody linked to Pseudomonas exotoxin. Nat. Med. 2, 350–353.

Papo, N. and Shai, Y. (2005). Host defense peptides as new weapons in cancer treatment. Cell. Mol. Life Sci. 62, 784–790.

Pastan, I., and FitzGerald, D. (1991). Recombinant toxins for cancer treatment. Science 254, 1173–1177.

Pérez, C.A., Eichwald, C., Burrone, O., and Mendoza, D. (2005). Rotavirus vp7 antigen produced by *Lactococcus lactis* induces neutralizing antibodies in mice. J. Appl. Microbiol. 99, 1158–1164.

Pérez-Cañadillas, J.M., Campos-Olivas, R., Lacadena, J., Martínez del Pozo, A., Gavilanes, J.G., Santoro, J., Rico, M., and Bruix, M. (1998). Characterization of pK(a) values and titration shifts in the cytotoxic ribonuclease α–sarcin by NMR. Relationship between electrostatic interactions, structure, and catalytic function. Biochemistry 37, 15865–15876.

Pérez-Cañadillas, J.M., Santoro, J., Campos-Olivas, R., Lacadena, J., Martínez del Pozo, A., Gavilanes, J.G., Rico, M., and Bruix, M. (2000). The highly refined solution structure of the cytotoxic ribonuclease α–sarcin reveals the structural requirements for substrate recognition and ribonucleolytic activity. J. Mol. Biol. 299, 1061–1073.

Pérez-Cañadillas, J.M., Guenneugues, M., Campos-Olivas, R., Santoro, J., Martínez del Pozo, A., Gavilanes, J.G., Rico, M., and Bruix, M. (2002). Backbone dynamics of the cytotoxic Ribonuclease α-sarcin by ^{15}N NMR relaxation methods. J. Biomol. NMR 24, 301–316.

Pérez-Cañadilllas, J.M., García-Mayoral, M.F., Laurents, D.V., Martínez del Pozo, A., Gavilanes, J.G., Rico, M., and Bruix, M. (2003). Tautomeric state of α-sarcin histidines. Nδ tautomers are a common feature in the active site of extracellular microbial ribonucleases. FEBS Lett. 534, 197–201.

Peumans, W.J., Hao, Q., and Van Damme, E.J. (2001). Ribosome-inactivating proteins from plants: more than RNA N-glycosidases? FASEB J. 15, 1493–1506.

Pfeiffer, S., Kariminejad, Y., and Ruterjans, H. (1997). Limits of NMR structure determination using variable target function calculations: Ribonuclease T1, a case study. J. Mol. Biol. 266, 400–423.

Piechura, J.E., Huang, C.J., Cohen, S.H., Kidd, J.M., Kurup, V.P., and Calvanico, N.J. (1983). Antigens of *Aspergillus fumigatus*. II. Electrophoretic and clinical studies. Immunology 49, 657–665.

Pouwels, P.H., Leer, R.J., and Boersma, W.J. (1996). The potential of *Lactobacillus* as a carrier for oral immunization: development and preliminary characterization of vector systems for targeted delivery of antigens. J. Biotechnol. 44, 183–192.

Ramakrishnan, V., and Moore, P.B. (2001). Atomic structures at last: the ribosome in 2000. Curr. Opin. Struct. Biol. 11, 144–154.

Ran, S., Downes, A., and Thorpe, P.E. (2002). Increased exposure of anionic phospholipids on the surface of tumor blood vessels. Cancer Res. 62, 6132–6140

Rathore, D., and Batra, J.K. (1996). Generation of active immunotoxins containing recombinant restrictocin. Biochem. Biophys. Res. Commun. 222, 58–63.

Rathore, D. and Batra, J.K. (1997a). Construction, expression and characterization of chimaeric toxins containing the ribonucleolytic toxin restrictocin: intracellular mechanism of action. Biochem. J. 324, 815–822.

Rathore, D., and Batra, J.K. (1997b). Cytotoxic activity of ribonucleolytic toxin restrictocin-based chimeric toxins targeted to epidermal growth factor receptor. FEBS Lett. 407, 275–279.

Rathore, D., Nayak, S.K., and Batra, J.K. (1997). Overproduction of fungal ribotoxin α-sarcin in *Escherichia coli*: Generation of an active immunotoxin. Gene 190, 31–35.

Robinson, K., Chamberlain, L.M., Schofield, K.M., Wells, J.M., and Le Page, R.W. (1997). Oral vaccination of mice against tetanus with recombinant *Lactococcus lactis*. Nat. Biotechnol. 15, 653–657.

Rodríguez, R., López-Otín, C., Barber, D., Fernández-Luna, J.L., González, G., and Méndez, E. (1982). Amino acid sequence homologies in α-sarcin, restrictocin and mitogillin. Biochem. Biophys. Res. Commun. 108, 315–321.

Roga, V., Hedeman, L.P., and Olson, B.H. (1971). Evaluation of mitogillin (NSC-69529) in the treatment of naturally occurring canine neoplasms. Cancer Chemother. Rep. 55, 101–113.

Sacco, G., Drickamer, K., and Wool, I.G. (1983). The primary structure of the cytotoxin α–sarcin. J. Biol. Chem. 258, 5811–5818.

Sandvig, K., and van Deurs, B. (2000). Entry of ricin and Shiga toxin into cells: molecular mechanisms and medical perspectives. EMBO J. 19, 5943–5950.

Schindler, D.G. and Davies, J.E. (1977). Specific cleavage of ribosomal RNA caused by α-sarcin. Nucleic Acids Res. 4, 1097–1110.

Schnell, R., Vitetta, E., Schindler, J., Barth, S., Winkler, U., Borchmann, P., Hansmann, M.L., Diehl, V., Ghetie, V., and Engert, A. (1998). Clinical trials with an anti-CD25 ricin A-chain experimental and immunotoxin (RFT5-SMPT-dgA) in Hodgkin's lymphoma. Leuk. Lymphoma 30, 525–537.

Seggerson, K. and Moore, P.B. (1998). Structure and stability of variants of the sarcin-ricin loop of 28S rRNA: NMR studies of the prokaryotic SRL and a functional mutant. RNA 4, 1203–1215.

Sevcik, J., Dodson, E.J., and Dodson, G.G. (1991). Determination and restrained least-squares refinement of the structures of ribonuclease Sa and its complex with 3'-guanylic acid at 1.8 A resolution. Acta Crystallogr. 47, 240–253.

Steidler, L., Hans, W., Schotte, L., Neirynck, S., Obermeier, F., Falk, W., Fiers, W., and Remaut, E. (2000). Treatment of murine colitis by *Lactococcus lactis* secreting interleukin-10. Science 289, 1352–1355.

Steyeaert, J. (1997). A decade of protein engineering on ribonuclease T1. Atomic dissection of the enzyme-substrate interactions. Eur. J. Biochem. 241, 1–11.

Steyaert, J., Hallenga, K., Wyns, L., Stanssens, P. (1990). Histidine 40 of ribonuclease T1 acts as base catalyst when the true catalytic base, glutamic acid 58, is replaced by alanine. Biochemistry 29, 9064–9072.

Stuart, A.D., and Brown, T.D. (2006). Entry of feline calicivirus is dependent on clathrin-mediated endocytosis and acidification in endosomes. J. Virol. 80, 7500–7509.

Sylvester, I.D., Roberts, L.M., and Lord, J.M. (1997). Characterization of prokaryotic recombinant Aspergillus ribotoxin α–sarcin. Biochim. Biophys. Acta 1358, 53–60.

Szewczak, A.A., and Moore, P.B. (1995). The sarcin/ricin loop, a modular RNA. J. Mol. Biol. 247, 81–98.

Szewczak, A.A., Moore, P.B., Chan, Y.L., and Wool, I.G. (1993). The conformation of the sarcin/ricin loop from 28S ribosomal RNA. Proc. Nat. Acad. Sci. USA 90, 9581–9585.

Turnay, J., Olmo, N., Jimenez, A., Lizarbe, M.A., and Gavilanes, J.G. (1993). Kinetic study of the cytotoxic effect of α–sarcin, a ribosome inactivating protein from Aspergillus giganteus, on tumour cell lines: protein biosynthesis inhibition and cell binding. Mol. Cell. Biochem. 122, 39–47.

Uchida, T., and Egami, F. (1971). Microbial ribonucleases with special reference to RNases T_1, T_2, N_1, and U_2. In The enzymes, P.D. Boyer, ed. (New York and London, Academic Press), pp. 205–250.

Uchiumi, T., Honma, S., Endo, Y., and Hachimori, A. (2002). Ribosomal proteins at the stalk region modulate functional rRNA structures in the GTPase Center. J. Biol. Chem. 277, 41401–41409.

Van Dyke, N., Xu, W., and Murgola, E.J. (2002). Limitation of ribosomal protein L11 availability in vivo affects translation termination. J. Mol. Biol. 319, 329–339.

Vatzaki, E.H., Allen, S.C., Leonidas, D.D., Trautwein-Fritz, K., Stackhouse, J., Benner, S.A., and Acharya, K.R. (1999). Crystal structure of a hybrid between ribonuclease A and bovine seminal ribonuclease – the basic surface, at 2.0 Å resolution. Eur. J. Biochem. 260, 176–182.

Ward, E.S., Gussow, D., Griffiths, A.D., Jones, P.T., and Winter, G. (1989). Binding activities of a repertoire of single immunoglobulin variable domains secreted from Escherichia coli. Nature 341, 544–546.

Wawrzynczak, E.J., Henry, R.V., Cumber, A.J., Parnell, G.D., Derbyshire, E.J., and Ulbrich, N. (1991). Biochemical, cytotoxic and pharmacokinetic properties of an immunotoxin composed of a mouse monoclonal antibody Fib75 and the ribosome-inactivating protein α–sarcin from Aspergillus giganteus. Eur. J. Biochem. 196, 203–209.

Wirth, J., Martínez del Pozo, A., Mancheño, J.M., Martínez-Ruiz, A., Lacadena, J., Oñaderra, M., and Gavilanes, J.G. (1997). Sequence determination and molecular characterization of gigantin, a cytotoxic protein produced by the mould Aspergillus giganteus IFO 5818. Arch. Biochem. Biophys. 343, 188–193.

Woo, J.H., Liu, Y.Y., Mathias, A., Stavrou, S., Wang, Z., Thompson, J., and Neville, D.M., Jr. (2002). Gene optimization is necessary to express a bivalent anti-human anti-T-cell immunotoxin in Pichia pastoris. Protein Expr. Purif. 25, 270–282.

Woo, J.H., Liu, Y.Y., Stavrou, S., and Neville, D.M., Jr. (2004). Increasing secretion of a bivalent anti-T-cell immunotoxin by Pichia pastoris. Appl. Environ. Microbiol. 70, 3370–3376.

Woo, J.H., Liu, Y.Y., and Neville, D.M., Jr. (2006). Minimization of aggregation of secreted bivalent anti-human T-cell immunotoxin in Pichia pastoris bioreactor culture by optimizing culture conditions for protein secretion. J. Biotechnol. 121, 75–85.

Woo, J.H., Liu, J.S., Kang, S.H., Singh, R., Park, S.K., Su, Y., Ortiz, J., Neville, D.M., Jr., Willingham, M.C., and Frankel, A.E. (2008). GMP production and characterization of the bivalent anti-human T-cell immunotoxin, A-dmDT390-bisFv(UCHT1) for phase I/II clinical trials. Protein Expr. Purif. 58, 1–11.

Wool, I.G., Gluck, A., and Endo, Y. (1992). Ribotoxin recognition of ribosomal RNA and a proposal for the mechanism of translocation. Trends Biochem. Sci. 17, 266–269.

Worn, A., and Pluckthun, A. (2001). Stability engineering of antibody single-chain Fv fragments. J. Mol. Biol. 305, 989–1010.

Xin, K.Q., Hoshino, Y., Toda, Y., Igimi, S., Kojima, Y., Jounai, N., Ohba, K., Kushiro, A., Kiwaki, M., Hamajima, K., Klinman, D., and Okuda, K. (2003). Immunogenicity and protective efficacy of orally administered recombinant Lactococcus lactis expressing surface-bound HIV Env. Blood 102, 223–228.

Yang, X., and Moffat, K. (1996). Insights into specificity of cleavage and mechanism of cell entry from the crystal structure of the highly specific Aspergillus ribotoxin, restrictocin. Structure 4, 837–852.

Yang, X., Gercei, T., Glover, L., and Correll, C.C. (2001). Crystal structures of restrictocin-inhibitor complexes with implications for RNA recognition and base flipping. Nat. Struct. Biol. 8, 968–973.

Yeung, T., Gilbert, G.E., Shi, J., Silvius, J., Kapus, A., and Grinstein, S. (2008). Membrane phosphatidylserine regulates surface charge and protein localization. Science 319, 210–213.

Yoshida, H. (2001). The ribonuclease T1 family. Meth. Enzymol. 341, 28–41.

Zegers, I., Haikal, A.F., Palmer, R., and Wyns, L. (1994). Crystal Structure of RNase T1 with 3′-Guanylic Acid and Guanosine. J. Biol. Chem. 269, 127–133.

Zegers, I., Loris, R., Dehollander, G., Haikal, A.F., Poortmans, F., Steyaert, J., and Wyns, L. (1998). Hydrolysis of a slow cyclic thiophosphate substrate of RNase T1 analysed by time-resolved crystallography. Nat. Struct. Biol. 5, 280–283.

Index